118257

Analytic Geometry with Vectors

Analytic Geometry with Vectors

Douglas F. Riddle

Wadsworth Publishing Company, Inc., Belmont, California

© 1972 by Wadsworth
Publishing Company, Inc.

© 1970 by Wadsworth
Publishing Company, Inc.,
Belmont, California 94002.
All rights reserved. No part of
this book may be reproduced,
stored in a retrieval system, or
transcribed, in any form or by
any means, electronic,
mechanical, photocopying,
recording, or otherwise,
without the prior written
permission of the publisher.

L. C. Cat. Card No.: 79-170679
ISBN: 0-534-00090-8

Printed in the United States
of America

2 3 4 5 6 7 8 9 10   76 75 74

The Table of Trigonometric
Functions in the appendix is
from The Calculus with
Analytic Geometry *by Louis
Leithold.*
Copyright © 1968 by Louis
Leithold. Reprinted by
permission of the publishers,
Harper & Row, Publishers,
Inc. The tables of Common
Logarithms and Squares,
Square Roots, and Prime
Factors are reprinted from
College Algebra, *2nd ed., by
Edwin F. Beckenbach, Irving
Drooyan, and William Wooton.*
Copyright © 1964, 1968 by
Wadsworth Publishing
Company, Inc., Belmont,
California.

Preface

Analytic Geometry with Vectors, designed for students with a reasonably sound background in algebra and trigonometry, contains more than enough material for a three-semester-hour or five-quarter-hour course in analytic geometry.

Vectors are used throughout. However, the approach is not one of "vectors for vectors sake"; rather, vectors are used only where they provide a simpler approach than traditional methods. Thus, for example, a vector approach is taken for determining the distance from a point to a line in the plane (page 56), while nonvector methods are used in finding the distance from a point to a plane in space (page 219).

It is traditional for many of the older texts on analytic geometry to include a separate chapter on curve fitting, while more recent texts tend to omit it entirely. I have adopted a middle course, considering curve fitting only for the line; this is by far the most important case and is, moreover, a simple application of analytic geometry that the student is likely to encounter in courses in engineering and the physical and social sciences.

A section relating conic sections to cones fills what I feel to be a gap in most students' knowledge of conic sections. Generally the student is told merely that the parabola, ellipse, and hyperbola are intersections of planes with a right circular cone, with no attempt being made to prove the statement. In Section 5.8 the student is shown this relationship, as well as the geometric significance, of the foci, directrices, and eccentricity. In addition, he is exposed to the elements of a synthetic approach to conic sections that he probably thought impossible.

My thanks go to Herbert C. Schmidt for his helpful suggestions in reviewing the manuscript.

Douglas F. Riddle

Contents

1 Plane Analytic Geometry

- 1.1 The Cartesian Plane 1
- 1.2 Distance Formula 4
- 1.3 Point-of-Division Formulas 7
- 1.4 Inclination and Slope 12
- 1.5 Parallel and Perpendicular Lines 14
- 1.6 Angle from One Line to Another 16
- 1.7 Analytic Proofs of Geometric Theorems 19
- 1.8 Graphs and Points of Intersection 23
- 1.9 An Equation of a Locus 25

2 Vectors in the Plane

- 2.1 Directed Line Segments and Vectors 29
- 2.2 Operations with Vectors 35
- 2.3 The Dot Product 42

3 The Line

3.1	Point-Slope and Two-Point Forms 48
3.2	Slope-Intercept and Intercept Forms 52
3.3	Distance from a Point to a Line 56
3.4	Families of Lines 61
3.5	Fitting a Line to Empirical Data 65

4 The Circle

4.1	The Standard Form for an Equation of a Circle 72
4.2	Conditions to Determine a Circle 77
4.3	Families of Circles 83

5 Conic Sections

5.1	Conic Sections 88
5.2	The Parabola 88
5.3	Parabola with Vertex at (h, k) 92
5.4	The Ellipse 96
5.5	Ellipse with Center (h, k) 102
5.6	The Hyperbola 106
5.7	Hyperbola with Center (h, k) 112
5.8	Conics and a Right Circular Cone 115

6 Transformation of Coordinates

6.1   Translation   119
6.2   Rotation   124
6.3   The General Equation of Second Degree   127

7 Curve Sketching

7.1   Intercepts and Asymptotes   132
7.2   Symmetry, Sketching   138
7.3   Radicals and the Domain of the Equation   144
7.4   Direct Sketching of Conics   148

8 Transcendental Curves

8.1   Trigonometric Functions   153
8.2   Inverse Trigonometric Functions   156
8.3   Exponential and Logarithmic Functions   160
8.4   Hyperbolic Functions   164

9 Polar Coordinates

9.1   Polar Coordinates   168
9.2   Graphs in Polar Coordinates   169
9.3   Points of Intersection   173

x **Contents**

	9.4	Relationships between Rectangular and Polar Coordinates 175
	9.5	Conics in Polar Coordinates 178

10 Parametric Equations

	10.1	Parametric Equations 183
	10.2	Parametric Equations of a Locus 187

11 Solid Analytic Geometry

	11.1	Introduction: The Distance and Point-of-Division Formulas 192
	11.2	Vectors in Space 196
	11.3	Direction Angles, Cosines, and Numbers 200
	11.4	The Line 204
	11.5	The Cross Product 209
	11.6	The Plane 214
	11.7	Distance between a Point and a Plane; Angles between Lines or Planes 219
	11.8	Cylinders and Spheres 223
	11.9	Quadric Surfaces 227
	11.10	Cylindrical and Spherical Coordinates 231

Table 1 Trigonometric Functions 237

Table 2 Common Logarithms 239

Table 3 Squares, Square Roots, and Prime Factors 240

Answers to Selected Problems 241

Index 297

Analytic Geometry with Vectors

1

Plane Analytic Geometry

1.1

The Cartesian Plane

Analytic geometry provides a bridge between algebra and geometry that makes it possible for geometric problems to be solved algebraically (or analytically). It also allows us to solve algebraic problems geometrically, but the former is far more important, especially when numbers are assigned to essentially geometric concepts. Consider, for instance, the length of a line segment or the angle between two lines. Even if the lines and points in question are accurately known, the number representing the length of a segment or the angle between two lines can be determined only approximately by measurement. Algebraic methods provide an exact determination of the number.

The association between the algebra and geometry is made by assigning numbers to points. Suppose we look at this assignment of numbers to the points on a line. First of all, we select a pair of points, O and P, on the line. The point O, which we call the origin, is assigned the number zero, and the point P is assigned the number one. Using OP as our unit of length, we assign numbers to all other points on the line in the following way: Q on the P side of the origin is assigned the positive number x if and only if its distance from the origin is x. A point Q on the opposite side of the origin is assigned the negative number $-x$ if and only if its distance from the origin is x units. In this way every point on the line is assigned a real number, and for each real number there corresponds a point on the line.

Thus a *scale* is established on the line, which we now call a *coordinate line*. The number representing a given point is called the *coordinate* of that point, and the point is called the *graph* of the number.   See   Figure 1.1.

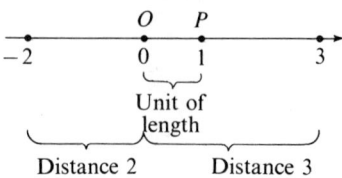

Figure 1.1

Just as points on a line (a one-dimensional space) are represented by single numbers, so points in a plane (a two-dimensional space) can be represented by pairs of numbers. Later we shall see that points in a three-dimensional space can be represented by triples of numbers.

In order to represent points in a plane by a pair of numbers, we select two intersecting lines and establish a scale on each line using the point of intersection as the origin. These two lines, called the axes, are distinguished by identifying symbols (usually by the letters x and y). For a given point P in the plane, there corresponds a point P_x on the x axis which is the point of intersection of the x axis and the line through P which is parallel to the y axis (if P is on the y axis, this line coincides with the y axis). Similarly, there exists a point P_y on the y axis which is the point of intersection of the y axis and the line through P which is parallel to (or is) the x axis. The

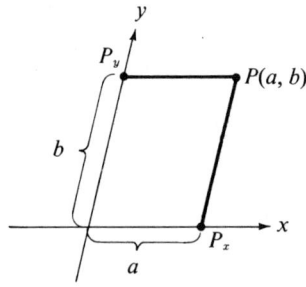

Figure 1.2

1.1 The Cartesian Plane

coordinates of these two points on the axes are the *coordinates* of P. If a is the coordinate of P_x and b is the coordinate of P_y, then the point P is represented by (a, b). In this example, a is called the *x coordinate*, or *abscissa*, of P and b is the *y coordinate*, or *ordinate*, of P. See Figure 1.2.

In a coordinate plane, the following conventions are normally followed:

(1) the axes are taken to be perpendicular to each other;
(2) the x axis is a horizontal line with the positive coordinates to the right of the origin, and the y axis is a vertical line with the positive coordinates above the origin;
(3) the same scale is used on both axes.

These are, of course, only conventions; they need not be followed when others are more convenient. One of these that we shall violate rather frequently is the third, because we shall sometimes consider figures that would be very difficult to sketch if we insisted upon using the same scale on both axes. In such cases, we shall feel free to use different scales, remembering that we have distorted the figure in the process. Unless a departure from convention is specifically stated or is obvious from the context, we shall always follow the other two conventions.

We can now identify the coordinates of the points in Figure 1.3. Note that all points on the x axis have the y coordinate zero, while those on the y axis have the x coordinate zero. The origin has both coordinates zero, since it is on both axes.

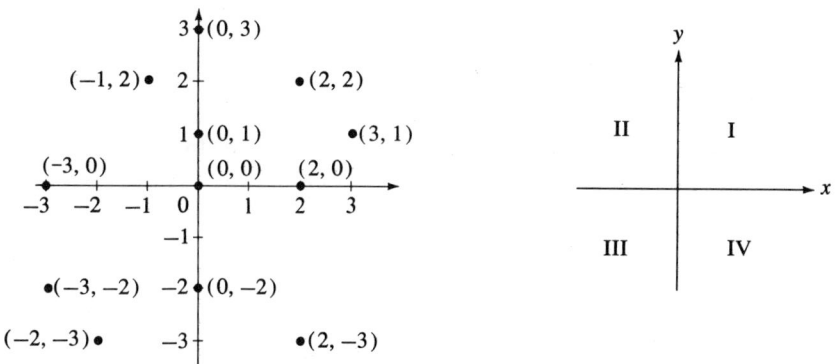

Figure 1.3   Figure 1.4

The axes separate the plane into four regions, called *quadrants*. It is convenient to identify them by the numbers shown in Figure 1.4. The points on the axes are not in any quadrant.

1.2

Distance Formula

Suppose we consider the distance between two points on a coordinate line. Let P_1 and P_2 be two points on a line, and let P_1 and P_2 have coordinates x_1 and x_2, respectively. If P_1 and P_2 are both to the right of the origin, with P_2 farther right than P_1 (as in Figure 1.5(a)), then*

$$\overline{P_1P_2} = \overline{OP_2} - \overline{OP_1} = x_2 - x_1.$$

Expressing the distance between two points is only slightly more complicated if one or both of the points are to the left of the origin. In Figure 1.5(b),

$$\overline{P_1P_2} = \overline{P_1O} - \overline{P_2O} = -x_1 - (-x_2) = x_2 - x_1,$$

and in Figure 1.5(c),

$$\overline{P_1P_2} = \overline{P_1O} + \overline{OP_2} = -x_1 + x_2 = x_2 - x_1.$$

Thus, we see that $\overline{P_1P_2} = x_2 - x_1$ in all three of these cases in which P_2 is to the right of P_1. If P_2 were to the left of P_1, then

$$\overline{P_1P_2} = x_1 - x_2,$$

as you can easily verify. Thus $\overline{P_1P_2}$ can always be represented as the larger coordinate minus the smaller. Since $x_2 - x_1$ and $x_1 - x_2$ differ only in that one is the negative of the other and since distance is always non-negative, we see that $\overline{P_1P_2}$ is the one of these two that is positive. Thus,

$$\overline{P_1P_2} = |x_2 - x_1|.$$

This form is especially convenient when the relative positions of P_1 and P_2 are unknown. However, since absolute values are sometimes rather bothersome, they will be avoided whenever possible.

Let us now turn our attention to the more difficult problem of finding the distance between two points in the plane. Suppose we are interested in the distance between $P_1: (x_1, y_1)$ and $P_2: (x_2, y_2)$ (see Figure 1.6). If a vertical line is drawn through P_1 and a horizontal line through P_2, they intersect at a point $Q: (x_1, y_2)$, and P_1P_2Q

---

* We shall use the notation AB for the line segment joining the points A and B, and $\overline{AB}$ for its length.

1.2 Distance Formula

forms a right triangle (assuming P_1 and P_2 are not on the same horizontal or vertical line) with the right angle at Q. Now we can use the theorem of Pythagoras to determine the length of P_1P_2. By the previous discussion,

$$\overline{QP_2} = |x_2 - x_1| \quad \text{and} \quad \overline{P_1Q} = |y_2 - y_1|$$

(the absolute values are retained here, since we want the resulting formula to hold for *any* choice of P_1 and P_2, not merely for the one shown in Figure 1.6). Now by the Pythagorean theorem,

$$\overline{P_1P_2} = \sqrt{|x_2 - x_1|^2 + |y_2 - y_1|^2}.$$

But, since $|x_2 - x_1|^2 = (x_2 - x_1)^2 = (x_1 - x_2)^2$, the absolute values may be dropped at this stage and we have

$$\overline{P_1P_2} = \sqrt{(x_2 - x_1)^2 + (y_2 - y_1)^2}.$$

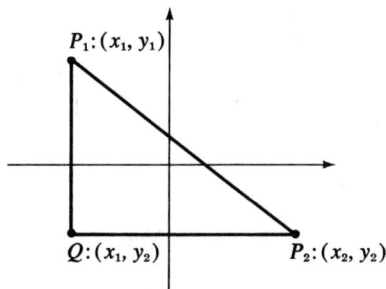

Figure 1.6

Thus we have proved the following theorem.

Theorem 1.1

The distance between two points $P_1: (x_1, y_1)$ and $P_2: (x_2, y_2)$ is

$$\overline{P_1P_2} = \sqrt{(x_2 - x_1)^2 + (y_2 - y_1)^2}.$$

In deriving this formula, we assumed that P_1 and P_2 are not on the same horizontal or vertical line; however, the above formula would hold even in these cases. For example, if P_1 and P_2 are on the same horizontal line, then $y_1 = y_2$ and $y_2 - y_1 = 0$. Thus,

$$\overline{P_1P_2} = \sqrt{(x_2 - x_1)^2} = |x_2 - x_1|.$$

Note that $\sqrt{(x_2 - x_1)^2}$ is *not always* $x_2 - x_1$. Since the symbol $\sqrt{}$ indicates the nonnegative square root, we see that if $x_2 - x_1$ is negative, then $\sqrt{(x_2 - x_1)^2}$ is not equal to $x_2 - x_1$ but, rather, equals $|x_2 - x_1|$.

Theorem 1.1 depends upon the convention that the axes are perpendicular. If this convention is not followed, Theorem 1.1 cannot be used. Another, more general formula, based upon the law of cosines can be derived; however, we shall not derive it here, since the convention of using perpendicular axes is so widely observed.

Example 1

Find the distance between $P_1: (1, 4)$ and $P_2: (-3, 2)$.

$$\overline{P_1 P_2} = \sqrt{(-3-1)^2 + (2-4)^2} = 2\sqrt{5}.$$

Example 2

Determine whether or not $A: (1, 7)$, $B: (0, 3)$, and $C: (-2, -5)$ are collinear.

$$\overline{AB} = \sqrt{(0-1)^2 + (3-7)^2} = \sqrt{17},$$
$$\overline{BC} = \sqrt{(-2-0)^2 + (-5-3)^2} = \sqrt{68} = 2\sqrt{17},$$
$$\overline{AC} = \sqrt{(-2-1)^2 + (-5-7)^2} = \sqrt{153} = 3\sqrt{17}.$$

Since $\overline{AC} = \overline{AB} + \overline{BC}$, the three points must be collinear (if they were not, they would form a triangle and any one side would be less than the sum of the other two).

Problems

In Problems 1–8, find the distance between the given points.

1. $(1, -3), (2, 5)$.
2. $(4, 13), (-1, 5)$.
3. $(3, -2), (3, -4)$.
4. $(-5, 1), (0, -10)$.
5. $(1/2, 3/2), (-5/2, 2)$.
6. $(2/3, 1/3), (-4/3, 4/3)$.
7. $(\sqrt{2}, 1), (2\sqrt{2}, 3)$.
8. $(\sqrt{3}, -\sqrt{2}), (-3\sqrt{3}, \sqrt{2})$.

In Problems 9–12, find the unknown quantity.

9. $P_1 = (1, 5)$, $P_2 = (x, 2)$, $\overline{P_1 P_2} = 5$.
10. $P_1 = (-3, y)$, $P_2 = (9, 2)$, $\overline{P_1 P_2} = 13$.
11. $P_1 = (x, x)$, $P_2 = (1, 4)$, $\overline{P_1 P_2} = \sqrt{5}$.
12. $P_1 = (x, 2x)$, $P_2 = (2x, 1)$, $\overline{P_1 P_2} = \sqrt{2}$.

In Problems 13–18, determine whether or not the three given points are collinear.

13. $(2, 1), (4, 3), (-1, -2)$.
14. $(3, 2), (4, 6), (0, -8)$.
15. $(-2, 3), (7, -2), (2, 5)$.
16. $(1, \sqrt{2}), (4, 3\sqrt{2}), (10, 6\sqrt{2})$.

1.3 Point-of-Division Formulas

17. $(-1/2, 2/3)$, $(1/4, 3/5)$, $(7/4, 7/15)$.
18. $(3/4, 1/8)$, $(2/3, 1/2)$, $(1/6, 11/4)$.

In Problems 19–22, determine whether or not the three given points are the vertices of a right triangle.

19. $(0, 2)$, $(-2, 4)$, $(1, 3)$.
20. $(-1, 3)$, $(4, 6)$, $(-3, 1)$.
21. $(\sqrt{3} - 2, 2\sqrt{3} + 1)$, $(\sqrt{3} + 2, -\sqrt{3} + 1)$, $(2\sqrt{3} - 2, 2\sqrt{3} + 2)$.
22. $(\sqrt{3} - 3, 2\sqrt{3} + 1)$, $(\sqrt{3} - 1, \sqrt{3} + 1)$, $(2\sqrt{3} - 1, \sqrt{3} + 2)$.
23. Show that $(5, 2)$ is on the perpendicular bisector of the segment AB where $A = (1, 3)$ and $B = (4, -2)$.
24. Show that $(-2, 4)$, $(2, 0)$, $(2, 8)$, and $(6, 4)$ are the vertices of a square.
25. Show that $(1, 1)$, $(4, 1)$, $(3, -2)$, and $(0, -2)$ are the vertices of a parallelogram.
26. Show that $(1, 2)$, $(4, 7)$, $(-6, 13)$, and $(-9, 8)$ are the vertices of a rectangle.
27. Graph the circle with center $(-2, 3)$ and radius 5. For each of the following points, indicate whether it is inside, on, or outside the circle: $(1, 7)$, $(-3, 8)$, $(2, 0)$, $(-5, 7)$, $(0, -1)$, $(-5, -1)$, $(-6, 6)$, $(4, 2)$.
28. Find the center and radius of the circle circumscribed about the triangle with vertices $(5, 1)$, $(6, 0)$, and $(-1, -7)$.

1.3

Point-of-Division Formulas

Suppose we are presented with the problem of finding the point which is some fraction of the way from A to B. Is it possible to express the coordinates of the point we want in terms of the coordinates of A and B? Let $A: (x_1, y_1)$ and $B: (x_2, y_2)$ be given and let $P: (x, y)$ be the point we are seeking. If we let

$$r = \frac{\overline{AP}}{\overline{AB}}$$

(see Figure 1.7), then P is $1/3$ of the way from A to B when $r = 1/3$, and P is $4/5$ of the way from A to B when $r = 4/5$, and so on. Thus we generalize the problem to one in which x and y are to be expressed in terms of x_1, y_1, x_2, y_2, and r. The problem can be simplified considerably by working with the x's and y's separately.

If A, B, and P are projected onto the x axis (see Figure 1.7) to give the points A_x, B_x, and P_x, respectively, we have, from elementary geometry,

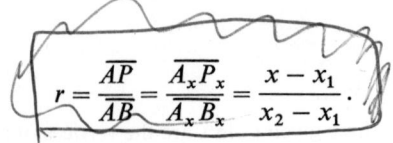

$$r = \frac{\overline{AP}}{\overline{AB}} = \frac{\overline{A_xP_x}}{\overline{A_xB_x}} = \frac{x - x_1}{x_2 - x_1}.$$

Solving for x gives

$$x = x_1 + r(x_2 - x_1).$$

By projecting onto the y axis, we have

$$y = y_1 + r(y_2 - y_1).$$

These two results, known as point-of-division formulas, are stated in the following theorem.

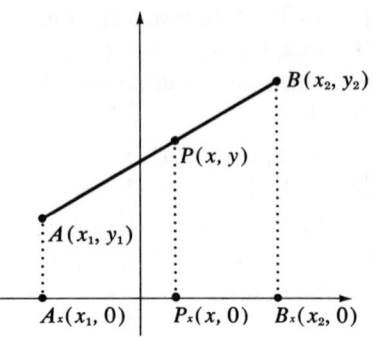

Figure 1.7

Theorem 1.2

If $A = (x_1, y_1)$, $B = (x_2, y_2)$, and P is a point such that $r = \overline{AP}/\overline{AB}$, then the coordinates of P are

$$x = x_1 + r(x_2 - x_1) \quad \text{and} \quad y = y_1 + r(y_2 - y_1).$$

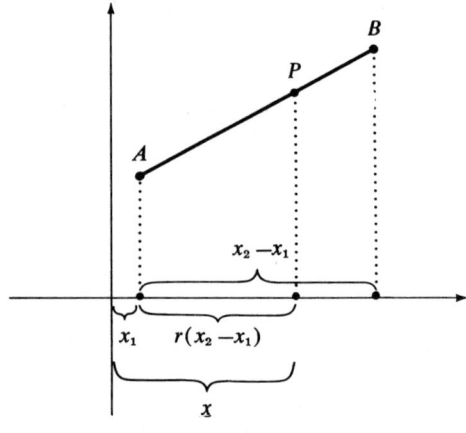

Figure 1.8

Figure 1.8 gives a geometric interpretation of the terms in the point-of-division formula for x. If you examine this figure carefully, it will help you to understand and remember the formulas. Note that if B is to the left of A, $x_2 - x_1$ is negative, but it has the same absolute value as the distance between the projections of A and B on the x axis. You might sketch this and compare your sketch with Figure 1.8. Of course, a similar figure can be used to interpret the y terms of the point-of-division formula.

Example 1

Find the point one-third of the way from $A: (2, 5)$ to $B: (8, -1)$.

1.3 Point-of-Division Formulas

$$r = \frac{\overline{AP}}{\overline{AB}} = \frac{1}{3}.$$

$$x = x_1 + r(x_2 - x_1) \qquad y = y_1 + r(y_2 - y_1)$$
$$= 2 + \frac{1}{3}(8 - 2) \qquad = 5 + \frac{1}{3}(-1 - 5)$$
$$= 4; \qquad = 3.$$

Thus the desired point is (4, 3).

So far we have tacitly assumed that r is between 0 and 1. If r is either 0 or 1, the point-of-division formulas would give us $P = A$ or $P = B$, respectively, a result that $r = \overline{AP}/\overline{AB}$ would lead us to expect. Similarly, if $r > 1$, then $r = \overline{AP}/\overline{AB}$ indicates $\overline{AP} > \overline{AB}$, which is exactly what the point-of-division formulas give. Thus if we wanted to extend the segment AB beyond B to a point P which is r times as far from A as B is, we could still use the point-of-division formulas.

Example 2

If the segment AB, where $A = (-3, 1)$ and $B = (2, 5)$, is extended beyond B to a point P twice as far from A as B is, find P.

$$r = \frac{\overline{AP}}{\overline{AB}} = 2.$$

$$x = x_1 + r(x_2 - x_1) \qquad y = y_1 + r(y_2 - y_1)$$
$$= -3 + 2[2 - (-3)] \qquad = 1 + 2(5 - 1)$$
$$= 7; \qquad = 9.$$

Thus $P = (7, 9)$.

While negative values of r do not make sense in $r = \overline{AP}/\overline{AB}$, we find that their use in the point-of-division formulas has the effect of extending the segment AB in the reverse direction—that is, from B through A to P. Suppose, for example, that $r = -2$. Then $\overline{AP}$ is twice $\overline{AB}$, and P and B are on opposite sides of A. However, we can get the same result by reversing the roles of A and B and using a positive value of r.

Example 3

Let the segment AB, where $A = (-3, 1)$ and $B = (2, 5)$, be extended beyond A to a point P twice as far from B as A is (see Figure 1.9); find P.

Suppose we use a negative value of r. Since $\overline{AP} = \overline{AB}$ with A between P and B,

$$r = -\frac{\overline{AP}}{\overline{AB}} = -1.$$

Therefore

$$x = x_1 + r(x_2 - x_1) \qquad y = y_1 + r(y_2 - y_1)$$
$$= -3 - 1[2 - (-3)] \qquad = 1 - 1(5 - 1)$$
$$= -8; \qquad = -3.$$

Thus $P = (-8, -3)$.

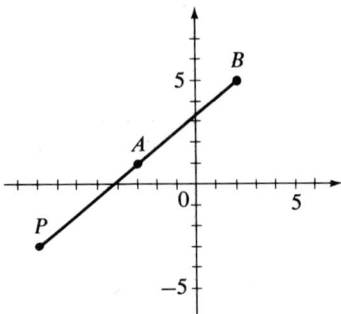

Figure 1.9

Reversing the roles of A and B, we have

$$r = \frac{\overline{BP}}{\overline{BA}} = 2, \quad B = (x_1, y_1) = (2, 5), \quad \text{and} \quad A = (x_2, y_2) = (-3, 1).$$

$$x = x_1 + r(x_2 - x_1) \qquad y = y_1 + r(y_2 - y_1)$$
$$= 2 + 2(-3 - 2) \qquad = 5 + 2(1 - 5)$$
$$= -8; \qquad = -3;$$

and $P = (-8, -3)$, as before.

One very important special case of the point-of-division formulas arises when $r = 1/2$, which gives the midpoint of the segment AB. Substituting into the point-of-division formulas, we have the following theorem.

Theorem 1.3

If P is the midpoint of AB, then the coordinates of P are:

$$x = \frac{x_1 + x_2}{2}, \qquad y = \frac{y_1 + y_2}{2}.$$

Thus, to find the midpoint of a segment AB, we merely average both the x and y coordinates of the given points. A moment of thought will reveal the reasonableness of this; the average of two grades is half-way between them, the average of two temperatures is half-way between them, and so forth.

1.3 Point-of-Division Formulas

Example 4

Find the midpoint of the segment AB, where $A = (1, 5)$ and $B = (-3, -1)$.

$$x = \frac{x_1 + x_2}{2} \qquad y = \frac{y_1 + y_2}{2}$$

$$= \frac{1 - 3}{2} \qquad = \frac{5 - 1}{2}$$

$$= -1; \qquad = 2.$$

Thus $P = (-1, 2)$.

Problems

In Problems 1–6, find the point P such that $\overline{AP}/\overline{AB} = r$.

1. $A = (3, 4)$, $B = (7, 0)$, $r = 1/4$.
2. $A = (4, -2)$, $B = (-2, -5)$, $r = 2/3$.
3. $A = (5, -1)$, $B = (-4, -5)$, $r = 1/5$.
4. $A = (2, 4)$, $B = (-5, 2)$, $r = 2/5$.
5. $A = (-4, 1)$, $B = (3, 8)$, $r = 3$.
6. $A = (-6, 2)$, $B = (4, 4)$, $r = 5/2$.

In Problems 7–10, find the midpoint of the segment AB.

7. $A = (5, -2)$, $B = (-1, 4)$.
8. $A = (-3, 3)$, $B = (1, 5)$.
9. $A = (4, -1)$, $B = (3, 3)$.
10. $A = (-1, 4)$, $B = (0, 2)$.
11. If $A = (3, 5)$, $P = (6, 2)$, and $\overline{AP}/\overline{AB} = 1/3$, find B.
12. If $P = (4, 7)$, $B = (2, -1)$, and $\overline{AP}/\overline{AB} = 2/5$, find A.
13. If $P = (2, -5)$, $B = (4, -3)$, and $\overline{AP}/\overline{AB} = 1/2$, find A.
14. If $A = (3, 3)$, $P = (5, 2)$, and $\overline{AP}/\overline{AB} = 3/5$, find B.

In Problems 15–18, find the point P between A and B such that AB is divided in the given ratio.

15. $A = (5, -3)$, $B = (-1, 6)$, $\overline{AP} : \overline{PB} = 1 : 2$.
16. $A = (-1, -3)$, $B = (-8, 11)$, $\overline{AP} : \overline{PB} = 3 : 4$.
17. $A = (2, -1)$, $B = (4, 5)$, $\overline{AP} : \overline{PB} = 2 : 3$.
18. $A = (5, 8)$, $B = (2, -1)$, $\overline{AP} : \overline{PB} = 5 : 1$.
19. If $P = (4, -1)$ is the midpoint of the segment AB, where $A = (2, 5)$, find B.
20. Find the point of intersection of the medians of the triangle with vertices $(5, 2)$, $(0, 4)$, and $(-1, -1)$.
21. Find the point of intersection of the diagonals of the parallelogram with vertices $(1, 1)$, $(4, 1)$, $(3, -2)$, and $(0, -2)$.
22. Find the center and radius of the circle circumscribed about the right triangle with vertices $(1, 1)$, $(1, 4)$, and $(7, 4)$.
23. The point $(1, 4)$ is at a distance 5 from the midpoint of the segment joining $(3, -2)$ and $(x, 4)$. Find x.
24. Prove Theorem 1.3.

1.4

Inclination and Slope

An important concept in the description of a line and one that is used quite extensively throughout calculus has to do with the inclination of a line. First let us recall the convention from trigonometry which states that angles measured in the counterclockwise direction are positive, those measured in the clockwise direction are negative. Thus we have the following definition.

Definition

*The **inclination** of a line that intersects the x axis is the measure of the smallest nonnegative angle which the line makes with the positive end of the x axis. The inclination of a line parallel to the x axis is 0.*

We shall use the symbol θ to represent an inclination. The inclination of a line is always less than 180°, or π radians, and every line has an inclination. Thus, for any line,

$$0° \leq \theta < 180° \quad \text{or} \quad 0 \leq \theta < \pi.$$

Figure 1.10 shows several lines with their inclinations. Note that the angular measure is given in both degrees and radians. Although there is no reason to show preference for one over the other at this time, radian measure is the preferred way of representing an angle in more advanced courses.

$m = \tan \theta = \dfrac{y_2 - y_1}{x_2 - x_1}$

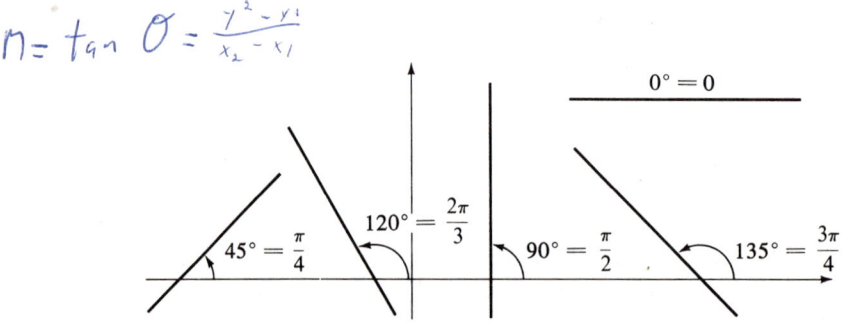

Figure 1.10

While the inclination of a line may seem like a simple representation, we cannot, in general, find a simple relationship between the inclination of a line and the coordinates of points on it without resorting to tables of trigonometric functions. Thus, we consider another expression related to the inclination—namely, the slope of a line.

1.4 Inclination and Slope

Definition

*The **slope** m of a line is the tangent of the inclination; thus,*

$$m = \tan \theta.$$

While it is possible for two different angles to have the same tangent, it is not possible for lines having two different inclinations to have the same slope. The reason for this is the restriction on the inclination, $0° \leq \theta < 180°$. Nevertheless, one minor problem does arise from the use of slope; not every line has a slope, since not every angle has a tangent; however, the only possible angle of inclination which does not have a tangent is 90°. Thus vertical lines have inclination 90° but no slope. *Do not confuse "no slope" with "zero slope."* A horizontal line definitely has a slope and that slope is the number 0, but there is no number at all (not even 0) which is the slope of a vertical line. Some might object to this nonexistence of tan 90° by saying that it is "infinity," or "∞." However, infinity is not a number. Also, while the symbol ∞ is quite useful in calculus when dealing with limits, its use in algebra or an algebraic development of trigonometry leads to trouble.

While the nonexistence of the slope of certain lines is somewhat bothersome, it is more than counterbalanced by the simple relationship between the slope and the

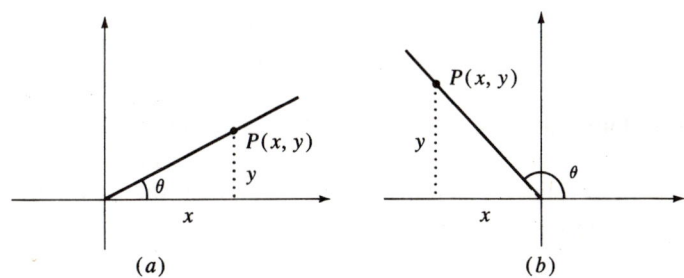

Figure 1.11

coordinates of a pair of points on the line. Recall that if θ is as shown in either of the two positions in Figure 1.11, then

$$\tan \theta = \frac{y}{x}.$$

Unfortunately, the lines with which we are dealing are not always so conveniently placed. Suppose we have a line with a pair of points, $P_1: (x_1, y_1)$ and $P_2: (x_2, y_2)$, on it (see Figure 1.12). If we place a pair of axes parallel to the old axes, with P_1 as the new origin, then the coordinates of P_2 with respect to this new coordinate system are $x = x_2 - x_1$ and $y = y_2 - y_1$. Now θ is situated in a position that allows us to use the definition of $\tan \theta$ and state the following theorem.

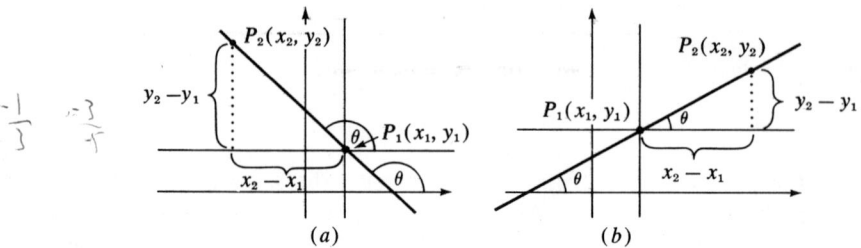

Figure 1.12

Theorem 1.4

A line through $P_1: (x_1, y_1)$ and $P_2: (x_2, y_2)$, where $x_1 \neq x_2$, has slope

$$m = \frac{y_2 - y_1}{x_2 - x_1} = \frac{y_1 - y_2}{x_1 - x_2}.$$

Example

Find the slope of the line containing $P_1: (1, 5)$ and $P_2: (7, -7)$.

$$m = \frac{y_2 - y_1}{x_2 - x_1} = \frac{-7 - 5}{7 - 1} = \frac{-12}{6} = -2.$$

Since a vertical line has no slope, Theorem 1.4 does not hold in that case; however, $x_1 = x_2$ for any pair of points on a vertical line, and the right-hand side of the slope formula is also nonexistent. Thus there is no slope when the right-hand side of the slope formula does not exist.

1.5

Parallel and Perpendicular Lines

If two nonvertical lines are parallel, they must have the same inclination and, thus, the same slope (see Figure 1.13). If two parallel lines are vertical, then neither one has slope. Similarly, if $m_1 = m_2$ or if neither line has slope, then the two lines are parallel. Thus, two lines are parallel if and only if $m_1 = m_2$ or neither line has slope.

If lines l_1 and l_2 with inclinations θ_1 and θ_2, respectively, are perpendicular (see Figure 1.14), then

1.5 Parallel and Perpendicular Lines

and
$$\theta_1 - \theta_2 = 90° \quad \text{or} \quad \theta_2 - \theta_1 = 90°,$$
$$\theta_1 = \theta_2 + 90° \quad \text{or} \quad \theta_1 = \theta_2 - 90°.$$

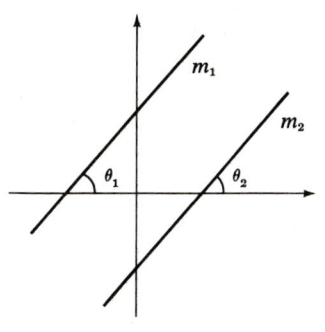

Figure 1.13

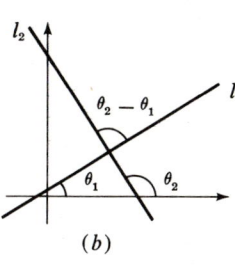

Figure 1.14

In either case (provided neither line is vertical),

$$\tan \theta_1 = -\cot \theta_2 = \frac{-1}{\tan \theta_2}$$

or

$$m_1 = \frac{-1}{m_2}.$$

Given $m_1 = -1/m_2$, the argument can be traced backward to prove that l_1 and l_2 are perpendicular. Thus if $m_1 = -1/m_2$, then l_1 and l_2 are perpendicular; and if l_1 and l_2 are perpendicular, then $m_1 = -1/m_2$ or one of the lines is horizontal and the other vertical.

Problems

In Problems 1–8, find the slope (if any) and the inclination of the line through the given points.

1. $(2, 3)$, $(5, 8)$.
2. $(-1, 4)$, $(4, 2)$.
3. $(-2, -2)$, $(4, 2)$.
4. $(3, -5)$, $(1, -1)$.
5. $(-4, 2)$, $(-4, 5)$.
6. $(2, 3)$, $(-4, 3)$.
7. (a, a), (b, b).
8. (a, a), $(-a, 2a)$.

In Problems 9–16, find the slopes of the lines through the two pairs of points; then determine whether the lines are parallel, coincident, perpendicular, or none of these.

9. $(1, -2)$, $(-2, -11)$; $(2, 8)$, $(0, 2)$.
10. $(1, 5)$, $(-2, -7)$; $(7, -1)$, $(3, 0)$.
11. $(1, 5)$, $(-1, -1)$; $(0, 3)$, $(2, 7)$.
12. $(1, 3)$, $(-1, -1)$; $(0, 2)$, $(4, -2)$.

13. $(1, 1), (4, -1); (-2, 3), (7, -3)$.
14. $(1, -4), (6, 1); (2, 3), (-1, 6)$.
15. $(1, 2), (3, 2); (4, 1), (4, -2)$.
16. $(1, 5), (1, 1); (-2, 2), (-2, 4)$.
17. If the line through $(x, 5)$ and $(4, 3)$ is parallel to a line with slope 3, find x.
18. If the line through $(x, 5)$ and $(4, 3)$ is perpendicular to a line with slope 3, find x.
19. If the line through $(x, 1)$ and $(0, y)$ is coincident with the line through $(1, 4)$ and $(2, -3)$, find x and y.
20. If the line through $(-2, 4)$ and $(1, y)$ is perpendicular to one through $(-2, 4)$ and $(x, 2)$, find a relationship between x and y.
21. If the line through $(x, 4)$ and $(3, 7)$ is parallel to one through $(x, -1)$ and $(5, 1)$, find x.
22. Show by means of slopes that $(1, 1), (4, 1), (3, -2)$, and $(0, -2)$ are the vertices of a parallelogram.
23. Show by means of slopes that $(-2, 4), (2, 0), (2, 8)$, and $(6, 4)$ are the vertices of a square.

1.6

Angle from One Line to Another

If l_1 and l_2 are two intersecting lines, then an angle from l_1 to l_2 is any angle measured from l_1 to l_2. If the measurement is in the counterclockwise direction, then the angle is positive; if it is in the clockwise direction, the angle is negative. While there are many angles from l_1 to l_2, all are related (see Figure 1.15) in that, if α is one of them,

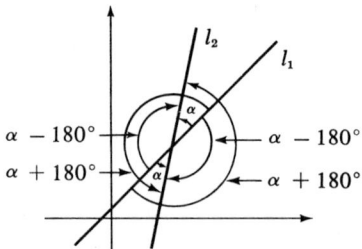

Figure 1.15

all can be expressed in the form

$$\alpha + n \cdot 180°,$$

where n is an integer (positive, negative, or zero). Since any two angles differ from each other by a multiple of 180°, they all have the same tangent.

1.6 Angle From One Line to Another

Theorem 1.5

If l_1 and l_2 are nonperpendicular lines with slopes m_1 and m_2, respectively, and α is any angle from l_1 to l_2, then

$$\tan \alpha = \frac{m_2 - m_1}{1 + m_1 m_2}.$$

Proof

Figure 1.16 shows that

$$\alpha = \theta_2 - \theta_1$$

for one of the angles α from l_1 to l_2. Thus

$$\tan \alpha = \frac{\tan \theta_2 - \tan \theta_1}{1 + \tan \theta_1 \tan \theta_2}.$$

But, since $m_1 = \tan \theta_1$ and $m_2 = \tan \theta_2$, we have

$$\tan \alpha = \frac{m_2 - m_1}{1 + m_1 m_2}.$$

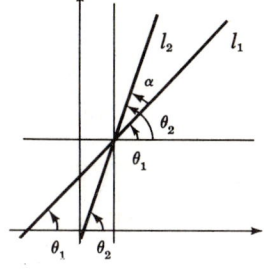

Figure 1.16

We have assumed in this argument that l_1 and l_2 intersect. If they do not, then $m_1 = m_2$. Using $m_1 = m_2$ in Theorem 1.5, we find that $\tan \alpha = 0$ and $\alpha = 0°$. Thus we shall use the convention that $\alpha = 0°$ if l_1 and l_2 are parallel. This is in agreement with the convention that $m = 0$ for horizontal lines.

The trigonometric identity used in this proof is, of course, true only when $\tan \alpha$ and $(m_2 - m_1)/(1 + m_1 m_2)$ both exist. Tan α does not exist if $\alpha = 90°$, but then $m_2 = -1/m_1$ and $1 + m_1 m_2 = 0$, which gives the one case in which $(m_2 - m_1)/(1 + m_1 m_2)$ does not exist. Thus Theorem 1.5 holds for all values of α except $\alpha = 90°$, for which case neither side of the equation exists.

Definition

The angle from l_1 to l_2 is the smallest nonnegative angle from l_1 to l_2.

Example 1

If l_1 and l_2 have slopes $m_1 = 3$ and $m_2 = -2$, respectively, find the angle from l_1 to l_2.

$$\tan \alpha = \frac{m_2 - m_1}{1 + m_1 m_2} = \frac{-2 - 3}{1 + 3(-2)} = 1.$$

Thus, from trigonometric tables, $\alpha = 45°$.

Example 2

Find the slope of the line bisecting the angle from l_1, with slope 7, to l_2, with slope 1.

18 Plane Analytic Geometry

Let m be the slope of the desired line. Since $\alpha_1 = \alpha_2$ (see Figure 1.17), we have

$$\tan \alpha_1 = \tan \alpha_2$$

and

$$\frac{m - m_1}{1 + m_1 m} = \frac{m_2 - m}{1 + m_2 m},$$

$$\frac{m - 7}{1 + 7m} = \frac{1 - m}{1 + m},$$

$$(m - 7)(1 + m) = (1 + 7m)(1 - m),$$

$$8m^2 - 12m - 8 = 0;$$

$$m = -1/2 \quad \text{or} \quad m = 2.$$

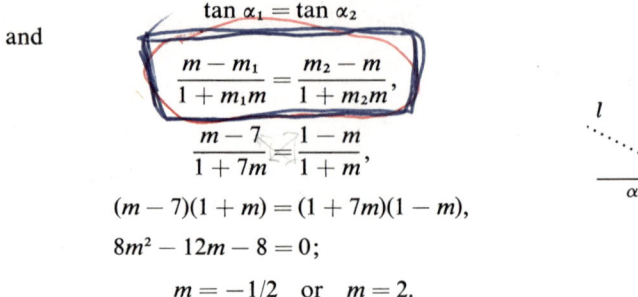

Figure 1.17

We have two answers, but obviously we want only one. Which one? Since one of them is the negative reciprocal of the other, they represent slopes of perpendicular lines, one of which is the bisector of the angle from l_1 to l_2, while the other bisects the angle from l_2 to l_1. An inspection of Figure 1.17 shows that the answer we want is $m = -1/2$.

Problems

In Problems 1–6, find the angle from l_1 to l_2 with slopes m_1 and m_2, respectively.

1. $m_1 = -2$, $m_2 = 3$.
2. $m_1 = 1$, $m_2 = 4$.
3. $m_1 = -3$, $m_2 = 2$.
4. $m_1 = 5$, $m_2 = -1$.
5. $m_1 = 10$, m_2 does not exist.
6. $m_1 = 0$, $m_2 = -1$.

In Problems 7–12, find the angle from l_1 to l_2, where l_1 and l_2 contain the points indicated.

7. l_1: (1, 4), (3, −1); l_2: (3, 2), (5, −1).
8. l_1: (2, 5), (−3, 10); l_2: (−1, −3), (3, 3).
9. l_1: (4, 5), (1, 1); l_2: (3, −3), (0, 4).
10. l_1: (1, 1), (0, 5); l_2: (4, 3), (−1, 2).
11. l_1: (2, 5), (4, 5); l_2: (1, 4), (3, −1).
12. l_1: (3, 4), (3, −1); l_2: (1, 5), (3, 5).

In Problems 13–18, find the slope of the line bisecting the angle from l_1 to l_2 with slope m_1 and m_2, respectively.

13. $m_1 = 3$, $m_2 = -2$.
14. $m_1 = 1$, $m_2 = -7$.
15. $m_1 = 2$, $m_2 = 3$.
16. $m_1 = -1$, $m_2 = 2$.
17. $m_1 = 10$, $m_2 = 13$.
18. $m_1 = 2$, $m_2 = 0$.
19. Find the interior angles of the triangle with vertices A: (1, 5), B: (3, −1), and C: (−1, −1).
20. Find the interior angles of the triangle with vertices A: (3, 2), B: (4, 5), and C: (−1, −1).
21. Find the slope of the line l_1 such that the angle from l_1 to l_2 is Arctan 2/3, where l_2 contains (2, 1) and (−4, −5).

1.7 Analytic Proofs of Geometric Theorems   19

22. Find the slope of the line l_1 such that the angle from l_1 to l_2 is 45°, where the slope of l_2 is -2.
23. Show by means of angles that $A:(1, 0)$, $B:(4, 4)$, and $C:(8, 1)$ are the vertices of an isosceles triangle.

1.7

Analytic Proofs of Geometric Theorems

Many geometric theorems can be proved quite easily using analytic geometry. The principal advantage of an analytic proof is that it rarely uses sophisticated methods; the disadvantage is that the algebraic manipulations sometimes become intolerable.

When proving a theorem analytically, remember that a plane does not come fully equipped with coordinate axes—the axes are artificially introduced to make possible the transition from geometry to algebra. Thus we may place them in *any* position we choose in order to make the coordinates of important points as simple as possible. For instance, one convenient placement of coordinate axes for dealing with theorems about triangles is shown in Figure 1.18(a). The placement used in this figure makes the algebra simpler and the proof of theorems no less general than the placement used in (b). Another convenient placement of axes is shown in Figure 1.18(c).

Although convenience may dictate the placement of the axes, we must be careful not to make simplifying assumptions about the figure which limit the generality of the argument. For example, unless we can be sure that angle A in Figure 1.18(a) is

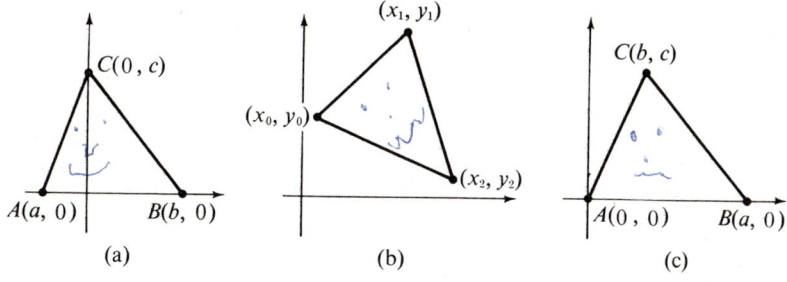

Figure 1.18

acute, we must not assume that a is negative—it may be zero or positive. Similarly, the point C of Figure 1.18(c) may be to the right of the y axis, as shown, or it may be on or to the left of the y axis. Thus b (the x coordinate of C) may be positive, negative, or zero.

Example 1

Prove analytically that the segment joining the midpoints of two sides of a triangle is parallel to and one-half the length of the third side.

The midpoints of AC and BC are D: $(a/2, c/2)$ and E: $(b/2, c/2)$, respectively (see Figure 1.19). Since the y coordinates of D and E are equal, DE is horizontal and parallel to AB. Furthermore,

$$\overline{DE} = \left|\frac{b}{2} - \frac{a}{2}\right| = \frac{1}{2}|b-a| = \frac{1}{2}\overline{AB}.$$

Of course the same argument can be used for the other two sides (with the axes placed in a similar manner).

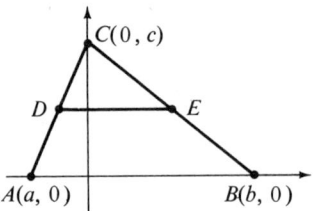

Figure 1.19

Note that the above argument is valid in any case, whether a is positive, negative, or zero. Furthermore, this generality is quite necessary. It might be argued that the axes can always be placed in such a way that the angle A is acute: choose the x axis to coincide with the side opposite the largest angle—angles A and B must then be acute. But, since the argument of Example 1 must be repeated three times using the same coordinate placement, this choice cannot be made. The argument must be valid for any angle A.

Example 2

Prove analytically that the midpoints of the sides of a quadrilateral are the vertices of a parallelogram.

1.7 Analytic Proofs of Geometric Theorems

Suppose the axes are placed in the position indicated in Figure 1.20. The midpoints are

$$E: \left(\frac{a+b}{2}, 0\right), \quad F: \left(\frac{a}{2}, \frac{c}{2}\right),$$

$$G: \left(\frac{d}{2}, \frac{c+e}{2}\right), \quad H: \left(\frac{b+d}{2}, \frac{e}{2}\right).$$

This gives the following slopes:

$$m_{EF} = \frac{c/2 - 0}{a/2 - (b+a)/2} = -\frac{c}{b}, \quad m_{GH} = \frac{(c+e)/2 - e/2}{d/2 - (b+d)/2} = -\frac{c}{b},$$

$$m_{FG} = \frac{(c+e)/2 - c/2}{d/2 - a/2} = \frac{e}{d-a}, \quad m_{EH} = \frac{e/2 - 0}{(b+d)/2 - (a+b)/2} = \frac{e}{d-a}.$$

Thus $EF \| GH$ and $FG \| EH$. $EFGH$ is a parallelogram.

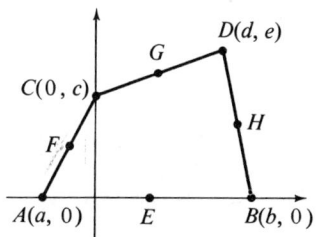

Figure 1.20   **Figure 1.21**

Example 3

Prove analytically that the diagonals of a rhombus are perpendicular.

Our first problem is that of labeling the vertices of a rhombus. Let us begin by noting that a rhombus is a special type of parallelogram. Thus we may label the vertices as shown in Figure 1.21. Now, since the sides are equal, we have

$$a = \sqrt{b^2 + c^2},$$

or

$$a^2 = b^2 + c^2.$$

From Figure 1.21 we see that the slopes of the diagonals are

$$m_{AD} = \frac{c - 0}{a + b - 0} = \frac{c}{a+b} \quad \text{and} \quad m_{BC} = \frac{c - 0}{b - a} = \frac{c}{b-a}.$$

This gives

$$m_{AD} \cdot m_{BC} = \frac{c}{a+b} \cdot \frac{c}{b-a} = \frac{c^2}{b^2-a^2} = \frac{c^2}{b^2-(b^2+c^2)} = -1,$$

which shows that the diagonals AD and BC are perpendicular.

Problems

Prove the following theorems analytically.

1. The diagonals of a rectangle are equal.
2. The diagonals of a parallelogram bisect each other.
3. The vertex and the midpoints of the three sides of an isosceles triangle are the vertices of a rhombus.
4. The sum of the squares of the four sides of a parallelogram is equal to the sum of the squares of the two diagonals.
5. One median of an isosceles triangle is an altitude.
6. The medians of an equilateral triangle are altitudes.
7. The midpoint of the hypotenuse of a right triangle is equidistant from the three vertices (and thus is the center of a circle circumscribed about the triangle).
8. The segment joining the midpoints of the nonparallel sides of a trapezoid is parallel to and one-half the sum of the lengths of the parallel sides.
9. The medians of a triangle are concurrent at a point two-thirds of the way from each vertex to the midpoint of the opposite side.
10. If the diagonals of a quadrilateral bisect each other, then the quadrilateral is a parallelogram.
11. The diagonals of an isosceles trapezoid are equal.
12. If the diagonals of a trapezoid are equal, then the trapezoid is isosceles.
13. The segments joining the midpoints of opposite sides of a quadrilateral bisect each other.
14. The segment joining the midpoints of the two diagonals of a quadrilateral is bisected by a segment joining the midpoints of a pair of opposite sides.
15. The lines joining the midpoints of the sides of a triangle subdivide it into four triangles of equal area.
16. The lines joining a vertex of a parallelogram and the midpoints of the opposite sides trisect a diagonal.
17. The base angles of an isosceles triangle are equal.
18. If the sum of the squares of two sides of a triangle equals the square of the third side, then the triangle is a right triangle.
19. If one of the parallel sides of a trapezoid is twice the length of the other side, then the diagonals intersect at a point of trisection of both of them.
20. If the lengths of the parallel sides of a trapezoid are in the ratio $1 : n$, then the point of intersection of the diagonals subdivides both of them in the ratio $1 : (n + 1)$.

1.8

Graphs and Points of Intersection

The graph of an equation in two variables x and y is simply the set of all points (x, y) in the plane whose coordinates satisfy the given equation. The determination of the graph of an equation is one of the principal problems of analytic geometry. Although we shall consider other methods in Chapter 7, we consider only point-by-point plotting here. To do this, we assign a value to either x or y, substitute the assigned value into the given equation, and solve for the other.

Example 1

Graph $x^2 + y^2 = 25$.

x	y
0	± 5
± 1	$\pm 2\sqrt{6}$
± 2	$\pm \sqrt{21}$
± 3	± 4
± 4	± 3
± 5	0

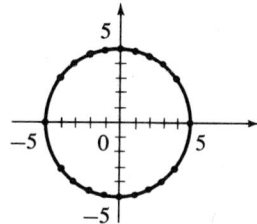

Figure 1.22

Example 2

Graph $y = |x| = \begin{cases} x, & \text{if } x \geq 0, \\ -x, & \text{if } x < 0. \end{cases}$

x	y
0	0
± 1	1
± 2	2
± 3	3
± 4	4

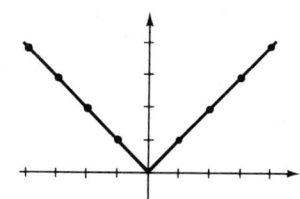

Figure 1.23

One obvious question that arises is: How many points must one plot before drawing the graph? There is no specific answer—just plot as many as are needed for a reasonable idea of what the graph looks like.

Since each point of a graph satisfies the given equation, a point of intersection of two graphs is simply a point that satisfies both equations. Thus, any such point can be found by solving the two equations simultaneously.

Example 3

Find all points of intersection of $x^2 + y^2 = 25$ and $x + y = 2$.

Solving the second equation for y and substituting into the first, we have

$$x^2 + (2 - x)^2 = 25,$$
$$2x^2 - 4x - 21 = 0,$$
$$x = \frac{4 \pm \sqrt{16 + 168}}{4} = \frac{2 \pm \sqrt{46}}{2} = 1 \pm \frac{1}{2}\sqrt{46},$$
$$y = 2 - x = 1 \mp \frac{1}{2}\sqrt{46}.$$

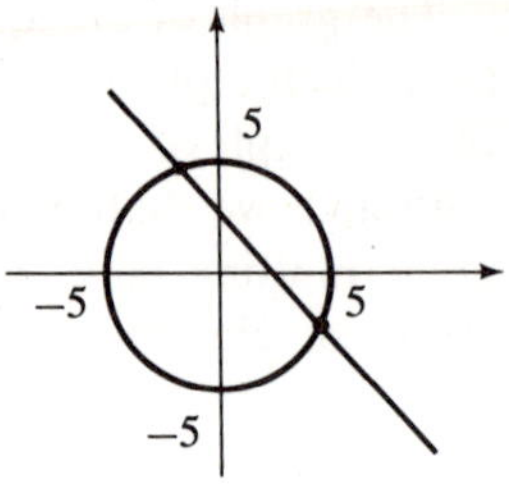

Thus, the two points of intersection are

$$\left(1 + \frac{1}{2}\sqrt{46},\ 1 - \frac{1}{2}\sqrt{46}\right) \quad \text{and} \quad \left(1 - \frac{1}{2}\sqrt{46},\ 1 + \frac{1}{2}\sqrt{46}\right).$$

Figure 1.24

Another interesting graph is represented by the equation $y = [x]$, where $[x]$ denotes the largest integer less than or equal to x. For instance, $[1] = 1$, $[3/2] = 1$, $[7/4] = 1$, $[-1/2] = -1$, and so on. The graph of $y = [x]$ is given in Figure 1.25. Note that each horizontal segment includes the left-hand end point but not the right-hand one. The relation between the weight of a letter and the amount of postage required has a similar graph.

$[x]$ lar. int $\leq x$

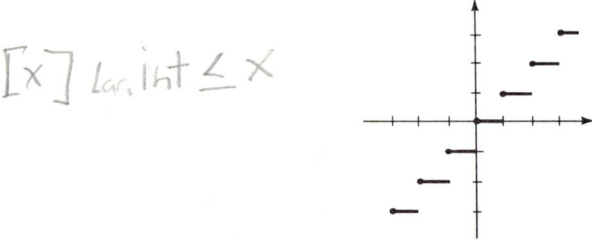

Figure 1.25

Problems

Plot the graphs of the equations in Problems 1–28.

1. $3x - 5y = 2$.
2. $y = 4x - 5$.
3. $y = 2x + 1$.
4. $x - y = 2$.
5. $x^2 + y^2 = 1$.   Circle
6. $x^2 - y^2 = 1$.
7. $x^2 - y^2 = -1$.
8. $x^2 + y^2 = 0$.

1.9 An Equation of a Locus

9. $4x^2 + y^2 = 4$. *(elipse)*
10. $4x^2 - y^2 = 4$.
11. $4x^2 - y^2 = -4$.
12. $x^3 = y^2$.
13. $y = |x| + 2$.
14. $y = |x + 2|$.
15. $y = |x| - 1$.
16. $y = |x - 1|$.
17. $|x| + |y| = 1$.
18. $|x + y| = 1$.
19. $y = [-x]$.
20. $y = [x + 1]$.
21. $y = [|x|]$.
22. $y = |[x]|$.
23. $y = x + [x]$.
24. $y = x - [x]$.
25. $y = \sqrt{x}$.
26. $\sqrt{x} + \sqrt{y} = 1$.
27. $y = \dfrac{x}{x+1}$.
28. $y = \dfrac{x+1}{x}$.

In Problems 29–38, find the points of intersection and sketch the graphs of the equations.

29. $3x - 5y = 2$
    $4x + 2y = 1$.
30. $4x + y = 3$
    $x - y = 1$.
31. $x + y = 2$
    $x^2 + y^2 = 1$.
32. $x + y = 2$
    $x^2 + y^2 = 2$.
33. $2x + y = 2$
    $x^2 - y^2 = 1$.
34. $y = x^2 + 1$
    $x + y = 1$.
35. $x^2 + y^2 = 4$
    $(x - 2)^2 + y^2 = 4$.
36. $y = \sqrt{4 - (x - 2)^2}$
    $x = \sqrt{4 - y^2}$.
37. $x^2 - y^2 = 1$
    $x^2 + y^2 = 7$.
38. $x^2 - y^2 = 3$
    $4y^2 - x^2 = 9$.
39. Graph $y = x$, $y = -x$, $y = |x|$, $y = \sqrt{x^2}$. Compare.
40. Graph $y = x$, $y = x^3$, $y = x^5$, using the same axes.
41. Graph $y = x^2$, $y = x^4$, $y = x^6$, using the same axes.
42. Sketch the "postage stamp graph." What is its equation?
43. Sketch $y = \{x\}$, where $\{x\} = \min(x - [x], 1 - x + [x])$. Interpret the values of y geometrically. *Note:* $\min(x - [x], 1 - x + [x])$ means the smaller of the two numbers $x - [x]$ and $1 - x + [x]$.

1.9

An Equation of a Locus

In the last section we considered one of the two principal problems of analytic geometry—finding the graph of an equation. Let us now consider the other—finding an equation of a locus. In other words, given a description of a curve, we want to find an equation representing that curve. Since an equation of a curve is a relationship between the x and y coordinates of every point on the curve (but no other point), we need merely consider an arbitrary point (x, y) on the curve and give the description of the curve in terms of x and y. Let us consider some examples.

Example 1

Find an equation for the set of all points in the xy plane which are equidistant from $(1, 3)$ and $(-2, 5)$.

Let (x, y) be one such point (see Figure 1.26). Thus

$$\sqrt{(x-1)^2 + (y-3)^2} = \sqrt{(x+2)^2 + (y-5)^2},$$
$$(x-1)^2 + (y-3)^2 = (x+2)^2 + (y-5)^2,$$
$$x^2 - 2x + 1 + y^2 - 6y + 9 = x^2 + 4x + 4 + y^2 - 10y + 25,$$
$$6x - 4y + 19 = 0.$$

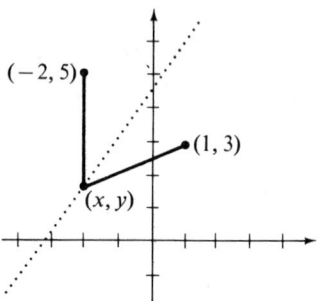

Figure 1.26

Note in the above example that the first equation given is an equation of the desired curve. Of course, we want the equation in as simple a form as possible. In carrying out the simplification, however, we must be sure that the final equation is equivalent to the original. One way to be sure of this is to simplify only by reversible steps.

It might also be noted in the example that the first step in simplifying was squaring both sides of the original equation. This is not normally a reversible operation. (If $x = 5$, then $x^2 = 25$; but if $x^2 = 25$, then $x = \pm 5$.) However, since we are dealing only with positive distances, we need only consider positive square roots in reversing the operation. Thus it is reversible. The remaining operations are clearly reversible; the first and last equations are equivalent.

There is a second way to verify that the first and last equations are equivalent. Clearly, any point (x, y) that satisfies the first equation must satisfy the subsequent ones. The only question in doubt is: Does any point satisfying the last equation also satisfy the first? In order to answer this question, let us consider a point (x, y) which satisfies the last equation. Solving for y, we have

$$y = \frac{6x + 19}{4}.$$

Let us now substitute this expression for y into both sides of the first equation. Since both sides simplify to

$$\frac{\sqrt{52x^2 + 52x + 65}}{4},$$

1.9 An Equation of a Locus

we see that the first equation is satisfied by any such point (x, y). In this way, we see that the first and last equations are equivalent.

Example 2

Find an equation for the set of all points (x, y) such that the sum of its distances from $(3, 0)$ and $(-3, 0)$ is 8.

$$\sqrt{(x-3)^2 + y^2} + \sqrt{(x+3)^2 + y^2} = 8,$$
$$\sqrt{(x-3)^2 + y^2} = 8 - \sqrt{(x+3)^2 + y^2},$$
$$x^2 - 6x + 9 + y^2 = 64 - 16\sqrt{(x+3)^2 + y^2} + x^2 + 6x + 9 + y^2,$$
$$16\sqrt{(x+3)^2 + y^2} = 64 + 12x,$$
$$4\sqrt{(x+3)^2 + y^2} = 16 + 3x,$$
$$16x^2 + 96x + 144 + 16y^2 = 256 + 96x + 9x^2,$$
$$7x^2 + 16y^2 = 112.$$

Again, the steps are reversible. Although we squared both sides twice, both sides of the equation had to be positive in each instance.

Problems

Find an equation for the set of all points (x, y) satisfying the given conditions.

1. It is equidistant from $(5, 8)$ and $(-2, 4)$.
2. It is equidistant from $(2, 3)$ and $(-4, 1)$.
3. It is on the line having slope 2 and containing the point $(3, -2)$.
4. It is on the line having slope -1 and containing the point $(-2, 5)$.
5. It is on the line containing $(3, -2)$ and $(5, 3)$.
6. It is on the line containing $(4, 2)$ and $(2, -1)$.
7. Its distance from $(5, 8)$ is 3.
8. Its distance from $(3, 1)$ is 4.
9. Its distance from $(0, 0)$ is three times its distance from $(4, 0)$.
10. Its distance from $(2, 5)$ is twice its distance from $(-3, 1)$.
11. It is the vertex of a right triangle with hypotenuse the segment from $(2, 5)$ to $(-1, 4)$.
12. It is the vertex of a right triangle with hypotenuse the segment from $(3, -2)$ to $(1, 4)$.
13. It is equidistant from $(4, 0)$ and the y axis.
14. It is equidistant from $(2, 3)$ and the x axis.
15. It is twice as far from $(3, 0)$ as it is from the y axis.

16. It is twice as far from the y axis as it is from $(3, 0)$.
17. The sum of its distances from $(0, 4)$ and $(0, -4)$ is 10.
18. The sum of its distances from $(4, -2)$ and $(2, 5)$ is 10.
19. The sum of the squares of its distances from $(3, 0)$ and $(-3, 0)$ is 50.
20. The sum of the squares of its distances from $(4, -2)$ and $(2, 5)$ is 20.
21. The difference of its distances from $(3, 0)$ and $(-3, 0)$ is 2.
22. The difference of its distances from $(4, 2)$ and $(1, -3)$ is 4.
23. The product of its distances from the coordinate axes is 4.
24. The product of its distances from $(4, 0)$ and the y axis is 4.

2

Vectors in the Plane

2.1

Directed Line Segments and Vectors

Vectors, which have direction as well as magnitude, are very important in physics and engineering. Furthermore, their use can considerably simplify geometric problems, especially in solid analytic geometry. One reason vectors are so useful is the wide range of interpretations they may be given. Since we are interested mainly in the geometric applications, vectors will be introduced geometrically by means of directed line segments.

Suppose A and B are points (not necessarily different) in space. The directed line segment from A to B is represented by $\overrightarrow{AB}$. B is called the *head* and A the *tail* of this segment. Two directed line segments $\overrightarrow{AB}$ and $\overrightarrow{CD}$ are *equivalent*, $\overrightarrow{AB} \equiv \overrightarrow{CD}$, (1) if both are of length zero or (2) if both have the same positive length, both lie on the same or parallel lines, and both are directed in the same way (see Figure 2.1 in which $\overrightarrow{AB} \equiv \overrightarrow{CD}$ and $\overrightarrow{EF} \equiv \overrightarrow{GH}$). With this information, you can easily prove the following theorem.

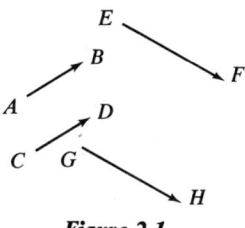

Figure 2.1

Theorem 2.1

(a) $\overrightarrow{AB} \equiv \overrightarrow{AB}$ for any directed line segment $\overrightarrow{AB}$;
(b) if $\overrightarrow{AB} \equiv \overrightarrow{CD}$, then $\overrightarrow{CD} \equiv \overrightarrow{AB}$;
(c) if $\overrightarrow{AB} \equiv \overrightarrow{CD}$ and $\overrightarrow{CD} \equiv \overrightarrow{EF}$, then $\overrightarrow{AB} \equiv \overrightarrow{EF}$.

Now let us choose an arbitrary directed line segment $\overrightarrow{AB}$. Let M_1 be the set of all directed line segments equivalent to $\overrightarrow{AB}$. Now let us choose another segment $\overrightarrow{CD}$ not in M_1 and let M_2 be the set of all directed line segments equivalent to $\overrightarrow{CD}$. Proceeding in this way, we can partition the set of all directed line segments into a collection of subsets no two of which have any element in common. These subsets are what we call *vectors*. Thus, a vector is a certain set of mutually equivalent directed line segments.

Definition

*The set of all directed line segments equivalent to a given directed line segment is a **vector v**. Any member of that set is a **representative** of v. The set of all directed line segments equivalent to one of length zero is called the **zero vector**, 0.*

It might be noted that a vector has magnitude (length) and direction, but not position. Any representative of a given vector has not only magnitude and direction but also position. Let us now consider how vectors may be combined.

Definition

*Suppose u and v are vectors. Let $\overrightarrow{AB}$ be a representative of u. Let $\overrightarrow{BC}$ be that representative of v with tail at B. The **sum u + v** of u and v is the vector w, having $\overrightarrow{AC}$ as a representative.*

Since the sum of two vectors is given in terms of representatives of those vectors, the question remains, "Is the sum well defined—that is, is it independent of the representatives used?" Theorem 2.1 and the congruence of triangles easily shows that the sum is well defined.

It might be noted that this definition is equivalent to the well-known parallelogram law for the addition of vectors (see Figure 2.2). Let us observe that the figures given here represent vectors graphically by means of representative directed line segments. In Figure 2.2 the vector **u** is represented by two equivalent directed line segments, both of which are labeled **u**.

The definition of addition of vectors can easily be extended to subtraction.

Definition

*If **u** and **v** are vectors, then **u** − **v** is the vector **w** such that **u** = **v** + **w**.*

2.1 Directed Line Segments and Vectors

Figure 2.2   Figure 2.3

This is represented graphically in Figure 2.3.

We shall use the word *scalar* for number. A scalar has magnitude but not direction.

Definition

*If **v** is a vector, then $|\mathbf{v}|$ is the length of any representative of **v**. It is called the **absolute value**, or **length**, of **v**.*

Note that the absolute value of a vector is not a vector, but a scalar.

Definition

*If k is a scalar and **v** a vector, then $k\mathbf{v}$ is a vector whose length is $|k| \cdot |\mathbf{v}|$ and whose direction is the same as or opposite to the direction of **v**, according to whether k is positive or negative. It is called a **scalar multiple** of **v**.*

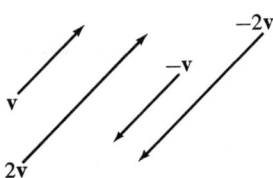

Figure 2.4

Figure 2.4 gives several examples of scalar multiples of the vector **v**.

Let us take note of the fact that we are not adding and multiplying ordinary numbers; thus it is not obvious that the rules of ordinary arithmetic hold—they must be proved from the given definitions.

Theorem 2.2

*The following properties hold for arbitrary vectors **u**, **v**, and **w** and arbitrary scalars a and b.*

(a) $\mathbf{u} + \mathbf{v} = \mathbf{v} + \mathbf{u}$.
(b) $\mathbf{u} + (\mathbf{v} + \mathbf{w}) = (\mathbf{u} + \mathbf{v}) + \mathbf{w}$.
(c) $(ab)\mathbf{v} = a(b\mathbf{v})$.
(d) $(a + b)\mathbf{v} = a\mathbf{v} + b\mathbf{v}$.
(e) $\mathbf{v} + \mathbf{0} = \mathbf{v}$.
(f) $0\mathbf{v} = \mathbf{0}$.
(g) $a\mathbf{0} = \mathbf{0}$.
(h) $|a\mathbf{v}| = |a| \cdot |\mathbf{v}|$.
(i) $|\mathbf{u} + \mathbf{v}| \leq |\mathbf{u}| + |\mathbf{v}|$.
(j) $a(\mathbf{u} + \mathbf{v}) = a\mathbf{u} + a\mathbf{v}$.

If we form the scalar multiple of the vector **v** and the scalar $1/|\mathbf{v}|$, the result is easily seen to be the unit vector (that is, the vector of length 1) in the direction of **v**. It is usually written

$$\frac{\mathbf{v}}{|\mathbf{v}|}.$$

Of special interest are the unit vectors along the axes.

Definition

If $O = (0, 0)$, $X = (1, 0)$, and $Y = (0, 1)$, then the vectors represented by $\overrightarrow{OX}$ and $\overrightarrow{OY}$ are denoted by **i** *and* **j**, *respectively, and are called* **basis vectors**.

Theorem 2.3

Every vector in the xy plane can be written in the form

$$a\mathbf{i} + b\mathbf{j}.$$

The numbers a and b are called the **components** *of the vector.*

Proof

Suppose we have a vector **v** in the plane. Let us consider the representative of **v** with its tail at the origin O (see Figure 2.5). The head is at $P:(a, b)$. Let us project P onto both axes, giving points $A:(a, 0)$ and $B:(0, b)$. Since $\overrightarrow{OA}$ represents a vector of length $|a|$ that is either in the direction of **i** or in the opposite direction, depending upon whether a is positive or negative, it represents $a\mathbf{i}$. Similarly $\overrightarrow{OB}$ represents $b\mathbf{j}$. It is clear that $\mathbf{v} = a\mathbf{i} + b\mathbf{j}$.

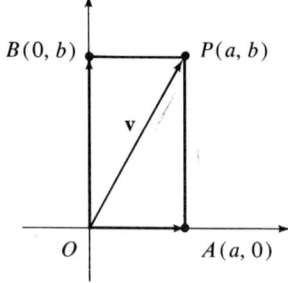

Figure 2.5

2.1 Directed Line Segments and Vectors

Theorem 2.4

If $\overrightarrow{AB}$, where $A = (x_1, y_1)$ and $B = (x_2, y_2)$, represents a vector **v** in the xy plane, then $\mathbf{v} = (x_2 - x_1)\mathbf{i} + (y_2 - y_1)\mathbf{j}$.

Proof

Let C be the point (x_2, y_1) (see Figure 2.6). Then

$$\mathbf{v} = \mathbf{u} + \mathbf{w},$$

where **u** is represented by $\overrightarrow{AC}$ and **w** by $\overrightarrow{CB}$. Since **u** is of length $|x_2 - x_1|$ and is in either the same or the opposite direction as **i**, depending upon whether $x_2 - x_1$ is positive or negative, it follows that $\mathbf{u} = (x_2 - x_1)\mathbf{i}$. Similarly $\mathbf{w} = (y_2 - y_1)\mathbf{j}$. Thus

$$\mathbf{v} = (x_2 - x_1)\mathbf{i} + (y_2 - y_1)\mathbf{j}.$$

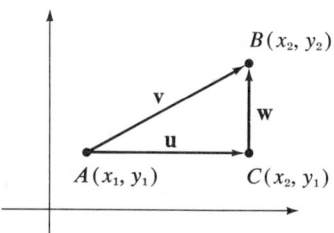

Figure 2.6

Example 1

Find the vector in the plane represented by the directed line segment from $(3, -2)$ to $(-1, 5)$.

$$\mathbf{v} = (x_2 - x_1)\mathbf{i} + (y_2 - y_1)\mathbf{j}$$
$$= (-1 - 3)\mathbf{i} + (5 + 2)\mathbf{j}$$
$$= -4\mathbf{i} + 7\mathbf{j}.$$

Theorem 2.5

$(a_1\mathbf{i} + b_1\mathbf{j}) + (a_2\mathbf{i} + b_2\mathbf{j}) = (a_1 + a_2)\mathbf{i} + (b_1 + b_2)\mathbf{j}$;
$(a_1\mathbf{i} + b_1\mathbf{j}) - (a_2\mathbf{i} + b_2\mathbf{j}) = (a_1 - a_2)\mathbf{i} + (b_1 - b_2)\mathbf{j}$;
$d(a\mathbf{i} + b\mathbf{j}) = da\mathbf{i} + db\mathbf{j}$;
$|a\mathbf{i} + b\mathbf{j}| = \sqrt{a^2 + b^2}$.

The proof is left to the student.

Example 2

If $u = 3i - j$ and $v = 4i + j$, find $u + v$, $u - v$, $3u$ and $|u|$.

$$u + v = (3 + 4)i + (-1 + 1)j = 7i,$$
$$u - v = (3 - 4)i + (-1 - 1)j = i - 2j,$$
$$3u = 3 \cdot 3i + 3(-1)j = 9i - 3j,$$
$$|u| = \sqrt{3^2 + (-1)^2} = \sqrt{10}.$$

Problems

In Problems 1–8, give in component form the vector v that is represented by $\overrightarrow{AB}$.

1. $A = (4, 3)$, $B = (-2, 1)$.
2. $A = (2, 5)$, $B = (0, 1)$.
3. $A = (-3, 2)$, $B = (4, 3)$.
4. $A = (-2, 4)$, $B = (0, 4)$.
5. $A = (1, -2)$, $B = (0, 3)$.
6. $A = (0, 2)$, $B = (1, 4)$.
7. $A = (-3, 2)$, $B = (1, -1)$.
8. $A = (4, -3)$, $B = (0, 5)$.

In Problems 9–16, give the unit vector in the direction of v.

9. $v = 3i - j$.
10. $v = 2i + 4j$.
11. $v = -i + 2j$.
12. $v = 3j$.
13. $v = i + 2j$.
14. $v = 3i - j$.
15. $v = -i + 2j$.
16. $v = 4j$.

In Problems 17–28, find the end points of the representative $\overrightarrow{AB}$ of v from the given information.

17. $v = 3i - j$, $A = (1, 4)$.
18. $v = 2i + 3j$, $A = (-1, 3)$.
19. $v = -i + 2j$, $B = (4, 2)$.
20. $v = 2i - 4j$, $B = (0, 3)$.
21. $v = 3i + 5j$, $(4, 1)$ is the midpoint of AB.
22. $v = 4i - 6j$, $(2, 5)$ is the midpoint of AB.
23. $v = i - j$, $A = (5, 1)$.
24. $v = 2i + j$, $A = (-2, 0)$.
25. $v = 3i - j$, $B = (4, 2)$.
26. $v = 2i + 2j$, $B = (1, 1)$.
27. $v = i + j$, $(2, 0)$ is the midpoint of AB.
28. $v = 2i - j$, $(0, -2)$ is the midpoint of AB.
29. If $u = 3i - j$ and $v = i + 2j$, find $u + v$. Draw a diagram showing u, v and $u + v$.
30. If $u = 2i + 3j$ and $v = 2i - j$, find $u + v$. Draw a diagram showing u, v and $u + v$.

31. If $u = i - j$ and $v = 2i + 2j$, find $u - v$. Draw a diagram showing u, v and $u - v$.
32. If $u = 3i + j$ and $v = 2i$, find $u - v$. Draw a diagram showing u, v and $u - v$.
33. If $u = i - 3j$ and $v = 2i + 4j$, find $2u + v$. Draw a diagram showing u, v and $2u + v$.
34. If $u = 4i + j$ and $v = i + 2j$, find $u + v$ and $u - v$.
35. If $u = 3i - j$ and $v = 2i - 2j$, find $u + v$ and $u - v$.
36. If $u = i - j$ and $v = 2i + j$, find $2u - v$.
37. If $u = 2i - 3j$ and $v = 4i - j$, find $3u + 2v$.
38. Prove Theorem 2.1.
39. Prove Theorem 2.2.
40. Prove Theorem 2.5.
41. Show that the sum of two vectors is well defined (see the paragraph following the definition of the sum).

2.2

Operations with Vectors

While the inclination and slope are quite convenient when considering lines, it is more convenient to use direction angles, cosines, and numbers when dealing with vectors. First let us consider the angle between two vectors.

Definition

The **angle** between two nonzero vectors u and v is the smallest angle between the representatives of u and v having their tails at the origin.

Note that the angle between two vectors is nondirected. That is, we do not consider the angle from one vector to another, which would imply a preferred direction, but rather the angle between two vectors. Thus there are no negative direction angles. If θ is the angle between two vectors, then

$$0° \leq \theta \leq 180°.$$

Definition

If v is a vector $\{\alpha, \beta\}$ is a pair of **direction angles** for v if α is the angle between v and i and β is the angle between v and j.

Vectors in the Plane

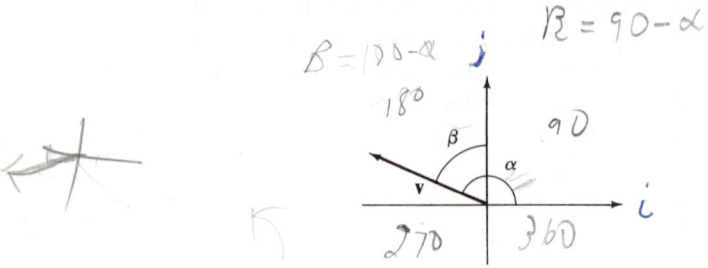

Figure 2.7

The direction angles for a vector **v** are illustrated in Figure 2.7. If a representative of **v** has its tail at the origin, then its head is in the first or fourth quadrant if α is acute and in the second or third if α is obtuse; similarly, the head is in the first or second quadrant if β is acute and in the third or fourth quadrant if β is obtuse. Furthermore, if **v** is directed upward (β is acute), then a line containing a representative of **v** has inclination α; if **v** is directed downward (β is obtuse), then such a line has inclination $180° - \alpha$.

Just as the inclination of a line is less convenient than the slope, so the direction angles of a vector are less convenient than other descriptive expressions, such as the direction cosine.

Definition

*If **v** is a vector, then $\{l, m\}$ is a pair of **direction cosines** for **v** if $l = \cos \alpha$ and $m = \cos \beta$, where $\{\alpha, \beta\}$ is a pair of direction angles.*

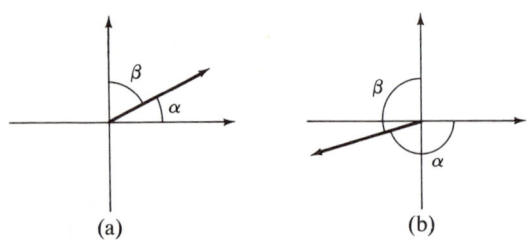

Figure 2.8

Let us first consider the relationship between the direction cosines of a vector and the slope of a line containing a representative of that vector. Suppose that a vector **v** is represented by a directed line segment with its tail at the origin and its head in the first quadrant (see Figure 2.8(a)). In this case

2.2 Operations with Vectors

or
$$\alpha + \beta = 90°,$$

$$\beta = 90° - \alpha.$$

Thus
$$\cos \beta = \sin \alpha$$

and
$$\frac{\cos \beta}{\cos \alpha} = \frac{\sin \alpha}{\cos \alpha} = \tan \alpha.$$

Since this vector is directed upward, α is the inclination of a line containing any representative of **v**. Therefore in this case,

$$m = \frac{\cos \beta}{\cos \alpha}.$$

Suppose now that the head of the vector is in the third quadrant (see Figure 2.8(b)). Then
$$\alpha + \beta = 270°,$$

or
$$\beta = 270° - \alpha.$$

Thus
$$\cos \beta = -\sin \alpha$$

and
$$\frac{\cos \beta}{\cos \alpha} = -\frac{\sin \alpha}{\cos \alpha} = -\tan \alpha.$$

But in this case the vector is directed downward; the inclination of a line containing a representative of **v** is $180° - \alpha$. Thus $m = \tan(180° - \alpha) = -\tan \alpha$. Again

$$m = \frac{\cos \beta}{\cos \alpha}.$$

The other two quadrants can be handled in the same way (see Problem 31) to give

$$m = \frac{\cos \beta}{\cos \alpha}$$

in any case.

In this discussion we found that either $\cos \beta = \sin \alpha$ or $\cos \beta = -\sin \alpha$. The fact that these results also hold in the other two quadrants leads to an important theorem that we shall use later.

Theorem 2.6

If $\{l, m\}$ is a pair of direction cosines for the vector **v**, then
$$l^2 + m^2 = 1.$$

Even more convenient for most purposes are direction numbers.

Definition

The pair $\{a, b\}$ is a pair of direction numbers for the vector **v** if there is a nonzero constant k such that $a = kl$ and $b = km$ where $\{l, m\}$ is a pair of direction cosines for **v**.

Of course, direction cosines are also direction numbers with $k = 1$. In any event, Theorem 2.6 allows us to find the direction cosines (except for the signs) from the direction numbers.
$$a^2 + b^2 = k^2 l^2 + k^2 m^2 = k^2(l^2 + m^2) = k^2.$$

Thus $k = \pm\sqrt{a^2 + b^2}$. The choice of the sign must be made on the basis of additional information.

Example 1

If the vector **v** has direction numbers $\{-4, 3\}$ and is directed downward, find the direction cosines.
$$k^2 = a^2 + b^2 = 16 + 9 = 25,$$
$$k = \pm 5.$$

Since **v** is directed downward, β is obtuse and m is negative. Thus k must be negative.
$$l = \frac{-4}{-5} = \frac{4}{5}, \quad m = \frac{3}{-5} = -\frac{3}{5}.$$

The relationship between the direction numbers $\{a, b\}$ of a vector **v** and the slope m of a line containing a representative of **v** is
$$m = \frac{\cos \beta}{\cos \alpha} = \frac{k \cos \beta}{k \cos \alpha} = \frac{b}{a}.$$

Direction numbers are particularly convenient because a vector may be represented directly using them.

Theorem 2.7

If $\mathbf{v} = a\mathbf{i} + b\mathbf{j}$ $(\mathbf{v} \neq 0)$, then $\{a, b\}$ is a pair of direction numbers for **v**.

2.2 Operations with Vectors

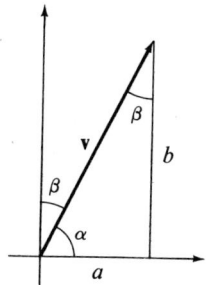

Figure 2.9

From Figure 2.9 we have

$$\cos \alpha = \frac{a}{|\mathbf{v}|}, \quad \cos \beta = \frac{b}{|\mathbf{v}|},$$

or

$$a = |\mathbf{v}|\cos \alpha, \quad b = |\mathbf{v}|\cos \beta.$$

Thus a and b are nonzero multiples ($k = |\mathbf{v}|$) of the direction cosines. Of course, Figure 2.9 considers only the case in which the head of the vector is in the first quadrant; however, the other cases can be handled similarly.

Example 2

Find a pair of direction numbers for the vector with head (3, 5) and tail (1, -2).

$$\mathbf{v} = (3 - 1)\mathbf{i} + (5 + 2)\mathbf{j} = 2\mathbf{i} + 7\mathbf{j}.$$

One pair of direction numbers is $\{2, 7\}$.

Example 3

The following forces are exerted on an object: 5 lb to the right, 10 lb upward, 2 lb upward and to the right, inclined to the horizontal at an angle of 30°. What single force is equivalent to them?

The three given forces are first represented by vectors; the equivalent force is their sum.

$$\mathbf{v}_1 = 5\mathbf{i} \quad \text{and} \quad \mathbf{v}_2 = 10\mathbf{j}.$$

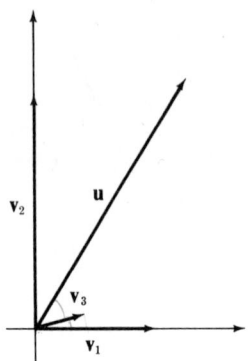

Figure 2.10

For the third vector we have the direction angles $\{30°, 60°\}$ and $|\mathbf{v}_3| = 2$ (see Figure 2.10). Thus

$$\mathbf{v}_3 = k\cos 30° \,\mathbf{i} + k\cos 60° \,\mathbf{j} = \frac{k\sqrt{3}}{2}\mathbf{i} + \frac{k}{2}\mathbf{j},$$

where k is chosen so that $|\mathbf{v}_3| = 2$ and k is positive.

$$|\mathbf{v}_3| = \sqrt{\frac{3k^2}{4} + \frac{k^2}{4}} = \sqrt{k^2} = k = 2.$$

This gives $\mathbf{v}_3 = \sqrt{3}\,\mathbf{i} + \mathbf{j}$.

$$\mathbf{u} = \mathbf{v}_1 + \mathbf{v}_2 + \mathbf{v}_3 = (5+\sqrt{3})\mathbf{i} + 11\mathbf{j},$$

$$|\mathbf{u}| = \sqrt{(5+\sqrt{3})^2 + 11^2} = \sqrt{149 + 10\sqrt{3}} = 12.9,$$

$$\cos\alpha = \frac{5+\sqrt{3}}{\sqrt{149+10\sqrt{3}}} = 0.5220, \quad \alpha = 58°30'.$$

Thus the three given forces are equivalent to a single force of 12.9 lb directed upward to the right at an angle of 58°30′ with the x axis.

Problems

In Problems 1–6, a pair of direction angles for a vector $\mathbf{v}$ is given. Find the inclination and slope of a line containing a representative of $\mathbf{v}$.

1. $\{30°, 120°\}$.
2. $\{45°, 45°\}$.

2.2 Operations with Vectors

3. $\{120°, 30°\}$.
4. $\{180°, 90°\}$.
5. $\{135°, 135°\}$.
6. $\{90°, 0°\}$.

In Problems 7–12, a pair of direction numbers for a vector v is given. Find the slope and inclination of a line containing a representative of v.

7. $\{4, 3\}$.
8. $\{\sqrt{3}, -1\}$.
9. $\{-2, -5\}$.
10. $\{4, 2\}$.
11. $\{-1, 3\}$.
12. $\{2, -6\}$.

In Problems 13–18, the slope of a line is given. A vector v with the given length and directed as indicated lies along the line (the line contains a representative of v). Find v.

13. $m = 3$; $|v| = 4$; v is directed upward.
14. $m = -1$; $|v| = \sqrt{2}$; v is directed downward.
15. $m = 0$; $|v| = 3$; v is directed to the left.
16. $m = -4$; $|v| = 2$; v is directed upward.
17. $m = 6$; $|v| = 5$; v is directed to the left.
18. $m = 1/2$; $|v| = 3$; v is directed downward.

In Problems 19–24, the given forces are acting on a body. What single force is equivalent to them?

19. 5 lb to the right; 10 lb upward.
20. 4 lb downward; 6 lb to the right; 2 lb upward.
21. 3 lb downward; 4 lb to the right and inclined upward at an angle of 45° with the horizontal.
22. 4 lb to the left; 5 lb to the right and inclined upward at an angle of 60° with the horizontal.
23. $\mathbf{f}_1 = 4\mathbf{i} + 3\mathbf{j}$, $\mathbf{f}_2 = \mathbf{i} - 2\mathbf{j}$, $\mathbf{f}_3 = \mathbf{i} + \mathbf{j}$.
24. $\mathbf{f}_1 = 2\mathbf{i} - \mathbf{j}$, $\mathbf{f}_2 = 3\mathbf{i} + 4\mathbf{j}$, $\mathbf{f}_3 = -\mathbf{i} + 2\mathbf{j}$.

In Problems 25–30, the given forces are acting on a body. What additional force will result in equilibrium? (A set of forces is in equilibrium if the sum of all of them is the zero vector.)

25. 2 lb to the left; 5 lb upward.
26. 4 lb to the right; 6 lb upward; 6 lb to the left.
27. 3 lb to the right; 5 lb to the right and upward inclined at an angle of 45° with the horizontal.
28. 3 lb upward; 2 lb to the right and inclined upward at an angle of 30° with the horizontal; 4 lb to the left and inclined downward at an angle of 60° with the horizontal.
29. $\mathbf{f}_1 = 2\mathbf{i} + \mathbf{j}$, $\mathbf{f}_2 = \mathbf{i} - 3\mathbf{j}$, $\mathbf{f}_3 = -3\mathbf{i} + \mathbf{j}$.
30. $\mathbf{f}_1 = 4\mathbf{i} - \mathbf{j}$, $\mathbf{f}_2 = \mathbf{i} + 3\mathbf{j}$, $\mathbf{f}_3 = 2\mathbf{i} - 5\mathbf{j}$.
31. Suppose l is a line containing a representative of a vector v, m is the slope of l and $\{\alpha, \beta\}$ is a pair of direction angles for v. Show that

$$m = \frac{\cos \beta}{\cos \alpha}.$$

32. Prove Theorem 2.6.
33. Prove that if $l^2 + m^2 = 1$, then $\{l, m\}$ is a pair of direction cosines for some vector.

2.3

The Dot Product

We have considered the sums and differences of pairs of vectors, but the only product considered so far is the scalar multiple—the product of a scalar and a vector. We now shall consider the product of two vectors. There are two different product operations for a pair of vectors, the dot product and the cross product. We shall take up the dot product in this section but defer a discussion of the cross product to Chapter 11 (since it requires three dimensions). First let us consider the angle between two vectors (which was defined in the previous section).

Theorem 2.8

If $\mathbf{u} = a_1\mathbf{i} + b_1\mathbf{j}$ and $\mathbf{v} = a_2\mathbf{i} + b_2\mathbf{j}$ ($\mathbf{u} \neq \mathbf{0}$ and $\mathbf{v} \neq \mathbf{0}$) and if θ is the angle between them, then

$$\cos \theta = \frac{a_1 a_2 + b_1 b_2}{|\mathbf{u}| \cdot |\mathbf{v}|}.$$

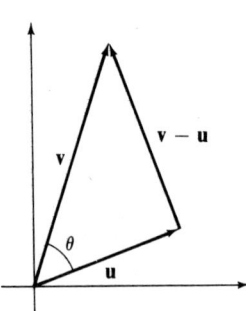

Figure 2.11

Proof

By the law of cosines (see Figure 2.11),

2.3 The Dot Product

$$|\mathbf{v}-\mathbf{u}|^2 = |\mathbf{u}|^2 + |\mathbf{v}|^2 - 2|\mathbf{u}|\cdot|\mathbf{v}|\cos\theta.$$

Since $\mathbf{v} - \mathbf{u} = (a_2 - a_1)\mathbf{i} + (b_2 - b_1)\mathbf{j}$, we have

$$(a_2 - a_1)^2 + (b_2 - b_1)^2 = a_1^2 + b_1^2 + a_2^2 + b_2^2 - 2|\mathbf{u}|\cdot|\mathbf{v}|\cos\theta,$$

$$|\mathbf{u}|\cdot|\mathbf{v}|\cos\theta = a_1 a_2 + b_1 b_2,$$

$$\cos\theta = \frac{a_1 a_2 + b_1 b_2}{|\mathbf{u}|\cdot|\mathbf{v}|}.$$

Example 1

Find the cosine of the angle between $\mathbf{u} = 3\mathbf{i} - 4\mathbf{j}$ and $\mathbf{v} = 5\mathbf{i} + 12\mathbf{j}$.

$$\cos\theta = \frac{a_1 a_2 + b_1 b_2}{|\mathbf{u}|\cdot|\mathbf{v}|}$$

$$= \frac{3\cdot 5 + (-4)\cdot 12}{\sqrt{9+16}\sqrt{25+144}}$$

$$= -\frac{33}{65}.$$

The dot product is closely related to $\cos\theta$.

Definition

If $\mathbf{u} = a_1\mathbf{i} + b_1\mathbf{j}$ and $\mathbf{v} = a_2\mathbf{i} + b_2\mathbf{j}$, then the **dot product (scaler product, inner product)** of $\mathbf{u}$ and $\mathbf{v}$ is

$$\mathbf{u}\cdot\mathbf{v} = a_1 a_2 + b_1 b_2.$$

Note that the dot product of two vectors is *not* another vector; it is a scalar.

Example 2

Find the dot product of $\mathbf{u} = 3\mathbf{i} - 2\mathbf{j}$ and $\mathbf{v} = \mathbf{i} + \mathbf{j}$.

$$\mathbf{u}\cdot\mathbf{v} = 3\cdot 1 - 2\cdot 1.$$
$$= 1.$$

Let us now consider some applications of the dot product.

Theorem 2.9

The vectors **u** *and* **v** *(not both* **0**) *are orthogonal (perpendicular) if and only if* **u** · **v** = 0 *(the zero vector is taken to be orthogonal to every other vector).*

This follows directly from Theorem 2.8 and the definition of the dot product. Thus we have a simple test for the orthogonality (perpendicularity) of two vectors. As we shall see later, orthogonality of vectors is an important concept.

Example 3

Determine whether or not **u** = 2**i** − **j** and **v** = **i** + 2**j** are orthogonal.

$$\mathbf{u} \cdot \mathbf{v} = 2 \cdot 1 - 1 \cdot 2,$$
$$= 0.$$

They are orthogonal, since **u** · **v** = 0.

Again there is the question of whether or not the dot product of vectors has the same properties as the product of numbers. The definition itself shows one difference, in that the dot product of two vectors is not itself a vector. While there are other differences, let us first note the similarities.

Theorem 2.10

If **u**, **v**, *and* **w** *are vectors, then*

$$\mathbf{u} \cdot \mathbf{v} = \mathbf{v} \cdot \mathbf{u},$$
$$(\mathbf{u} + \mathbf{v}) \cdot \mathbf{w} = \mathbf{u} \cdot \mathbf{w} + \mathbf{v} \cdot \mathbf{w}.$$

This is easily proved from the definitions. The proof is left to the student. It might be noted that the dot product of three vectors **u** · **v** · **w** is meaningless, since the dot product of any pair of them is a scalar.

Theorem 2.11

If **u** *and* **v** *are vectors and* θ *is the angle between them, then*

$$\mathbf{u} \cdot \mathbf{v} = |\mathbf{u}|\,|\mathbf{v}| \cos \theta,$$
$$\mathbf{v} \cdot \mathbf{v} = |\mathbf{v}|^2.$$

The proof is left to the student. We might note some special cases of this theorem.

2.3 The Dot Product

If **u** and **v** are orthogonal, $\theta = 90°$ and $\mathbf{u} \cdot \mathbf{v} = 0$, as we have seen before. If **u** and **v** are parallel, $\theta = 0°$ or $\theta = 180°$ and $\mathbf{u} \cdot \mathbf{v} = \pm |\mathbf{u}| \cdot |\mathbf{v}|$.

The projection of one vector upon another is determined by the angle between them or the dot product.

Definition

The **projection** of **u** on **v** ($\mathbf{v} \neq 0$) is a vector **w** such that if $\overrightarrow{AB}$ is a representative of **u** and $\overrightarrow{CD}$ is a representative of **v**, then a representative of **w** is a directed line segment $\overrightarrow{EF}$ such that $EF \perp AE$, $AE \perp CD$, and $BF \perp CD$ (see Figure 2.12).

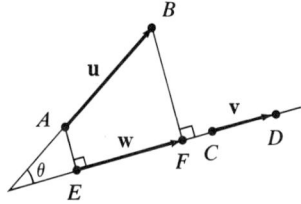

Figure 2.12

The projection of **u** on **v** is defined in terms of representatives of these vectors. Thus we again have the question of whether or not the projection is well defined. Theorem 2.1 and the congruence of triangles show that the projection of **u** on **v** is independent of the representatives considered.

Theorem 2.12

If **w** is the projection of **u** on **v** and θ is the angle between **u** and **v**, then

$$|\mathbf{w}| = \frac{|\mathbf{u} \cdot \mathbf{v}|}{|\mathbf{v}|}.$$

Proof

We can see from Figure 2.11, that

$$|\mathbf{w}| = |\mathbf{u}| |\cos \theta|$$

$$= \frac{|\mathbf{u}| |\mathbf{v}| |\cos \theta|}{|\mathbf{v}|}$$

$$= \frac{|\mathbf{u} \cdot \mathbf{v}|}{|\mathbf{v}|} \quad \text{(by Theorem 2.11)}.$$

Example 4

Find the projection **w** of $\mathbf{u} = 3\mathbf{i} - \mathbf{j}$ on $\mathbf{v} = \mathbf{i} + \mathbf{j}$.

By Theorem 2.12,

$$|w| = \frac{|u \cdot v|}{|v|}$$

$$= \frac{|2|}{\sqrt{2}}$$

$$= \sqrt{2}.$$

Since w and v have the same (or opposite) directions, there is a number c such that

$$w = cv$$
$$= ci + cj.$$

Thus $|w| = \sqrt{2} = \sqrt{2c^2}$ and

$$c = \pm 1.$$

Since $u \cdot v$ is positive, $\cos \theta$ is positive. Thus $\theta < 90°$ and w and v have the same direction. Thus $c = 1$ and $w = i + j$.

Problems

In Problems 1–8, find the angle θ between the given vectors.

1. $u = 3i - j$, $v = i + 2j$.
2. $u = 4i + j$, $v = i + 2j$.
3. $u = -i + 2j$, $v = 2i + j$.
4. $u = i + j$, $v = 2i - j$.
5. $u = 2i - j$, $v = i + 2j$.
6. $u = 3i + 2j$, $v = i - j$.
7. $u = 2i - 2j$, $v = 4i + j$.
8. $u = i + j$, $v = 2i + 4j$.

In Problems 9–16, find $u \cdot v$ and indicate whether or not u and v are orthogonal.

9. $u = i - j$, $v = 2i + j$.
10. $u = 2i + j$, $v = i - 2j$.
11. $u = 3i + 2j$, $v = 2i - j$.
12. $u = 2i - 4j$, $v = 2i + j$.
13. $u = i - j$, $v = 3i + 4j$.
14. $u = i + j$, $v = 2i - 3j$.
15. $u = 2i - 3j$, $v = 3i + j$.
16. $u = 4i$, $v = i + j$.

In Problems 17–24, find the projection of u on v.

17. $u = 2i - j$, $v = i + j$.
18. $u = i - 3j$, $v = 2i + j$.
19. $u = 2i + 4j$, $v = i - 2j$.
20. $u = 4i + j$, $v = 2i + j$.
21. $u = i - j$, $v = 2i + j$.
22. $u = 2i - 3j$, $v = 3i + 2j$.
23. $u = 2i + j$, $v = 4i - 2j$.
24. $u = 3i - j$, $v = 2i + 2j$.

In Problems 25–36, determine the value(s) of a so that the given conditions are satisfied.

25. $u = 3i - j$, $v = i + aj$, u and v are perpendicular.
26. $u = i + j$, $v = 3i - aj$, u and v are perpendicular.
27. $u = i - 2j$, $v = ai + j$, u and v are parallel.
28. $u = ai - j$, $v = 2i + aj$, u and v are parallel.

2.3 The Dot Product

29. $\mathbf{u} = a\mathbf{i} + 2\mathbf{j}$, $\mathbf{v} = \mathbf{i} - \mathbf{j}$, the angle between $\mathbf{u}$ and $\mathbf{v}$ is $\pi/3$.
30. $\mathbf{u} = 3\mathbf{i} - a\mathbf{j}$, $\mathbf{v} = 2\mathbf{i} + \mathbf{j}$, the angle between $\mathbf{u}$ and $\mathbf{v} = \pi/4$.
31. $\mathbf{u} = 4\mathbf{i} + \mathbf{j}$, $\mathbf{v} = 2\mathbf{i} + a\mathbf{j}$, $\mathbf{u}$ and $\mathbf{v}$ are perpendicular.
32. $\mathbf{u} = 2\mathbf{i} - \mathbf{j}$, $\mathbf{v} = a\mathbf{i} + \mathbf{j}$, $\mathbf{u}$ and $\mathbf{v}$ are perpendicular.
33. $\mathbf{u} = \mathbf{i} + \mathbf{j}$, $\mathbf{v} = a\mathbf{i} - \mathbf{j}$, $\mathbf{u}$ and $\mathbf{v}$ are parallel.
34. $\mathbf{u} = a\mathbf{i} + 3\mathbf{j}$, $\mathbf{v} = 2\mathbf{i} + \mathbf{j}$, $\mathbf{u}$ and $\mathbf{v}$ are parallel.
35. $\mathbf{u} = 2\mathbf{i} + \mathbf{j}$, $\mathbf{v} = a\mathbf{i} - \mathbf{j}$, the angle between $\mathbf{u}$ and $\mathbf{v}$ is $2\pi/3$.
36. $\mathbf{u} = \mathbf{i} - \mathbf{j}$, $\mathbf{v} = 4\mathbf{i} + a\mathbf{j}$, the angle between $\mathbf{u}$ and $\mathbf{v}$ is $\pi/6$.

In Problems 37–40, let $\mathbf{u}$ be represented by $\overrightarrow{AB}$, $\mathbf{v}$ by $\overrightarrow{AC}$ and $\mathbf{w}$ by $\overrightarrow{BC}$. Find the projections of $\mathbf{v}$ and $\mathbf{w}$ on $\mathbf{u}$.

37. $A = (0, 0)$, $B = (1, 4)$, $C = (2, -1)$.
38. $A = (2, 3)$, $B = (-3, -1)$, $C = (4, 2)$.
39. $A = (4, 1)$, $B = (3, -1)$, $C = (0, 2)$.
40. $A = (2, 0)$, $B = (5, 5)$, $C = (3, 5)$.
41. Prove Theorem 2.9.
42. Prove Theorem 2.10.
43. Prove Theorem 2.11.

3

The Line

3.1

Point-Slope and Two-Point Forms

The last section of Chapter 1 dealt with the problem of finding an equation of a curve from a description of it. In this chapter, as well as the next two, we shall consider this problem in more detail. Let us begin with a consideration of the line. The two simplest ways of determining a line are by a pair of points and by one point and the slope. Thus, if a line is described in either of these ways, we should be able to give an equation for it. We begin with a line described by its slope and a point on it.

Theorem 3.1

(*Point-slope form of a line.*) A line that has slope m and contains the point (x_1, y_1) has equation
$$y - y_1 = m(x - x_1).$$

Proof

Let (x, y) be any point different from (x_1, y_1) on the given line (see Figure 3.1).

3.1 Point-Slope and Two-Point Forms

Since the line has slope, it is not vertical. Thus $x \neq x_1$, which gives

$$m = \frac{y - y_1}{x - x_1}$$

and

$$y - y_1 = m(x - x_1).$$

Although the formula was derived only for points on the line different from the given point (x_1, y_1), it is easily seen that (x_1, y_1) also satisfies the equation. Thus, every point on the line satisfies the equation. Suppose now that the point (x_2, y_2) satisfies the equation—that is,

$$y_2 - y_1 = m(x_2 - x_1).$$

If $x_2 = x_1$, then $y_2 - y_1 = 0$, or $y_2 = y_1$. In this case, $(x_2, y_2) = (x_1, y_1)$, which is on the line. If $x_2 \neq x_1$, then

$$\frac{y_2 - y_1}{x_2 - x_1} = m.$$

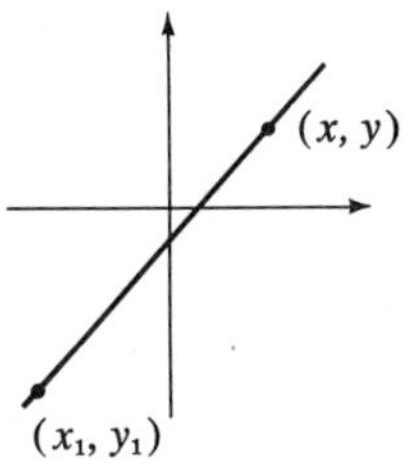

Figure 3.1

Thus, the slope of the line joining (x_1, y_1) and (x_2, y_2) is m, and this line has the point (x_1, y_1) in common with the given line. Thus, (x_2, y_2) is on the given line since there can be only one line with slope m containing (x_1, y_1).

Example 1

Find an equation of the line through $(3, -2)$ with slope 4.

$$y - y_1 = m(x - x_1),$$
$$y - (-2) = 4(x - 3),$$
$$4x - y - 14 = 0.$$

Of course vertical lines cannot be represented by the point-slope form, since they have no slope. Again, remember that "no slope" does not mean "zero slope." A horizontal line has $m = 0$, and it can be represented by the point-slope form, which gives $y - y_1 = 0$. There is no x in the resulting equation! But the points on a horizontal line satisfy the condition that they all have the same y coordinate, no matter what the x coordinate is. Similarly, the points on a vertical line satisfy the condition that all have the same x coordinate. Thus, if (x_1, y_1) is one point on a vertical line, then $x = x_1$, or $x - x_1 = 0$ for every point (x, y) on it.

Example 2

Find an equation of the vertical line through $(5, -2)$.

Since the x coordinate of the given point is 5, all points on the line have x coordinates 5. Thus,

$$x = 5 \quad \text{or} \quad x - 5 = 0.$$

Theorem 3.2

(*Two-point form of a line.*) A line through (x_1, y_1) and (x_2, y_2), $x_1 \neq x_2$, has equation

$$y - y_1 = \frac{y_2 - y_1}{x_2 - x_1}(x - x_1).$$

It might be noted that this result is often stated in the form

$$\frac{y - y_1}{x - x_1} = \frac{y_2 - y_1}{x_2 - x_1}.$$

While the symmetry of this form is appealing, the form has one serious defect—the point (x_1, y_1) is on the desired line, but it does not satisfy this equation. It does satisfy the equation of Theorem 3.2.

The proof of Theorem 3.2 follows directly from Theorem 3.1 and the fact that $m = (y_2 - y_1)/(x_2 - x_1)$, provided $x_1 \neq x_2$. Actually this follows so easily from Theorem 3.1 that you may prefer to use the earlier theorem after finding the slope from the two given points. Of course, the designation of the two points as "point 1" and "point 2" is quite arbitrary.

Example 3

Find an equation of the line through $(4, 1)$ and $(-2, 3)$.

$$y - y_1 = \frac{y_2 - y_1}{x_2 - x_1}(x - x_1),$$

$$y - 1 = \frac{3 - 1}{-2 - 4}(x - 4),$$

$$x + 3y - 7 = 0.$$

Problems

In Problems 1–16, find an equation of the line indicated and sketch the graph.

1. Through $(2, -4)$; $m = -2$.
2. Through $(5, 3)$; $m = 4$.
3. Through $(2, 2)$; $m = 1$.
4. Through $(-4, 6)$; $m = 5$.
5. Through $(9, 0)$; $m = 1$.
6. Through $(0, 3)$; $m = 2$.
7. Through $(4, -2)$; $m = 0$.
8. Through $(2, 5)$; no slope.
9. Through $(1, 4)$ and $(3, 5)$.
10. Through $(2, -1)$ and $(4, 4)$.
11. Through $(3, 3)$ and $(1, 1)$.
12. Through $(2, 1)$ and $(-3, 3)$.
13. Through $(0, 0)$ and $(1, 5)$.
14. Through $(0, 1)$ and $(-2, 0)$.
15. Through $(2, 3)$ and $(5, 3)$.
16. Through $(5, 1)$ and $(5, 3)$.
17. Find equations of the three sides of the triangle with vertices $(1, 4)$, $(3, 0)$, and $(-1, -2)$.
18. Find equations of the medians of the triangle of Problem 17.
19. Find equations of the altitudes of the triangle of Problem 17.
20. Find the vertices of the triangle with sides $x - 5y + 8 = 0$, $4x - y - 6 = 0$, and $3x + 4y + 5 = 0$.
21. Find equations of the medians of the triangle of Problem 20.

3.1 Point-Slope and Two-Point Forms

22. Find equations of the altitudes of the triangle of Problem 20.
23. Find an equation of the chord of the circle $x^2 + y^2 = 25$ which joins $(-3, 4)$ and $(5, 0)$. Sketch the circle and its chord.
24. Find an equation of the chord of the parabola $y = x^2$ which joins $(-1, 1)$ and $(2, 4)$. Sketch the curve and its chord.
25. Find an equation of the perpendicular bisector of the segment joining $(4, 2)$ and $(-2, 6)$.
26. Find an equation of the line through the points of intersection of the circles

$$x^2 + y^2 + 2x - 19 = 0 \quad \text{and} \quad x^2 + y^2 - 10x - 12y + 41 = 0.$$

    Look over your work. Is there any easier way?
27. Repeat Problem 26 for the circles

$$x^2 + y^2 + 4x + 2y + 3 = 0 \quad \text{and} \quad x^2 + y^2 - 6x - 8y + 21 = 0.$$

    What is wrong?
28. Find an equation of the line through the centers of the two circles of Problem 26.
29. What condition must the coordinates of a point satisfy in order that it be equidistant from $(2, 5)$ and $(4, -1)$?
30. Find the center and radius of the circle through the points $(1, 3)$, $(4, -6)$, and $(-3, 1)$.
31. Consider the triangle with vertices $A : (3, 1)$, $B : (0, 5)$ and $C : (7, 4)$. Find equations of the altitude and the median from A. What do your results tell us about the triangle?
32. The pressure within a partially evacuated container is being measured by means of an open end manometer. This gives the difference between the pressure in the container and atmospheric pressure. It is known that a difference of 0 mm of mercury corresponds to a pressure of 1 atmosphere and that if the pressure in the container were reduced to 0 atmospheres, a difference of 760 mm of mercury would be observed. Assuming that the difference D in mm of mercury and the pressure P in atmospheres are related by a linear relation, determine what such a relation is.
33. Knowing that water freezes at 0°C, or 32°F, that it boils at 100°C, or 212°F, and that the relation between the temperature in degrees centigrade C and in degrees fahrenheit F is linear, find that relation.
34. Show that a line through points (x_1, y_1) and (x_2, y_2) can be represented by

$$\begin{vmatrix} x & y & 1 \\ x_1 & y_1 & 1 \\ x_2 & y_2 & 1 \end{vmatrix} = 0.$$

35. Show that the points (x_1, y_1), (x_2, y_2), (x_3, y_3) are collinear if and only if

$$\begin{vmatrix} x_1 & y_1 & 1 \\ x_2 & y_2 & 1 \\ x_3 & y_3 & 1 \end{vmatrix} = 0.$$

36. Show that if no pair of the equations

$$A_1 x + B_1 y + C_1 = 0$$
$$A_2 x + B_2 y + C_2 = 0$$
$$A_3 x + B_3 y + C_3 = 0$$

    represent parallel lines, then the lines are concurrent if and only if

$$\begin{vmatrix} A_1 & B_1 & C_1 \\ A_2 & B_2 & C_2 \\ A_3 & B_3 & C_3 \end{vmatrix} = 0.$$

3.2

Slope-Intercept and Intercept Forms

The x and y intercepts of a line are the points at which the line crosses the x and y axes, respectively. These points are of the form $(a, 0)$ and $(0, b)$ (see Figure 3.2), but they are usually represented simply by a and b, since the 0's are understood by their position on the axes. We shall continue using the convention that the x and y intercepts of a line are represented by the symbols a and b, respectively. It might be noted that lines parallel to the x axis have no x intercept and those parallel to the y axis have no y intercept. While the x axis has infinitely many points in common with the x axis, we shall adopt the convention that it has no x intercept. Similarly, the y axis has no y intercept. Thus no horizontal line has an x intercept and no vertical line has a y intercept. One other special case is that of a line through the origin which is neither horizontal nor vertical; it has a single point (the origin) which is both its x and y intercept. In this case $a = b = 0$. With these special points defined, we now introduce two more forms of a line.

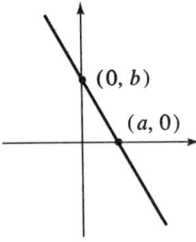

Figure 3.2

Theorem 3.3

(*Slope-intercept form of a line.*) *A line with slope m and y intercept b has equation*

$$y = mx + b.$$

Proof

Since the y intercept is really the point $(0, b)$, the use of the point-slope form gives

$$y - b = m(x - 0) \quad \text{or} \quad y = mx + b.$$

Theorem 3.4

(*Intercept form of a line.*) *A line with nonzero intercepts a and b has equation*

$$\frac{x}{a} + \frac{y}{b} = 1.$$

Proof

Since the intercepts are the points $(a, 0)$ and $(0, b)$, the line has slope

$$m = -\frac{b}{a}.$$

3.2 Slope-Intercept and Intercept Forms

Using the slope-intercept form, we have

$$y = -\frac{b}{a}x + b.$$

Dividing through by b gives

$$\frac{y}{b} = \frac{-x}{a} + 1, \quad \text{or} \quad \frac{x}{a} + \frac{y}{b} = 1.$$

It might be noted that these two forms are merely special cases of the point-slope and two-point forms; thus, the earlier forms may be used in place of these at any time. However, these forms, especially the slope-intercept form, are so convenient to use that it is well to remember them. We shall see an example of their use shortly.

Example 1

Find an equation of the line with slope 2 and y intercept 5.

$$y = mx + b,$$
$$y = 2x + 5,$$
$$2x - y + 5 = 0.$$

There is no commonly used special form for a line with a given slope and x intercept. Although one can easily be derived, it has not proved as convenient as the slope-intercept form. If you know the slope and the x intercept, simply use the point-slope form, the point being $(a, 0)$.

Example 2

Find an equation of the line with x and y intercepts 5 and -2, respectively.

$$\frac{x}{a} + \frac{y}{b} = 1,$$
$$\frac{x}{5} + \frac{y}{-2} = 1,$$
$$-2x + 5y = -10,$$
$$2x - 5y - 10 = 0.$$

Just as it was true that vertical lines could not be represented by the point-slope form, we see that vertical lines cannot be represented by the slope-intercept form, since vertical lines have neither slope nor y intercept. The intercept form is even more restrictive. Neither horizontal nor vertical lines can be put into the intercept form, since horizontal lines have no x intercept and vertical lines have no y intercept.

Furthermore, no line through the origin can be put into the intercept form, since (except for the horizontal or vertical ones) $a = b = 0$, giving 0's in the denominators.

In all of the examples we have considered so far, we used the special forms only as a starting point; the final form was always $Ax + By + C = 0$. The question arises, Can every equation representing a line be put into such a form and does every equation in such a form represent a line?

Theorem 3.5

(*General form of a line.*) *Every line can be represented by an equation of the form*

$$Ax + By + C = 0,$$

where A and B are not both zero, and any such equation represents a line.

Proof

Any line we consider is either vertical or can be put into slope-intercept form. Thus any line can be represented by either

$$x = k \quad \text{or} \quad y = mx + b.$$

Thus any line is in the form

$$x - k = 0 \quad \text{or} \quad mx - y + b = 0.$$

Both are special cases of $Ax + By + C = 0$.

Suppose we have an equation of the form $Ax + By + C = 0$, where A and B are not both 0. Let us consider two cases.

Case I: $B = 0$. Then

$$Ax + C = 0 \quad \text{and} \quad x = -\frac{C}{A}$$

(since $B = 0$ and A and B are not both 0, we know that $A \neq 0$ and we may divide by A). This represents an equation of a vertical line.

Case II: $B \neq 0$. Solving $Ax + By + C = 0$ for y, we have

$$y = -\frac{A}{B}x - \frac{C}{B}$$

(since $B \neq 0$, we may divide by B). This represents an equation of a line with slope $-A/B$ and y intercept $-C/B$.

Theorem 3.5 has the following implication for graphing: any equation of the form $Ax + By + C = 0$ represents a line, and its graph can be determined by two of its points. Since the intercepts are so easily found, finding the line through these two points (if there are two) is the quickest way of sketching a line. Of course, vertical or horizontal lines do not have two intercepts, but these are easily sketched. The only problem comes from lines through the origin. The origin is both the x and y intercept; so just find a second point in any convenient way.

3.2 Slope-Intercept and Intercept Forms

Example 3

Sketch the line $2x - 3y - 6 = 0$.

When $y = 0$, $x = 3$, and when $x = 0$, $y = -2$. We did not put the equation into intercept form in order to determine the intercepts, although we might have done so; however, we can find the intercepts by inspection by setting y and x equal to zero in turn and solving for the other. Actually this represents a convenient way of putting the line into intercept form. Since $a = 3$ and $b = -2$, the intercept form of $2x - 3y - 6 = 0$ is

$$\frac{x}{3} + \frac{y}{-2} = 1.$$

The graph of this equation is given in Figure 3.3.

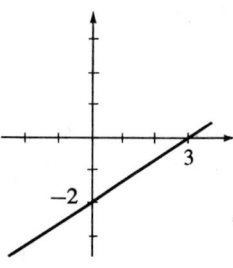

Figure 3.3

Problems

In Problems 1–18, find an equation of the line described and express it in general form with integer coefficients. Sketch the line.

1. $m = 4$, $b = 2$.
2. $m = -1$, $b = 3$.
3. $m = 5$, $b = 1/2$.
4. $m = 2/3$, $b = -1/3$.
5. $m = 3/4$, $b = 2/3$.
6. $m = -1/6$, $b = -5/4$.
7. $m = 5$, $a = -2$.
8. $m = 6$, $a = 3$.
9. $a = 4$, $b = 2$.
10. $a = -1$, $b = 3$.
11. $a = 2$, $b = 1/2$.
12. $a = 1/2$, $b = 1/2$.
13. $a = 2/3$, $b = -2/5$.
14. $a = -3/4$, $b = 2/3$.
15. $a = b = 0$, through $(2, 5)$.
16. $a = b = 0$, through $(-2, -3)$.
17. $a = 4$, no b.
18. No a, $b = -3$.
19. Find an equation of the line parallel to $2x - 5y + 1 = 0$ and containing the point $(2, 3)$.
20. Find an equation of the line parallel to $4x + y + 2 = 0$ with y intercept 3.
21. Find an equation of the line perpendicular to $x + 2y - 5 = 0$ and containing the point $(4, 1)$.
22. Find an equation of the line perpendicular to $4x - y - 3 = 0$ with x intercept 4.
23. Find the center of the circle circumscribed about the triangle with vertices $(1, 3)$, $(4, -2)$, $(-2, 1)$.
24. Find the center of the circle circumscribed about the triangle with sides $x + y = 2$, $x - y = 0$, $2x - y = 4$.
25. Find the orthocenter (points of concurrency of the altitudes) of the triangle with vertices $(1, 4)$, $(7, 3)$, $(2, -3)$.
26. Prove analytically that the altitudes of a triangle are concurrent.
27. For what value(s) of m does the line $y = mx - 5$ have x intercept 2?
28. For what value(s) of m does the line $y = mx + 2$ contain the point $(4, 5)$?

29. For what value(s) of a does the line $(x/a) - (y/2) = 1$ have slope 2?
30. For what value(s) of b does the line $(x/3) + (y/b) = 1$ have slope -4?
31. Plot the graph of $x^2 - y^2 = 0$.
32. Plot the graph of $xy = 0$.
33. Plot the graph of $x^2 - 5x + 6 = 0$.
34. Plot the graph of $(x + y - 1)(3x - y + 2) = 0$.
35. Show that $\mathbf{v} = A\mathbf{i} + B\mathbf{j}$ is perpendicular to $Ax + By + C = 0$.
36. Show that $\mathbf{v} = B\mathbf{i} - A\mathbf{j}$ is parallel to $Ax + By + C = 0$.
37. Work Problem 34 of the previous section without expanding the determinant. (*Hint:* Use Theorem 3.5.)

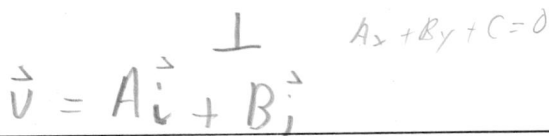

3.3

Distance from a Point to a Line

Before considering the distance from a point to a line, let us note the result of Problem 35 of the previous section: the vector $\mathbf{v} = A\mathbf{i} + B\mathbf{j}$ is perpendicular to $Ax + By + C = 0$. This perpendicularity allows us to find the distance from any point to a given line.

Theorem 3.6

The distance from the point (x_1, y_1) to the line $Ax + By + C = 0$ is

$$d = \frac{|Ax_1 + By_1 + C|}{\sqrt{A^2 + B^2}}.$$

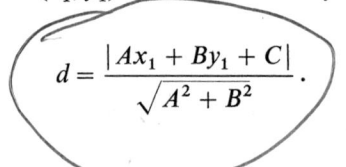

Proof

The distance we are considering here is the shortest, or perpendicular, distance. As noted above, the vector $\mathbf{v} = A\mathbf{i} + B\mathbf{j}$ is perpendicular to $Ax + By + C = 0$. Let (x, y) be a point on $Ax + By + C = 0$ and $\mathbf{u}$ the vector represented by the segment from (x, y) to (x_1, y_1) (see Figure 3.4). Thus

$$\mathbf{u} = (x_1 - x)\mathbf{i} + (y_1 - y)\mathbf{j}.$$

3.3 Distance From a Point to a Line

The length we seek is the length of the projection **w** of **u** upon **v**. By Theorem 2.12,

$$d = |\mathbf{w}| = \frac{|\mathbf{v} \cdot \mathbf{u}|}{|\mathbf{v}|} = \frac{|A(x_1 - x) + B(y_1 - y)|}{\sqrt{A^2 + B^2}}$$

$$= \frac{|Ax_1 + By_1 - (Ax + By)|}{\sqrt{A^2 + B^2}}$$

$$= \frac{|Ax_1 + By_1 + C|}{\sqrt{A^2 + B^2}}.$$

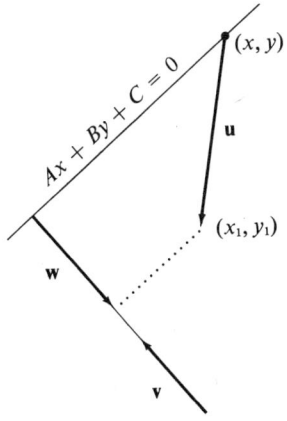

Figure 3.4

Example 1

Find the distance from the point $(1, 4)$ to the line $3x - 5y + 2 = 0$.

$$d = \frac{|Ax_1 + By_1 + C|}{\sqrt{A^2 + B^2}}$$

$$= \frac{|3 \cdot 1 - 5 \cdot 4 + 2|}{\sqrt{3^2 + (-5)^2}}$$

$$= \frac{15}{\sqrt{34}}.$$

The absolute value in the distance formula is sometimes very inconvenient in practice. We could get rid of it if we knew whether $Ax_1 + By_1 + C$ were positive or negative. The following theorem gives us a method of determining this.

Theorem 3.7

If $P(x_1, y_1)$ is a point not on the line $Ax + By + C = 0$ ($B \neq 0$), then
(a) B and $Ax_1 + By_1 + C$ agree in sign if P is above the line;
(b) B and $Ax_1 + By_1 + C$ have opposite signs if P is below the line.

Proof

Case I: $B > 0$. Let Q be the point on the given line with abscissa x_1 (see Figure 3.5). If P is above the line, then $y_1 > y$. $By_1 > By$, since $B > 0$. Therefore,

$$Ax_1 + By_1 + C > Ax_1 + By + C.$$

Since (x_1, y) is on the line,

$$Ax_1 + By + C = 0 \quad \text{and} \quad Ax_1 + By_1 + C > 0.$$

If P is below the line, all of the above inequalities are reversed and

$$Ax_1 + By_1 + C < 0.$$

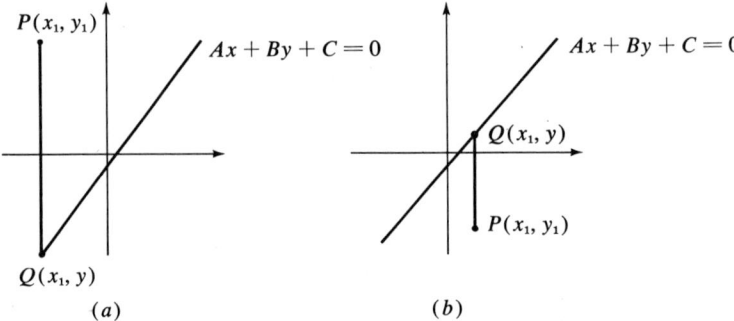

Figure 3.5

Case II: $B < 0$. If P is above the line, then $y_1 > y$. $By_1 < By$, since $B < 0$. Thus,

$$Ax_1 + By_1 + C < Ax_1 + By + C.$$

Again

$$Ax_1 + By + C = 0 \quad \text{and} \quad Ax_1 + By_1 + C < 0.$$

As with Case I, all of these inequalities are reversed if P is below the line, and

$$Ax_1 + By_1 + C > 0.$$

If $B = 0$, the line is vertical and there is no "above" nor "below." Theorem 3.7 does not apply to this case, but the distance from a point to a vertical line is easily found without using Theorem 3.6. Other methods of determining the sign of $Ax_1 + By_1 + C$ are given in Problems 32 and 33.

3.3 Distance From a Point to a Line

Example 2

Find an equation of the line bisecting the angle from $3x - 4y - 3 = 0$ to $5x + 12y + 1 = 0$.

If (x, y) is any point on the desired line (see Figure 3.6), then it is equidistant from the two given lines. By Theorem 3.6,

$$\frac{|5x + 12y + 1|}{\sqrt{5^2 + 12^2}} = \frac{|3x - 4y - 3|}{\sqrt{3^2 + (-4)^2}},$$

$$5|5x + 12y + 1| = 13|3x - 4y - 3|.$$

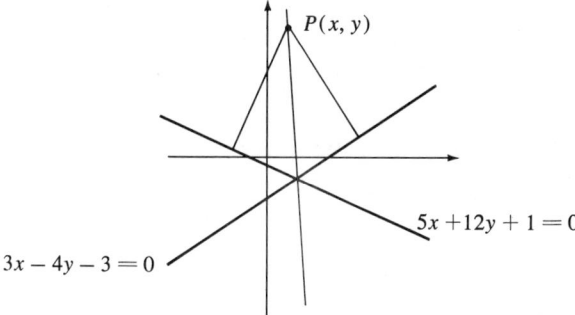

Figure 3.6

Now let us apply Theorem 3.7. Since P is above $5x + 12y + 1 = 0$ and the coefficient of y is positive, $5x + 12y + 1$ is also positive. Similarly, since P is above $3x - 4y - 3 = 0$ and B is negative,

$$3x - 4y - 3 < 0.$$

Thus

$$5(5x + 12y + 1) = -13(3x - 4y - 3) \quad \text{or} \quad 32x + 4y - 17 = 0.$$

Perhaps you object to the designation of P above both lines. Not every point on the bisector is above them. While this is true, the points on the bisector that are not above both are below both. Thus, we still have one expression positive and the other negative, and the result is the same.

It might be noted that we can avoid the use of Theorem 3.7 by considering both cases; that is, $5x + 12y + 1$ and $3x - 4y - 3$ either agree in signs or have opposite signs. We then get two answers, and Figure 3.6 indicates which is correct.

Problems

In Problems 1–10, find the distance from the given point to the given line.

1. $x + y - 5 = 0$, $(2, 5)$.
2. $2x - 4y + 2 = 0$, $(1, 3)$.

$\sqrt{2}$

3. $4x + 5y - 3 = 0$, $(-2, 4)$.
4. $x - 3y + 5 = 0$, $(1, 2)$.
5. $3x + 4y - 5 = 0$, $(1, 1)$.
6. $5x + 12y + 13 = 0$, $(0, 2)$.
7. $2x - 5y = 3$, $(3, -3)$.
8. $2x + y = 5$, $(4, -1)$.
9. $3x + 4 = 0$, $(2, 4)$.
10. $y = 3$, $(1, 5)$.
11. Find the altitudes of the triangle with vertices $(1, 2)$, $(5, 5)$, $(-1, 7)$.
12. Find the altitudes of the triangle with sides $x + y - 3 = 0$, $x - 2y + 4 = 0$, $2x + 3y = 5$.

In Problems 13–18, find an equation of the line bisecting the angle from the first line to the second.

13. $3x - 4y - 2 = 0$, $4x - 3y + 4 = 0$.
14. $8x + 15y - 5 = 0$, $5x - 12y + 1 = 0$.
15. $24x - 7y + 1 = 0$, $3x + 4y - 5 = 0$.
16. $12x + 35y - 4 = 0$, $15y - 8x + 3 = 0$.
17. $x + y - 2 = 0$, $2x - 3 = 0$.
18. $2x + y + 3 = 0$, $y + 5 = 0$.

In Problems 19–24, find the distance between the given parallel lines.

19. $2x - 5y + 3 = 0$, $2x - 5y + 7 = 0$.
20. $x + 2y - 2 = 0$, $x + 2y + 5 = 0$.
21. $2x + y + 2 = 0$, $4x + 2y - 3 = 0$.
22. $4x - y + 2 = 0$, $12x - 3y + 1 = 0$.
23. $2x - y + 1 = 0$, $2x - y - 7 = 0$.
24. $3x + 2y = 0$, $6x + 4y - 5 = 0$.

25. Find the area of the triangle of Problem 11.
26. Find the area of the triangle of Problem 12.
27. The center of the circle inscribed in a triangle is the incenter of the triangle. The center of a circle which is tangent to one side and the extensions of the other two sides is an excenter of the triangle. Find the incenter and the three excenters of the triangle with vertices $(3, 1)$, $(5, 6)$, $(-9/4, 31/10)$.
28. For what value(s) of m is the line $y = mx + 5$ at a distance 4 from the origin?
29. For what value(s) of m is the line $y = mx + 1$ at a distance 3 from $(4, 1)$?
30. For what value(s) of a is the line $(x/a) + (y/2) = 1$ at a distance 2 from the point $(5, 4)$?
31. For what value(s) of b is the line $(x/3) + (y/b) = 1$ at a distance 1 from the origin?
32. Prove that if $P(x_1, y_1)$ is a point not on the line $Ax + By + C = 0$ $(A \neq 0)$, then
    (a) A and $Ax_1 + By_1 + C$ agree in sign if P is to the right of the line;
    (b) A and $Ax_1 + By_1 + C$ have opposite signs if P is to the left of the line.
33. Prove that if $P(x_1, y_1)$ is a point not on the line $Ax + By + C = 0$ $(C \neq 0)$, then
    (a) C and $Ax_1 + By_1 + C$ agree in sign if P and the origin are on the same side of the line;
    (b) C and $Ax_1 + By_1 + C$ have opposite signs if P and the origin are on opposite sides of the line.
34. Find the center of the circle inscribed in the triangle with vertices $(0, 0)$, $(4, 0)$, $(0, 3)$.
35. Show that the line through (x_1, y_1) and parallel to $Ax + By + C = 0$ is $Ax + By - (Ax_1 + By_1) = 0$. Show that $Bx - Ay = 0$ is perpendicular to $Ax + By + C = 0$.
36. Use the results of Problem 35 to derive the distance from a point to a line without using vectors.

3.4 Families of Lines

37. Suppose that α is the inclination of a line perpendicular (or normal) to the line l and p is the directed distance of l from the origin, p being positive if l is above the origin and

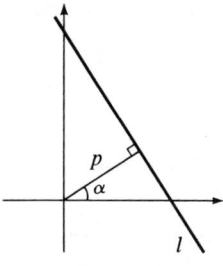

Figure 3.7

negative if l is below. Show that l can be put into the form

$$x \cos \alpha + y \sin \alpha - p = 0.$$

This is called the normal form of the line.

3.4

Families of Lines

The equation

$$y = 2x + b$$

is in the form $y = mx + b$, with $m = 2$, and thus it represents a line with slope 2 and y intercept b. But what is b? Clearly we could substitute many different values for b and get equations of many different lines. It is of interest then to consider the following set, or family, of equations representing lines:

$$M = \{y = 2x + b \mid b \text{ real}\}.$$

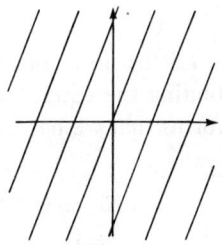

M represents a set of parallel lines all having slope 2; in fact, it represents the set of *all* lines having slope 2 (see Figure 3.8). The b in $y = 2x + b$ is called a parameter. Since the equation has a single parameter, M is called a one-parameter family of lines. Let us consider a few more examples.

Figure 3.8

Example 1

$\{y - 2 = m(x - 1) \mid m \text{ real}\}$ represents a family of lines through the point $(1, 2)$; however, it does not represent all such lines. The vertical line $x = 1$ (which has no slope) is not a member of this family. The set of *all* lines through the point $(1, 2)$ is $\{y - 2 = m(x - 1) \mid m \text{ real}\} \cup \{x = 1\}$.

Example 2

$\left\{\dfrac{x}{2} + \dfrac{y}{b} = 1 \mid b \text{ real}, b \neq 0\right\}$ represents a family of lines, all having x intercept 2 and some y intercept. It represents all such lines. However, it does not represent all lines having x intercept 2, since the line $x = 2$ is not represented, nor does it represent all lines through $(2, 0)$, since $x = 2$ and $y = 0$ are not included.

Example 3

$\{y = mx + b \mid m, b \text{ real}\}$ is a two-parameter family of lines representing all non-vertical lines.

Example 4

$\{x = k \mid k \text{ real}\}$ is the family of all vertical lines.

Example 5

$\{2x + 3y - 6 + k(4x - y + 2) = 0 \mid k \text{ real}\}$ represents a family of lines (no matter what value we choose for k, the resulting equation is linear) all containing the point of intersection of

$$2x + 3y - 6 = 0 \quad \text{and} \quad 4x - y + 2 = 0$$

(because any point satisfying $2x + 3y - 6 = 0$ and $4x - y + 2 = 0$ must satisfy

$$2x + 3y - 6 + k(4x - y + 2) = 0$$

no matter what value of k we choose). Again, it does not represent *all* such lines; the line $4x - y + 2 = 0$ is not a member of this family.

Example 6

$\{Ax + By + C = 0 \mid A, B, C \text{ real}\}$ is a three-parameter family representing all lines in the plane.

Let us now consider the use of families of lines. This concept is most useful in finding the equation of a line which cannot be represented in any of the standard forms that we have seen. Suppose we consider the following example.

Example 7

Find an equation(s) of a line(s) that contains the point $(6, 0)$ and is a distance 5 from the point $(1, 3)$.

3.4 Families of Lines

$\{y = m(x - 6) \mid m \text{ real}\}$ represents a family of lines all containing the point (6, 0). Note that it does not represent all lines containing the point (6, 0); the only one not represented is the vertical line with equation $x = 6$. Thus, the family of all lines containing (6, 0) is (see Figure 3.9)

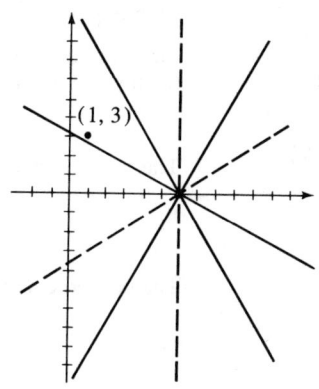

Figure 3.9

$$\{y = m(x - 6) \mid m \text{ real}\} \cup \{x = 6\}.$$

Now we must choose those members of the family that are at a distance 5 from (1, 3). We first consider those lines of the form $y = m(x - 6)$, which can be rewritten in the form

$$mx - y - 6m = 0.$$

The distance from this line to the point (1, 3) is

$$\frac{|m - 3 - 6m|}{\sqrt{m^2 + 1}} = 5.$$

Thus,

$$|-3 - 5m| = 5\sqrt{m^2 + 1},$$
$$9 + 30m + 25m^2 = 25m^2 + 25,$$
$$m = \frac{8}{15}.$$

Substituting this value back into the original equation, we get

$$y = \frac{8}{15}(x - 6),$$
$$8x - 15y - 48 = 0.$$

Now we must consider the line $x = 6$, which is a distance 5 from the point (1, 3). Thus, the two lines we want are

$$8x - 15y - 48 = 0 \quad \text{and} \quad x - 6 = 0.$$

Example 8

Find an equation(s) of the line(s) parallel to $3x - 5y + 2 = 0$ and containing the point (3, 8).

$\{3x - 5y = k \mid k \text{ real}\}$ is the family of all lines parallel to $3x - 5y + 2 = 0$ (including the given line). The member of the family which contains (3, 8) satisfies the condition

$$3 \cdot 3 - 5 \cdot 8 = k,$$
$$k = -31.$$

The equation desired is $3x - 5y + 31 = 0$. The above procedure is simple enough to carry out entirely in one's head, and a similar procedure can be used for perpendicular lines.

Example 9

Find an equation(s) of the line(s) perpendicular to $3x - 5y + 2 = 0$ and containing the point (3, 8).

$\{5x + 3y = k \mid k \text{ real}\}$ is the family of all lines perpendicular to $3x - 5y + 2 = 0$. The member that contains (3, 8) satisfies the conditions

$$5 \cdot 3 + 3 \cdot 8 = k,$$
$$k = 39.$$

The desired equation is $5x + 3y - 39 = 0$.

Problems

In Problems 1–14, describe the family of lines given. Indicate whether or not it contains *every* line of that description, and, if not, give all the lines with that description which are not included in the family.

1. $\{y - 4 = m(x + 1) \mid m \text{ real}\}$.
2. $\{y = mx - 5 \mid m \text{ real}\}$.
3. $\left\{ \dfrac{x}{2} + \dfrac{y}{b} = 1 \mid b \text{ real}, b \neq 0 \right\}$.
4. $\{x = ky \mid k \text{ real}\}$.
5. $\{Ax + By = 0 \mid A, B \text{ real}\}$.
6. $\{2x - 3y = k \mid k \text{ real}\}$.
7. $\left\{ \dfrac{x}{a} + \dfrac{y}{b} = 1 \mid a, b, \text{ real}, a \neq 0, b \neq 0 \right\}$.
8. $\{y = mx + b \mid m, b \text{ real}\}$.
9. $\{2x + 3y + 1 + k(4x + 2y - 5) = 0 \mid k \text{ real}\}$.
10. $\{x = k \mid k \text{ real}\}$.
11. $\left\{ \dfrac{x}{a} + \dfrac{y}{2a} = 1 \mid a \text{ real}, a \neq 0 \right\}$.
12. $\left\{ \dfrac{x}{a} + \dfrac{y}{3-a} = 1 \mid a \text{ real}, a \neq 0, a \neq 3 \right\}$.
13. $\{y = mx + m \mid m \text{ real}\}$.
14. $\{y - a = m(x - a) \mid a, m \text{ real}\}$.

In Problems 15–24, give, in set notation, the family described.

15. All lines parallel to $3x - 5y - 7 = 0$.
16. All lines perpendicular to $3x - 5y - 7 = 0$.
17. All lines containing (2, 5).
18. All lines with x intercept twice the y intercept.
19. All lines containing the point of intersection of $3x - 5y + 1 = 0$ and $2x + 3y - 7 = 0$.
20. All horizontal lines.
21. All lines containing the origin.

3.5 Fitting a Line to Empirical Data

22. All lines at a distance 3 from the origin.
23. All lines at a distance 5 from (6, 0).
24. All lines which form with the coordinate axes a triangle of area 4.

In Problems 25–28, find the lines satisfying the given condition that are (a) parallel and (b) perpendicular, respectively, to the given line.

25. Containing (5, 8); $3x - 5y + 1 = 0$.
26. Containing (3, 2); $2x + 3y - 7 = 0$.
27. y intercept 5; $4x + 2y - 5 = 0$.
28. x intercept 2; $3x + y + 2 = 0$.
29. Find an equation(s) of the line(s) with slope 5 at a distance 3 from the origin.
30. Find an equation(s) of the line(s) perpendicular to $3x - 4y + 1 = 0$ and at a distance 4 from (2, 3).
31. Find an equation(s) of the line(s) containing (5, 4) and at a distance 2 from $(-1, -3)$.
32. Find an equation(s) of the line(s) containing $(3, -1)$ and at a distance 4 from $(-1, 3)$.
33. Find an equation(s) of the line(s) containing (7, 1) and at a distance 5 from $(2, -5)$.
34. Find an equation(s) of the line(s) containing $(-4, 3)$ and at a distance 5 from $(-2, 2)$.
35. Find an equation(s) of the line(s) containing the point of intersection of $3x - y - 5 = 0$ and $2x + 2y - 3 = 0$ and having slope 2.
36. Find an equation(s) of the line(s) containing the point of intersection of $4x + 5y - 1 = 0$ and $3x - 2y + 1 = 0$ and the point (1, 1).
37. Find an equation(s) of the line(s) containing $(4, -3)$ such that the sum of the intercepts is 5.
38. Find an equation(s) of the line(s) with slope 3 such that the sum of the intercepts is 12.
39. Find an equation(s) of the line(s) containing (2, 3) and forming with the coordinate axes a triangle of area 16.
40. Prove analytically that the bisector of an exterior angle determined by the two equal sides of an isosceles triangle is parallel to the third side.
41. An isosceles right triangle is circumscribed about the circle with center (2, 2) and radius 2. The coordinate axes are two of the sides. What is the third?
42. An isosceles right triangle is circumscribed about the circle with center (4, 2) and radius 2. The x axis is the hypotenuse. What are the other two sides?
43. An equilateral triangle is circumscribed about the circle with center (4, 2) and radius 2. The x axis is one side. What are the other two?

3.5

Fitting a Line to Empirical Data

In experimental work one is often called upon to fit a line to a given set of empirical data. For example, the electrical resistance of a wire of a given material and diameter is directly proportional to its length—in symbols,

$$R = kL.$$

Now suppose we have found the following data from a laboratory experiment.

L (cm)	1.3	4.2	7.0	10.1	14.2
R (ohm)	11.8	32.7	58.4	81.4	115.2

It is easily seen (see Figure 3.10) that there is no line containing all of the points determined by the above data. In fact, it is unrealistic to expect a line to contain all of the points, since there must be some experimental error involved. Our problem, then, is to find the line which most nearly fits the given data or, equivalently, to find the best value of k from the given data.

One way of doing so is by the method of selected points. This is a graphical method. The points determined by the data are graphed and the line which seems to fit the data best is drawn using a transparent straightedge. Then k is determined by randomly selecting a pair of points on the line (these will not normally coincide with any of the data points) and using them to find the slope of the line (which is k).

Example 1

Determine the value of k for the data given above using the method of selected points.

The points determined by the data, as well as a line which seems to fit the data best are given in Figure 3.10. Note that the point $(0, 0)$, which was not given in the data, must be on the line because of the form of the equation. Let us choose the points $(0, 0)$ and $(10, 82.0)$ to determine k.

$$k = \frac{82.0 - 0}{10 - 0} = 8.20.$$

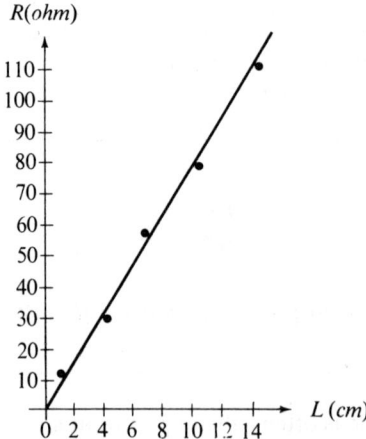

Figure 3.10

3.5 Fitting a Line to Empirical Data

The method just used has some serious drawbacks. The selection of the line which best fits the data is strictly a matter of guesswork. As such it is purely subjective—two people might differ on which line fits the data best. Another method, which eliminates the guessing, is the method of averages. Before going into this method let us define a term used in it.

Definition

*The **residual** for a given value of x is the observed y coordinate minus the computed y coordinate at that value of x.*

For example, if $y = 4x$ were found from data which included the point $(2, 8.2)$, then the residual at $x = 2$ is $r = 8.2 - 8 = 0.2$, since 8.2 is the observed value at $x = 2$ and 8 is the value computed from $y = 4x$. For Example 1, the residual at $L = 7.0$ is

$$r = 58.4 - (8.2)(7.0) = 1.0.$$

The method of averages simply directs that the value of m in $y = mx$ be chosen in such a way that the sum of the residuals is zero. Suppose we are given the data $(x_1, y_1), (x_2, y_2), \ldots, (x_n, y_n)$. Then

$$r_1 = y_1 - mx_1,$$
$$r_2 = y_2 - mx_2,$$
$$\vdots$$
$$r_n = y_n - mx_n.$$

Adding, we have

$$r_1 + r_2 + \cdots + r_n = y_1 + y_2 + \cdots + y_n - m(x_1 + x_2 + \cdots + x_n),$$

or, using abbreviations,

$$\sum r_i = r_1 + r_2 + \cdots + r_n$$
$$\sum y_i = y_1 + y_2 + \cdots + y_n$$
$$\sum x_i = x_1 + x_2 + \cdots + x_n,$$

and so we have

$$\sum r_i = \sum y_i - m \sum x_i.$$

Since $\sum r_i = 0$,

$$m = \frac{\sum y_i}{\sum x_i}.$$

Example 2

Use the method of averages to find the value of k in Example 1.

$$\sum L_i = 1.3 + 4.2 + 7.0 + 10.1 + 14.2 = 36.8,$$
$$\sum R_i = 11.8 + 32.7 + 58.4 + 81.4 + 115.2 = 299.5.$$

Thus,
$$k = \frac{\sum R_i}{\sum L_i} = \frac{299.5}{36.8} = 8.14.$$

In the preceding examples the situation was relatively simple, since we merely wanted to determine m in $y = mx$. Suppose now we consider the problem of determining m and b in $y = mx + b$. The method of selected points is essentially unchanged. Of course the origin is not necessarily on the line and both the slope and the y intercept are to be found. The method of averages must be altered somewhat when dealing with the equation $y = mx + b$. If we have n points given, then the addition of n equations of the form $r_i = y_i - (mx_i + b)$ leads to

$$\sum r_i = \sum y_i - m \sum x_i - nb = 0.$$

Since there is only one equation in two unknowns (m and b), we cannot solve for either. This is easily remedied by dividing the given data into two sets. Each set leads to an equation of the form just shown, and the two resulting equations can be solved simultaneously for m and b.

Example 3

Find m and b of $y = mx + b$ by both methods of this section from the data given.

x	-1	0	1	2	3	4
y	-3.6	-1.4	1.3	3.1	5.4	8.5

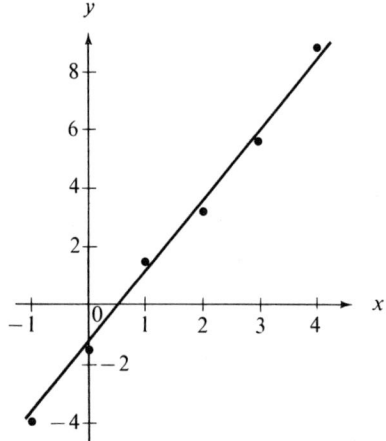

Figure 3.11

3.5 Fitting a Line to Empirical Data

Figure 3.11 shows the data points and a line that is a reasonable fit. Now $b = -1.25$, as can be read from the graph directly; and, using $(0, -1.25)$ and $(4, 8.25)$, we have $m = 2.38$.

For the method of averages, let us divide the data into two sets using the first three points for the first set and the last three points for the other. The first set gives

$$-3.7 - 3b = 0$$

and the second gives

$$17 - 9m - 3b = 0.$$

Solving simultaneously, we have $b = -1.23$ and $m = 2.30$.

In many cases a nonlinear equation can be handled by the above methods we have used, as the following example illustrates.

Example 4

Find k of $y = kx^3$ from the given data.

x	0	1	2	3	4
y	0	1.35	10.40	36.01	84.52

Since the given equation is nonlinear, our first problem is to convert it to a linear equation. This is easily done by the substitution $z = x^3$. Thus the equation becomes $y = kz$ and the data are as follows:

z	0	1	8	27	64
y	0	1.35	10.40	36.01	84.52

Now, by the method of averages,

$$k = \frac{\sum y_i}{\sum z_i} = \frac{132.28}{100} = 1.323.$$

In addition to the methods given here, there are other, more sophisticated methods which yield somewhat better results.*

Problems

In Problems 1–10, find the unknown constant(s) by both methods of this section from the data given.

1. Find m of $y = mx$.

x	0	1	2	3	4
y	0	4.1	8.5	12.8	16.3

---

*For a discussion of the method of least squares and the method of moments, see Ivan S. Sokolnikoff and Elizabeth S. Sokolnikoff, *Higher Mathematics for Engineers and Physicists*, 2d ed., (New York, McGraw-Hill Book Company, Inc., 1941), pp. 536–545.

2. Find m of $y = mx$.

x	-3	-2	-1	0	1	2
y	7.2	4.5	2.4	0	-2.3	-4.7

3. Find k of $P = kT$.

T	50	71	102	140	182
P	20.8	29.4	42.0	58.0	74.2

4. Find k of $C = kt$.

t	5	10	20	40	80
C	2.0	4.4	8.3	16.4	33.9

5. Find m and b of $y = mx + b$.

x	1	3	5	7	10	15
y	6.4	10.9	14.5	19.8	26.0	35.2

6. Find m and b of $y = mx + b$.

x	1	2	5	7	10	15
y	-3.3	1.0	14.1	21.8	34.7	57.0

7. Find p and q of $y = px + q$.

x	2	10	14	22	31	45
y	8.6	29.0	37.2	56.8	78.0	112.9

8. Find k and c of $y = kt + c$.

t	0	20	40	60	80	100
y	75.6	72.8	69.6	66.2	62.6	58.9

9. Find k of $P = k(1/V)$.

V	9.0	4.8	2.5	1.3	0.7
P	1.2	2.3	4.2	8.8	16.2

10. Find k of $y = kx^2$.

x	1	2	3	4	5
y	5.3	20.5	47.0	82.9	131.0

11. The relationship between the vapor pressure P of a liquid and its absolute temperature, T, is given by the Clausius-Clapeyron equation,

$$2.303 \log_{10} P = \frac{-\Delta H}{R} \cdot \frac{1}{T} + C,$$

where ΔH is the molar heat of vaporization of the liquid and R is the ideal gas constant, 1.987 calories degree^{-1} mole^{-1}. The following data were found.

$1/T$	0.00364	0.00357	0.00341	0.00328	0.00319
$\log_{10} P$	0.0000218	0.0000230	0.0000250	0.0000272	0.0000287

What is the molar heat of vaporization of the liquid?

3.5 Fitting a Line to Empirical Data

12. The Freundlich equation for adsorption is

$$y = kC^{1/n},$$

where y represents the weight in grams of substance adsorbed, C the concentration in moles/liter of the solute. In logarithmic form, the equation is

$$\log_{10} y = \log_{10} k + \frac{1}{n}\log_{10} C.$$

Experimentation with the adsorption of acetic acid from water solutions by charcoal gave the following results.

C	0.079	0.036	0.019	0.0097	0.0045
y	0.054	0.038	0.029	0.022	0.016

What are k and n?

4

The Circle

4.1

The Standard Form for an Equation of a Circle

The standard form for an equation of a circle is a direct consequence of the definition and the length formula.

Definition

A **circle** is the set of all points in a plane at a fixed positive distance (**radius**) from a fixed point (**center**).

Theorem 4.1

A circle with center (h, k) and radius r has equation
$$(x - h)^2 + (y - k)^2 = r^2.$$

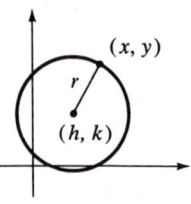

Figure 4.1

Proof

If (x, y) is any point on the circle, then the distance from the center (h, k) to (x, y) is r (see Figure 4.1):
$$r = \sqrt{(x - h)^2 + (y - k)^2}.$$

4.1 The Standard Form for an Equation of a Circle

Squaring, we have

$$(x - h)^2 + (y - k)^2 = r^2.$$

Since the steps above are reversible, we see that every point satisfying the equation of Theorem 4.1 is on the circle described.

Example 1

Give an equation for the circle with center $(3, -5)$ and radius 2.

From Theorem 4.1, an equation is

$$(x - 3)^2 + [y - (-5)]^2 = 2^2,$$

or

$$(x - 3)^2 + (y + 5)^2 = 4.$$

Although the above form is a convenient one, in that it shows at a glance the center and radius of the circle, another form is usually used. It is called the general form and is comparable to the general form of a line. Before giving the generalized equation for this form, let us illustrate it with the result of Example 1. Squaring the two binomials and combining similar terms, we have

$$(x - 3)^2 + (y + 5)^2 = 4,$$
$$x^2 - 6x + 9 + y^2 + 10y + 25 = 4,$$
$$x^2 + y^2 - 6x + 10y + 30 = 0.$$

Normally an equation of a circle will be given in this form. Let us now repeat the manipulation, starting with the standard form of Theorem 4.1.

$$(x - h)^2 + (y - k)^2 = r^2,$$
$$x^2 - 2hx + h^2 + y^2 - 2ky + k^2 = r^2,$$
$$x^2 + y^2 - 2hx - 2ky + (h^2 + k^2 - r^2) = 0,$$

which is in the form

$$x^2 + y^2 + D'x + E'y + F' = 0.$$

Upon multiplication by a nonzero constant, A, we have

$$Ax^2 + Ay^2 + Dx + Ey + F = 0 \quad (A \neq 0),$$

as the following theorem states.

Theorem 4.2

Every circle can be represented in the general form

$$\boxed{Ax^2 + Ay^2 + Dx + Ey + F = 0 \quad (A \neq 0).}$$

It is a simple matter to take an equation of a circle in the standard form and reduce it to the general form. We have already seen an example of this. However, it is somewhat more difficult to go from the general form to the standard form. The latter is accomplished by the process of "completing the square." To see how this is accomplished, suppose we consider

$$(x + a)^2 = x^2 + 2ax + a^2.$$

The constant term a^2 and the coefficient of x have a definite relationship; namely, the constant term is the square of one-half the coefficient of x. Thus,

$$a^2 = \left[\frac{1}{2}(2a)\right]^2.$$

Note, however, that this relationship holds only when the coefficient of x^2 is 1.

This relationship suggests the following procedure. If the coefficients of x^2 and y^2 are not one, make them one by division. Group the x terms and the y terms on one side of the equation and take the constant to the other side. Then complete the square on both the x and the y terms. Remember that whatever is added to one side of an equation must be added to the other in order to maintain equality.

Example 2

Express $2x^2 + 2y^2 - 2x + 6y - 3 = 0$ in the standard form.

$$2x^2 + 2y^2 - 2x + 6y - 3 = 0,$$

$$x^2 + y^2 - x + 3y - \frac{3}{2} = 0,$$

$$(x^2 - x \quad) + (y^2 + 3y \quad) = \frac{3}{2},$$

$$\left(x^2 - x + \frac{1}{4}\right) + \left(y^2 + 3y + \frac{9}{4}\right) = \frac{3}{2} + \frac{1}{4} + \frac{9}{4},$$

$$\left(x - \frac{1}{2}\right)^2 + \left(y + \frac{3}{2}\right)^2 = 4.$$

Thus, the original equation represents a circle with center $(1/2, -3/2)$ and radius 2.

The next two examples show that the converse of Theorem 4.2 is not true: that is, an equation of the form

$$Ax^2 + Ay^2 + Dx + Ey + F = 0$$

does not necessarily represent a circle.

4.1 The Standard Form for an Equation of a Circle

Example 3

Express $x^2 + y^2 + 4x - 6y + 13 = 0$ in standard form.

$$x^2 + y^2 + 4x - 6y + 13 = 0,$$
$$(x^2 + 4x \quad) + (y^2 - 6y \quad) = -13,$$
$$(x^2 + 4x + 4) + (y^2 - 6y + 9) = -13 + 4 + 9,$$
$$(x + 2)^2 + (y - 3)^2 = 0.$$

Since neither of the two expressions on the left-hand side of the last equation can be negative, their sum can be zero only if both expressions are zero. This is possible only when $x = -2$ and $y = 3$. Thus, the point $(-2, 3)$ is the only point in the plane that satisfies the original equation.

Example 4

Express $x^2 + y^2 + 2x + 8y + 19 = 0$ in standard form.

$$x^2 + y^2 + 2x + 8y + 19 = 0,$$
$$(x^2 + 2x \quad) + (y^2 + 8y \quad) = -19,$$
$$(x^2 + 2x + 1) + (y^2 + 8y + 16) = -19 + 1 + 16,$$
$$(x + 1)^2 + (y + 4)^2 = -2.$$

Again, since neither expression on the left-hand side of the last equation can be negative, their sum cannot possibly be negative. There is no point in the plane satisfying this equation. Its graph is the empty set.

The results illustrated by the last three examples are stated in the next theorem.

Theorem 4.3

Every equation of the form

$$Ax^2 + Ay^2 + Dx + Ey + F = 0 \quad (A \neq 0)$$

represents either a circle, a point, or the empty set. (The last two cases are called the degenerate cases of a circle.)

Problems

In Problems 1–16, write an equation of the circle described in both the standard form and the general form. Sketch.

1. Center $(1, 3)$; radius 5.
2. Center $(0, 0)$; radius 1.
3. Center $(5, -2)$; radius 2.
4. Center $(0, 3)$; radius $1/2$.
5. Center $(1/2, -3/2)$; radius 2.
6. Center $(-2/3, -1/2)$; radius $3/2$.

7. Center $(4, -2)$; $(3, 3)$ on the circle. 8. Center $(-1, 0)$; $(4, -3)$ on the circle.
9. $(2, -3)$ and $(-2, 0)$ are the end points of a diameter.
10. $(-3, 5)$ and $(2, 4)$ are the end points of a diameter.
11. Radius 3; in the first quadrant and tangent to both axes.
12. Radius 5; in the fourth quadrant and tangent to both axes.
13. Radius 2; tangent to $x = 2$ and $y = -1$ and above and to the right of these lines.
14. Radius 3; tangent to $x = -3$ and $y = 4$ and below and to the left of these lines.
15. Tangent to both axes at $(4, 0)$ and $(0, -4)$.
16. Tangent to $x = -2$ and $y = 2$ at $(-2, 0)$ and $(-4, 2)$.

In Problems 17–28, express the equation in standard form. Sketch if the graph is nonempty.

17. $x^2 + y^2 - 2x - 4y + 1 = 0$.
18. $x^2 + y^2 + 4x - 6y - 3 = 0$.
19. $x^2 + y^2 + 6x - 16 = 0$.
20. $x^2 + y^2 - 10x + 4y + 29 = 0$.
21. $4x^2 + 4y^2 - 4x - 12y + 1 = 0$.
22. $9x^2 + 9y^2 - 12x - 24y - 13 = 0$.
23. $5x^2 + 5y^2 - 8x - 4y - 121 = 0$.
24. $9x^2 + 9y^2 - 18x - 12y - 23 = 0$.
25. $9x^2 + 9y^2 - 6x + 18y + 11 = 0$.
26. $36x^2 + 36y^2 - 36x + 24y - 23 = 0$.
27. $36x^2 + 36y^2 - 48x - 36y + 25 = 0$.
28. $8x^2 + 8y^2 + 24x - 4y + 19 = 0$.
29. Find the point(s) of intersection of

$$x^2 + y^2 - x - 3y - 6 = 0 \quad \text{and} \quad 4x - y - 9 = 0.$$

30. Find the point(s) of intersection of

$$x^2 + y^2 + 4x - 12y + 6 = 0 \quad \text{and} \quad 3x - 5y + 2 = 0.$$

31. Find the point(s) of intersection of

$$x^2 + y^2 + 5x + y - 26 = 0 \quad \text{and} \quad x^2 + y^2 + 2x - y - 15 = 0.$$

32. Find the point(s) of intersection of

$$x^2 + y^2 + x + 12y + 8 = 0 \quad \text{and} \quad 2x^2 + 2y^2 - 4x + 9y + 4 = 0.$$

33. What happens when we try to solve simultaneously

$$x^2 + y^2 - 2x + 4y + 1 = 0 \quad \text{and} \quad x - 2y + 2 = 0?$$

Interpret geometrically.

34. What happens when we try to solve simultaneously

$$x^2 + y^2 - 4x - 2y + 1 = 0 \quad \text{and} \quad x^2 + y^2 + 6x - 6y + 14 = 0?$$

Interpret geometrically.

35. Find the line through the points of intersection of

$$x^2 + y^2 - x + 3y - 10 = 0 \quad \text{and} \quad x^2 + y^2 - 2x + 2y - 11 = 0.$$

36. For what value(s) of k is the line $x + 2y + k = 0$ tangent to the circle

$$x^2 + y^2 - 2x + 4y + 1 = 0?$$

37. Prove analytically that if P_1 and P_2 are the ends of a diameter of a circle and Q is any point on the circle, then $\angle P_1 Q P_2$ is a right angle.
38. A set of points in the plane has the property that every point in it is twice as far from $(1, 1)$ as it is from $(5, 3)$. What equation must be satisfied by every point (x, y) in the set?

39. Find the relation between A, D, E, and F of Theorem 4.2 in order that the equation represent (a) a circle, (b) a point, (c) the empty set. If the equation represents a circle, find h, k, and r in terms of A, D, E, and F.
40. In general, squaring both sides of an equation is not reversible (if $x = 2$, then $x^2 = 4$; but if $x^2 = 4$, then $x = \pm 2$). Yet, in the proof of Theorem 4.1, the argument was declared to be reversible even though both sides of an equation were squared. Why?

4.2

Conditions to Determine a Circle

We have seen two forms for equations of a circle: the standard form,

$$(x - h)^2 + (y - k)^2 = r^2,$$

with the three parameters h, k, and r, and the general form,

$$Ax^2 + Ay^2 + Dx + Ey + F = 0 \quad (A \neq 0),$$

with the parameters A, D, E, and F. However, since $A \neq 0$, we can divide through by A to obtain

$$x^2 + y^2 + D'x + E'y + F' = 0,$$

which, like the standard form, has only three parameters. Thus we need three equations in h, k, and r or in D', E', and F' in order to determine these parameters and give the equation desired. Since each condition on a circle determines one such equation, three conditions are required to determine a circle.

Example 1

Find an equation of the circle through points $(1, 5)$, $(-2, 3)$, and $(2, -1)$.

The desired equation is

$$x^2 + y^2 + D'x + E'y + F' = 0$$

for suitable choices of D', E', and F'. Since the three given points are on the circle, they satisfy this equation. Thus

$$1 + 25 + D' + 5E' + F' = 0,$$
$$4 + 9 - 2D' + 3E' + F' = 0,$$
$$4 + 1 + 2D' - E' + F' = 0,$$

or

$$D' + 5E' + F' = -26,$$
$$-2D' + 3E' + F' = -13,$$
$$2D' - E' + F' = -5.$$

Solving simultaneously, we have $D' = -9/5$, $E' = -19/5$, and $F' = -26/5$. Thus the circle is

$$x^2 + y^2 - \frac{9}{5}x - \frac{19}{5}y - \frac{26}{5} = 0,$$

or

$$5x^2 + 5y^2 - 9x - 19y - 26 = 0.$$

Example 2

Find an equation(s) of the circle(s) of radius 4 with center on the line $4x + 3y + 7 = 0$ and tangent to $3x + 4y + 34 = 0$.

The three conditions lead to the following three relations involving h, k, and r (see Figure 4.2).

$$r = 4, \tag{1}$$
$$4h + 3k + 7 = 0, \tag{2}$$
$$\frac{|3h + 4k + 34|}{\sqrt{3^2 + 4^2}} = r. \tag{3}$$

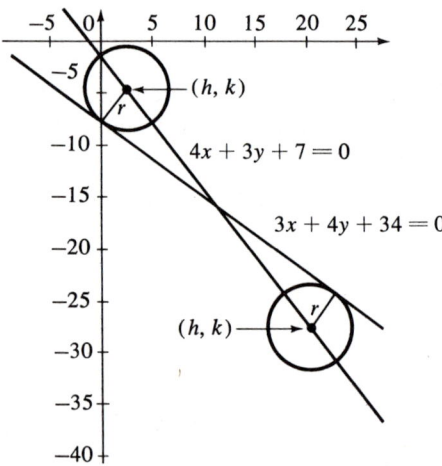

Figure 4.2

The first and third give

$$|3h + 4k + 34| = 20.$$

4.2 Conditions to Determine a Circle

Solving the second for k, we have

$$k = -\frac{4h+7}{3},$$

and substituting into $|3h + 4k + 34| = 20$, we have

$$\left|3h - \frac{16h+28}{3} + 34\right| = 20,$$

$$|74 - 7h| = 60,$$

$$74 - 7h = \pm 60,$$

$$h = 2 \quad \text{or} \quad h = \frac{134}{7};$$

and $k = -5$ or $k = -195/7$, respectively. Thus the two solutions are

$$(x-2)^2 + (y+5)^2 = 16 \quad \text{and} \quad \left(x - \frac{134}{7}\right)^2 + \left(y + \frac{195}{7}\right)^2 = 16,$$

or

$$x^2 + y^2 - 4x + 10y + 13 = 0 \quad \text{and} \quad 49x^2 + 49y^2 - 1876x + 2730y + 55{,}197 = 0.$$

This problem can also be solved in the following way. Since the desired circle has radius 4 and is tangent to $3x + 4y + 34 = 0$, its center is on a line parallel to $3x + 4y + 34 = 0$ and at a distance 4 from it. There are two such lines (see Figure 4.3) given by

$$\frac{|3x + 4y + 34|}{5} = 4,$$

$$3x + 4y + 34 = \pm 20,$$

$$3x + 4y + 14 = 0 \quad \text{or} \quad 3x + 4y + 54 = 0.$$

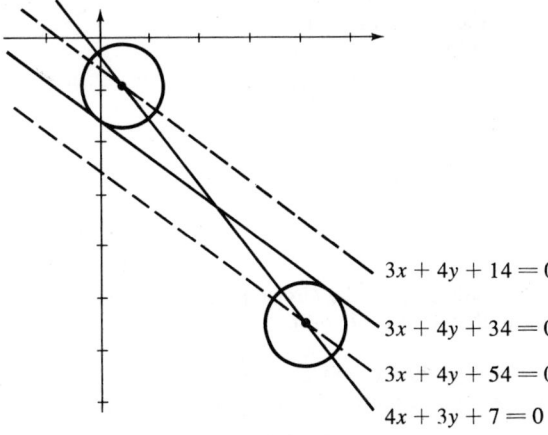

Figure 4.3

Since the center is also on $4x + 3y + 7 = 0$, we can find its coordinates by solving this equation simultaneously with each of the two equations above. From

$$3x + 4y + 14 = 0 \quad \text{and} \quad 4x + 3y + 7 = 0,$$

we get center $(2, -5)$; from

$$3x + 4y + 54 = 0 \quad \text{and} \quad 4x + 3y + 7 = 0,$$

we get center $(134/7, -195/7)$. Using these centers with the given radius, 4, we have the desired circles.

Example 3

Find an equation(s) of the circle(s) tangent to both axes and containing the point $(-8, -1)$.

The three conditions give

$$h = -r, \tag{1}$$
$$k = -r, \tag{2}$$
$$(-8 - h)^2 + (-1 - k)^2 = r^2 \tag{3}$$

(see Figure 4.4). Substituting (1) and (2) into (3), we have

$$(-8 + r)^2 + (-1 + r)^2 = r^2,$$
$$r^2 - 18r + 65 = 0,$$
$$(r - 5)(r - 13) = 0,$$
$$r = 5 \quad \text{or} \quad r = 13.$$

Thus we have the circle with radius 5 and center $(-5, -5)$ with equation

$$(x + 5)^2 + (y + 5)^2 = 25,$$

or

$$x^2 + y^2 + 10x + 10y + 25 = 0;$$

or we have the circle with radius 13 and center $(-13, -13)$ with equation

$$(x + 13)^2 + (y + 13)^2 = 169,$$

or

$$x^2 + y^2 + 26x + 26y + 169 = 0.$$

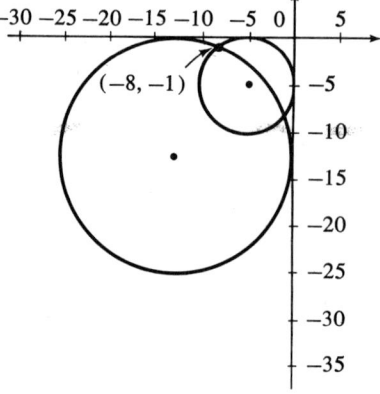

Figure 4.4

Example 4

Find an equation(s) of the circle(s) tangent to $3x - 4y - 4 = 0$ at $(0, -1)$ and containing the point $(-1, -8)$.

The center of the desired circle is on the line perpendicular to the tangent line at $(0, -1)$ (see Figure 4.5). An equation of this perpendicular is

$$4x + 3y = 4 \cdot 0 + 3(-1),$$

4.2 Conditions to Determine a Circle

or
$$4x + 3y + 3 = 0.$$

Thus, for center (h, k), we have
$$4h + 3k + 3 = 0. \qquad (1)$$

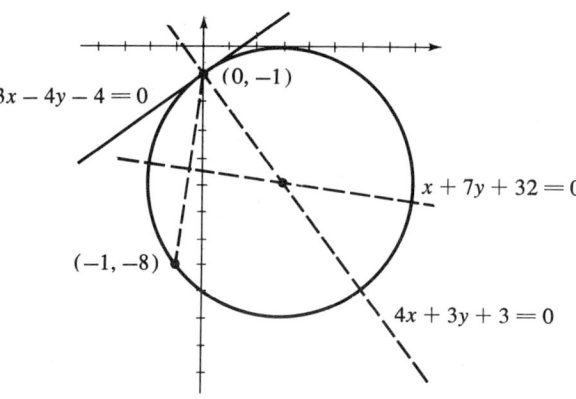

Figure 4.5

The center is also on the perpendicular bisector of the line joining $(0, -1)$ and $(-1, -8)$ (see Figure 4.5). The slope of the line joining $(0, -1)$ and $(-1, -8)$ is 7; thus the slope of a perpendicular line is $-1/7$. The midpoint of the segment from $(0, -1)$ to $(-1, -8)$ is $(-1/2, -9/2)$. By the point-slope formula, the perpendicular bisector is
$$y + \frac{9}{2} = -\frac{1}{7}\left(x + \frac{1}{2}\right),$$

or
$$x + 7y + 32 = 0.$$

Thus,
$$h + 7k + 32 = 0. \qquad (2)$$

Solving (1) and (2) simultaneously, we have
$$h = 3 \quad \text{and} \quad k = -5.$$

Using this point with $(0, -1)$, we find the radius:
$$r = \sqrt{(3-0)^2 + (-5+1)^2} = 5.$$

Thus, the desired equation is
$$(x - 3)^2 + (y + 5)^2 = 25,$$

or
$$x^2 + y^2 - 6x + 10y + 9 = 0.$$

Problems

In Problems 1–23, find an equation(s) of the circle(s) described.
1. Through $(-1, 2)$, $(3, 4)$, and $(2, -1)$.
2. Through $(-2, -1)$, $(0, 3)$, and $(2, 0)$.
3. Circumscribed about the triangle with vertices $(2, 3)$, $(0, 5)$, and $(1, -1)$.
4. Circumscribed about the triangle with vertices $(1, 1)$, $(-2, 1)$, and $(1, 4)$.
5. Circumscribed about the triangle with sides $x - y = 0$, $x + 2y = 0$, and $4x + y = 35$.
6. Through $(2, 1)$, $(-4, 4)$, and $(6, -1)$. (*Watch out!*)
7. Tangent to the x axis; center on $2x + y - 1 = 0$; radius 5.
8. Tangent to $2x + 3y + 13 = 0$ and $2x - 3y - 1 = 0$; contains $(0, 4)$.
9. Tangent to $3x + 4y - 15 = 0$ at $(5, 0)$; contains $(-2, -1)$.
10. Tangent to $5x - 12y + 89 = 0$ at $(-1, 7)$; contains $(16, 0)$.
11. Tangent to $x + y = 0$ and $x - y - 6 = 0$; center on $3x - y + 3 = 0$.
12. Tangent to $x - 3y - 7 = 0$ and $3x + y - 21 = 0$; center on $x - 3y + 3 = 0$.
13. Tangent to $x - 3y = 0$ at $(0, 0)$; center on $2x + y + 1 = 0$.
14. Tangent to $x - y = 0$ at $(2, 2)$; center on $2x + 3y - 7 = 0$.
15. Contains $(-1, 4)$ and $(3, 2)$; center on $3x - y + 3 = 0$.
16. Contains $(5, 2)$ and $(-1, 6)$; center on $x = y$.
17. Tangent to $2x + 3y - 5 = 0$ at $(1, 1)$; tangent to $2x + 3y + 10 = 0$.
18. Tangent to $y = 0$ at $(4, 0)$; tangent to $3x - 4y - 17 = 0$.
19. Tangent to both axes; radius 3.
20. Tangent to $x = 0$; center on $x + y = 10$; contains $(2, 9)$.
21. Tangent to $3x - 4y + 3 = 0$ at $(-1, 0)$; radius 7.
22. Tangent to $x^2 + y^2 - 22x + 20y + 77 = 0$ at $(91/17, 10/17)$; containing $(0, 1)$.
23. Tangent to $x^2 + y^2 - 8x - 22y + 112 = 0$ and $3x + 4y + 19 = 0$; radius 5.
24. Show that if (x_1, y_1), (x_2, y_2), and (x_3, y_3) are three noncollinear points, then the circle containing these three points has equation

$$\begin{vmatrix} x^2 + y^2 & x & y & 1 \\ x_1^2 + y_1^2 & x_1 & y_1 & 1 \\ x_2^2 + y_2^2 & x_2 & y_2 & 1 \\ x_3^2 + y_3^2 & x_3 & y_3 & 1 \end{vmatrix} = 0.$$

25. Show that if (x_1, y_1), (x_2, y_2), and (x_3, y_3) are three collinear points, then the determinant of Problem 24 is linear.
26. Find an equation(s) of the line(s) tangent to $x^2 + y^2 + 4x - 10y + 4 = 0$ from the point $(3, 2)$.
27. Find an equation(s) of the line(s) tangent to $x^2 + y^2 - 8x + 2y - 152 = 0$ and having slope $1/3$.

4.3

Families of Circles

We can, of course, consider families of circles just as we did families of lines. In particular, let us suppose that

$$Ax^2 + Ay^2 + Dx + Ey + F = 0 \quad \text{and} \quad A'x^2 + A'y^2 + D'x + E'y + F' = 0$$

represent two circles which intersect at the points P_1 and P_2. Now let us consider the family

$$M = \{Ax^2 + Ay^2 + Dx + Ey + F + k(A'x^2 + A'y^2 + D'x + E'y + F') = 0 \,|\, k \text{ real}\}.$$

Since the coordinates of P_1 satisfy the equations of both of the given circles, they must satisfy the equation of the family M no matter what value k might have. Thus the point P_1 belongs to every member of the family. Similarly P_2 belongs to every member. Since the coefficients of x^2 and y^2 are the same, the members of M consist of circles containing P_1 and P_2 together with the line containing these two points (when $k = -A/A'$). Note, however, that the family does not include *all* circles through P_1 and P_2; no value of k gives

$$A'x^2 + A'y^2 + D'x + E'y + F' = 0.$$

Example 1

Find an equation of the circle containing $(2, -1)$ and the points of intersection of

$$x^2 + y^2 - 2x - 4y + 1 = 0 \quad \text{and} \quad x^2 + y^2 - 6x - 2y + 9 = 0.$$

Since neither circle contains $(2, -1)$, the desired circle must be one of the family

$$\{x^2 + y^2 - 2x - 4y + 1 + k(x^2 + y^2 - 6x - 2y + 9) = 0 \,|\, k \text{ real}\}.$$

If $(2, -1)$ is on the desired circle, then, for the proper choice of k, $(2, -1)$ satisfies the family equation.

$$4 + 1 - 4 + 4 + 1 + k(4 + 1 - 12 + 2 + 9) = 0,$$
$$6 + 4k = 0,$$
$$k = -\frac{3}{2}.$$

Thus the circle is

$$x^2 + y^2 - 2x - 4y + 1 - \frac{3}{2}(x^2 + y^2 - 6x - 2y + 9) = 0$$

or
$$x^2 + y^2 - 14x + 2y + 25 = 0.$$

We might note the procedure necessary to solve this problem without using families of circles. First we would have to find the points of intersection of the given circles. These are $(3, 2)$ and $(11/5, 2/5)$. Then, using these two points together with the point $(2, -1)$, we would have to find the circle containing them all (see Example 1, page 77). This would be considerably more difficult than the method above. Furthermore, if the points of intersection had been irrational, this method would be even more tedious. Thus the use of families of circles is quite convenient here.

Example 2

Find the line through the points of intersection of the circles
$$x^2 + y^2 + 4x - 2y - 4 = 0 \quad \text{and} \quad x^2 + y^2 - 4x = 0.$$

The line desired is the one line of the family
$$\{x^2 + y^2 + 4x - 2y - 4 + k(x^2 + y^2 - 4x) = 0 \,|\, k \text{ real}\}.$$

We get the line by choosing $k = -1$ so that the second-degree terms cancel. Thus we have
$$x^2 + y^2 - 4x - 2y - 4 - (x^2 + y^2 - 4x) = 0,$$
$$-2y - 4 = 0,$$
$$y + 2 = 0.$$

We must be careful when using the above family to be sure that the given circles actually intersect at two points. If they are tangent or have no point in common, the set discussed above is still a family of circles (including degenerate circles) together with one line. If the given circles are tangent, the family consists of circles and a line tangent to both of the given circles. If the given circles have no point in common and are not concentric, the family consists of circles with their centers on the line of centers of the given circles together with one line perpendicular to the line of centers. If the given circles are concentric, the family consists of circles concentric with each other and the given circles (there is no line in the family in this case).

Whenever the given circles are not concentric the family contains one line, called the *radical axis* of the two circles. This line has some interesting properties. From the discussion above it follows that the radical axis must be perpendicular to the line of centers of the given circles. Furthermore, if P is any point on the radical axis of two circles C_1 and C_2 and P is outside of both C_1 and C_2, and if T_1 and T_2 are points of C_1 and C_2, respectively, such that PT_1 and PT_2 are tangent to C_1 and C_2, respectively, then $PT_1 = PT_2$ (see Figure 4.6). The proof of this statement is left to the student (see Problems 23, 28, and 29).

Of course other families of circles may also be considered. Some of these are given in the problems that follow (see Problems 15–20).

4.3 Families of Circles

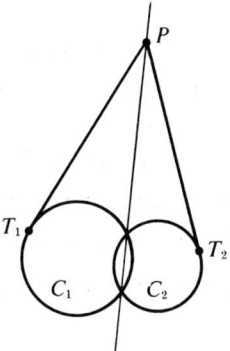

Figure 4.6

Problems

1. Find an equation(s) of the circle(s) containing $(1, -4)$ and the points of intersection of
$$x^2 + y^2 + 2x - 4y + 1 = 0 \quad \text{and} \quad x^2 + y^2 + 4x + 6y - 3 = 0.$$

2. Find an equation(s) of the circle(s) containing $(2, 0)$ and the points of intersection of
$$x^2 + y^2 - 2x + 6y + 6 = 0 \quad \text{and} \quad x^2 + y^2 + 2x - 2y - 7 = 0.$$

3. Find an equation(s) of the circle(s) containing $(-1, -2)$ and the points of intersection of
$$x^2 + y^2 + 3x - 4y - 7 = 0 \quad \text{and} \quad x^2 + y^2 - 2x - 5y - 5 = 0.$$

4. Find an equation(s) of the circle(s) containing $(2, 1)$ and the points of intersection of
$$2x^2 + 2y^2 + 3x - y - 8 = 0 \quad \text{and} \quad x^2 + y^2 - 2x + y - 5 = 0.$$

5. Find an equation(s) of the circle(s) with center on $x + y - 2 = 0$ and containing the points of intersection of
$$x^2 + y^2 + 4x + 6y - 3 = 0 \quad \text{and} \quad x^2 + y^2 + 2x + 2y - 2 = 0.$$

6. Find an equation(s) of the circle(s) with center on $x - 2y + 5 = 0$ and containing the points of intersection of
$$x^2 + y^2 - 4x + 2y + 2 = 0 \quad \text{and} \quad x^2 + y^2 - 10x + 4 = 0.$$

7. Find an equation(s) of the circle(s) with center $(3, -1)$ and containing the points of intersection of
$$x^2 + y^2 - 4x - 6y + 9 = 0 \quad \text{and} \quad x^2 + y^2 - 2x - 14y + 15 = 0.$$

8. Find an equation(s) of the circle(s) with center $(4, 0)$ and containing the points of intersection of
$$x^2 + y^2 - 2x - 2y - 2 = 0 \quad \text{and} \quad x^2 + y^2 - 5x - y + 4 = 0.$$

9. Find an equation(s) of the circle(s) with radius 2 and containing the points of inter-

section of
$$x^2 + y^2 + 2x - 2y - 3 = 0 \quad \text{and} \quad x^2 + y^2 - x - 1 = 0.$$

10. Find an equation(s) of the circle(s) with radius 3 and containing the points of intersection of
$$x^2 + y^2 - 4x - 2y - 1 = 0 \quad \text{and} \quad 2x^2 + 2y^2 - 8y + 3 = 0.$$

11. Find an equation(s) of the line(s) containing the points of intersection of
$$x^2 + y^2 - 2x - 8y + 8 = 0 \quad \text{and} \quad x^2 + y^2 + 2x - 3 = 0.$$

12. Find an equation(s) of the line(s) containing the points of intersection of
$$x^2 + y^2 - 2x + 4y = 0 \quad \text{and} \quad x^2 + y^2 - 6x - 2y + 6 = 0.$$

13. Suppose you are asked to use the method of this section to find an equation(s) of the circle(s) containing (3, 0) and the points of intersection of
$$x^2 + y^2 + 2x + 2y - 7 = 0 \quad \text{and} \quad x^2 + y^2 - 4x - 6y + 9 = 0.$$
What is the result? Does this represent all possible circles satisfying the given conditions? If not, why not? Sketch the given circles and the result.

14. Suppose you are asked to use the method of this section to find an equation(s) of the circle(s) containing $(-2, -1)$ and the points of intersection of
$$x^2 + y^2 + 4x - 8y + 16 = 0 \quad \text{and} \quad x^2 + y^2 - 4x - 2y + 1 = 0.$$
What is the result? Does this represent all possible circles satisfying the given conditions? If not, why not? Sketch the given circles and the result.

In Problems 15–20, describe the family of circles given. Indicate whether or not it contains *every* circle of that description, and, if not, give all circles with that description which are not included in the family.

15. $\{(x - h)^2 + (y - 1)^2 = 1 \mid h \text{ real}\}$.
16. $\{(x - h)^2 + (y - k)^2 = k^2 \mid h, k \text{ real}\}$.
17. $\{(x - h)^2 + (y - h)^2 = h^2 \mid h \text{ real}\}$.
18. $\{x^2 + y^2 + Dx + Ey = 0 \mid D, E \text{ real}\}$.
19. $\{(x - h)^2 + (y - k)^2 = 1 \mid h, k \text{ real}, \ h^2 + k^2 = 1\}$.
20. $\{(x - h)^2 + (y - k)^2 = 1 \mid h, k \text{ real}, \ |h| = 2, |k| = 2\}$.
21. Given two points P_1 and P_2 such that three different circles all contain P_1 and P_2, show that the centers of the circles are collinear.
22. Show that the radical axis of a pair of circles is perpendicular to their line of centers.
23. Show that if $P : (x_1, y_1)$ is outside the circle $(x - h)^2 + (y - k)^2 = r^2$ and T is a point of the circle such that PT is tangent to the circle, then the length of PT is
$$\sqrt{(x - h)^2 + (y - k)^2 - r^2} \quad \text{(see Figure 4.7).}$$

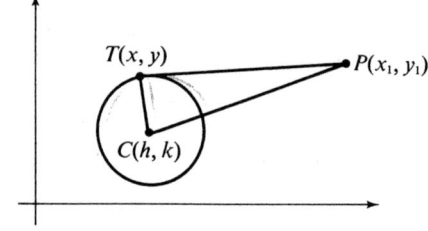

Figure 4.7

In Problems 24–27, use the result of Problem 23 to find the length of the tangent from the given point to the given circle (PT).

4.3 Families of Circle

24. $(x-1)^2 + (y-2)^2 = 4$; (5, 3).
25. $(x+3)^2 + (y-1)^2 = 9$; (4, 3).
26. $x^2 + y^2 - 4x + 2y + 1 = 0$; (3, 5).
27. $x^2 + y^2 + 8x - 4y + 11 = 0$; (1, −3).
28. Use the result of Problem 23 to show that, if P has equal tangents to the circles

$$x^2 + y^2 + Dx + Ey + F = 0 \quad \text{and} \quad x^2 + y^2 + D'x + E'y + F' = 0,$$

then P is on the radical axis of the two circles.
29. Show that the argument of Problem 28 is reversible, so that if P is on the radical axis and there exist tangents from P to the two circles, then the tangents are equal.

5

Conic Sections

5.1

Conic Sections

Up to this point the only second-degree equations that we have considered systematically have been equations of circles. We shall see that equations of the second degree represent (with two trivial exceptions) conic sections—that is, curves formed by the intersection of a plane with a right circular cone. There are three general types of curves formed in this way, the parabola, the ellipse, and the hyperbola.

5.2

The Parabola

Definition

*A **parabola** is the set of all points in a plane equidistant from a fixed point (focus) and a fixed line (directrix) not containing the focus.*

5.2 The Parabola

Suppose we choose the focus to be the point $(c, 0)$ and the directrix to be $x = -c$, $c \neq 0$ (see Figure 5.1). Let us choose a point (x, y) on the parabola and see what condition must be satisfied by x and y. From the definition, we have

$$\overline{PF} = \overline{PD},$$
$$\sqrt{(x-c)^2 + y^2} = |x + c|, \qquad \text{(See Note 1)}$$
$$(x-c)^2 + y^2 = (x+c)^2, \qquad \text{(See Note 2)}$$
$$x^2 - 2cx + c^2 + y^2 = x^2 + 2cx + c^2,$$
$$y^2 = 4cx. \quad \text{Focus } (c,0) \quad D \quad x=-c$$

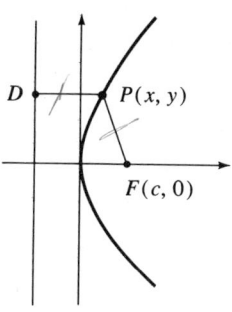

Note 1: Since $\overline{PD}$ is a horizontal distance,

$$\overline{PD} = |x - (-c)| = |x + c|.$$

Figure 5.1

You might feel that we should drop the absolute-value signs, since it is clear from Figure 5.1 that $x + c$ must be positive. However, we did not insist that c be positive (although Figure 5.1 is given for a positive value of c). If c is negative, x is also negative and $x + c$ is negative.

Note 2: When we square both sides of an equation, there is a possibility of introducing extraneous roots. For instance, $(0, 1)$ is not a root of $x + y = x - y$, but it is a root of $(x + y)^2 = (x - y)^2$. The reason is that $x + y = 1$, while $x - y = -1$ for $(0, 1)$, and $1^2 = (-1)^2 = 1$. In any case in which $(x + y)^2 = (x - y)^2$ and $x + y$ and $x - y$ are either both positive, both negative, or both zero, the above situation cannot occur. Since $\sqrt{(x-c)^2 + y^2}$ and $|x + c|$ must both be positive in any case, we have introduced no extraneous roots; that is, any point satisfying

$$(x - c)^2 + y^2 = (x + c)^2$$

must also satisfy

$$\sqrt{(x-c)^2 + y^2} = |x + c|.$$

We see then that if a point is on the parabola with focus $(c, 0)$ and directrix $x = -c$, it must satisfy the equation $y^2 = 4cx$. Furthermore, since Note 2 indicates that all steps in the above argument are reversible, any point satisfying the equation $y^2 = 4cx$ is on the given parabola.

Theorem 5.1

A point (x, y) is on the parabola with focus $(c, 0)$ and directrix $x = -c$ if and only if it satisfies the equation

$$y^2 = 4cx.$$

Let us observe some properties of this parabola before considering others. First of all, the x axis is a line of symmetry: that is, the portion below the x axis is the mirror image of the portion above. This line is called the *axis* of the parabola. It is perpendicular to the directrix and contains the focus (see Figure 5.2). The point of intersection of the axis and the parabola is the *vertex*. The vertex of the parabola $y^2 = 4cx$ is the origin. Finally, the line segment through the focus, perpendicular to the axis and having both ends on the parabola is the *latus rectum* (literally, straight side). Since the latus rectum of $y^2 = 4cx$ must be vertical and since it contains $(c, 0)$, the x coordinate of both ends is c. Substituting $x = c$ into $y^2 = 4cx$, we have

$$y^2 = 4c^2,$$
$$y = \pm 2c.$$

Thus one end of the latus rectum is $(c, 2c)$ and the other $(c, -2c)$. Its length is $4|c|$.

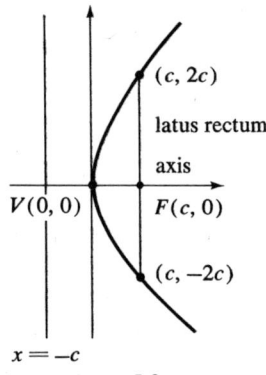

Figure 5.2

Finally the role of the x and y may be reversed throughout, as the next theorem states.

Theorem 5.2

A point (x, y) is on the parabola with focus $(0, c)$ and directrix $y = -c$ if and only if it satisfies the equation

$$x^2 = 4cy.$$

Example 1

Sketch and discuss $y^2 = 8x$.

The equation is of the form

$$y^2 = 4cx,$$

with $c = 2$. Thus, it represents a parabola with vertex at the origin and axis on the x axis. The focus is at $(2, 0)$, and the directrix is $x = -2$. Finally, the length of the latus rectum is 8. This length may be used to determine the ends, $(2, \pm 4)$,

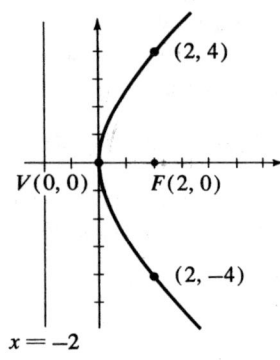

Figure 5.3

5.2 The Parabola

of the latus rectum, which helps in sketching the curve (see Figure 5.3).

Example 2

Sketch and discuss $x^2 = -12y$.

This equation is in the form $x^2 = 4cy$, with $c = -3$. Thus, it is a parabola with vertex at the origin and axis on the y axis. The focus is $(0, -3)$, the length of the latus rectum is 12, and the equation of the directrix is $y = 3$ (see Figure 5.4).

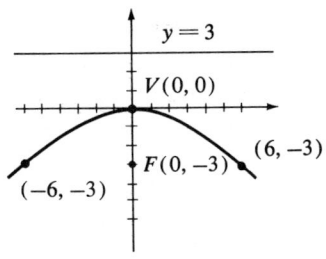

Figure 5.4

Example 3

Find an equation(s) of the parabola(s) with vertex at the origin and focus $(-4, 0)$.

Since the focus and vertex are on the x axis, the x axis is the axis of the parabola. Thus the equation is in the form $y^2 = 4cx$. Since the focus is $(-4, 0)$, $c = -4$ and the equation is $y^2 = -16x$.

Problems

In Problems 1–12, sketch and discuss the given parabola.

1. $y^2 = 16x$.
2. $y^2 = -12x$.
3. $x^2 = 4y$.
4. $x^2 = -8y$.
5. $y^2 = 10x$.
6. $x^2 = -7y$.
7. $x^2 = 5y$.
8. $y^2 = -9x$.
9. $x^2 = -2y$.
10. $y^2 = 3x$.
11. $x^2 = 6y$.
12. $y^2 = -5x$.

In Problems 13–20, find an equation(s) of the parabola(s) described.

13. Vertex: $(0, 0)$; axis: x axis; contains $(1, 5)$.
14. Vertex: $(0, 0)$; axis: y axis; contains $(1, 5)$.
15. Vertex: $(0, 0)$; axis: x axis; length of latus rectum: 5.
16. Vertex: $(0, 0)$; focus: $(0, 5)$.
17. Focus: $(-3, 0)$; directrix: $x = 3$.
18. Focus: $(0, 8)$; directrix: $y = -8$.
19. Vertex: $(0, 0)$; contains $(2, 3)$ and $(-2, 3)$.
20. Vertex: $(0, 0)$; contains $(-3, -4)$ and $(-3, 4)$.
21. Find an equation of the parabola with focus $(4, 0)$ and directrix $x = 0$.
22. Find an equation of the parabola with focus $(2, 4)$ and directrix $y = -2$.
23. Find an equation of the parabola with focus $(0, 0)$ and directrix $x + y = 4$.
24. Find an equation of the parabola with focus $(1, 1)$ and directrix $x + y = 0$.

A line is tangent to a parabola if it has one and only one point in common with the parabola and is not on or parallel to its axis. Use this fact to work problems 25–28.

25. Find an equation of the line tangent to $y = x^2$ at $(1, 1)$.
26. Find an equation of the line tangent to $x^2 = -5y$ at $(5, -5)$.
27. Find an equation of the line tangent to $y^2 = -16x$ and parallel to $x + y = 1$.
28. Find equations of the lines tangent to $y^2 = 4x$ and containing $(-2, 1)$.
29. Prove Theorem 5.2.
30. Prove that the ordinate of any point P of the parabola $y^2 = 4cx$ is the mean proportional between the length of the latus rectum and the abscissa of P.

5.3

Parabola with Vertex at (h, k)

Suppose we have a parabola with vertex at (h, k) and axis $y = k$ (see Figure 5.5). Let us put in a new pair of axes, the x' and y' axes, which are parallel to and in the same directions as the original axes and have their origin at the point (h, k) of the original system. Since the parabola's vertex is now at the origin of this coordinate system, its equation is

$$y'^2 = 4cx',$$

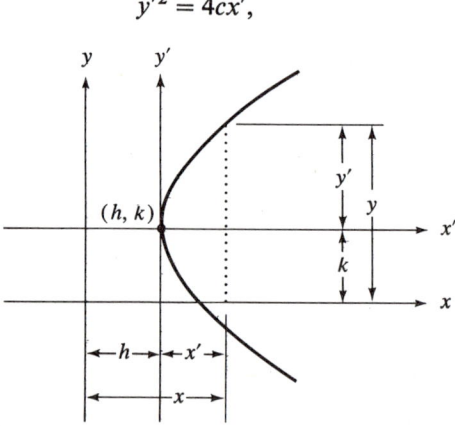

Figure 5.5

where $|c|$ is the distance from vertex to focus. Now the relationship between the old and new coordinates is (see Figure 5.5)

$$x = x' + h, \qquad y = y' + k$$

or

5.3 Parabola with Vertex at (h, k)

$$x' = x - h, \quad y' = y - k.$$

Thus the equation of the parabola in the original coordinate system is

$$(y - k)^2 = 4c(x - h).$$

Theorem 5.3

A point (x, y) is on the parabola with focus $(h + c, k)$ and directrix $x = h - c$ if and only if it satisfies the equation

$$(y - k)^2 = 4c(x - h).$$

Theorem 5.4

A point (x, y) is on the parabola with focus $(h, k + c)$ and directrix $y = k - c$ if and only if it satisfies the equation

$$(x - h)^2 = 4c(y - k).$$

Theorem 5.4 can be proved by an argument similar to that for Theorem 5.3. Of course Theorems 5.1 and 5.2 are special cases of these two, with h and k both 0. The equation of Theorem 5.3 can be put into the form

$$y^2 - 4cx - 2ky + (k^2 + 4ch) = 0,$$

or

$$y^2 + D'x + E'y + F' = 0,$$

where $D' = -4c$, $E' = -2k$, and $F' = k^2 + 4ch$. Finally, if we multiply through by some number C, we have

$$Cy^2 + Dx + Ey + F = 0.$$

Similarly, the equation of Theorem 5.4 can be put into the form

$$Ax^2 + Dx + Ey + F = 0.$$

Theorem 5.5

A parabola with axis parallel to (or on) a coordinate axis can be represented by an equation in one of the two forms:

$$Ax^2 + Dx + Ey + F = 0, \quad A \neq 0,$$
$$Cy^2 + Dx + Ey + F = 0, \quad C \neq 0.$$

An equation of the form given in Theorem 5.5 is referred to as a *general form* for a parabola, while those of Theorems 5.3 and 5.4 are called *standard forms*. We have seen similar forms for equations of lines and circles.

Example 1

Sketch and discuss the parabola
$$(x-1)^2 = -8(y+2).$$

First of all, the vertex is $(1, -2)$. Since the square is on the x term, the axis is parallel to the y axis. Finally $c = -2$, giving focus $(1, -4)$ and directrix $y = 0$, and the length of the latus rectum is 8 (see Figure 5.6).

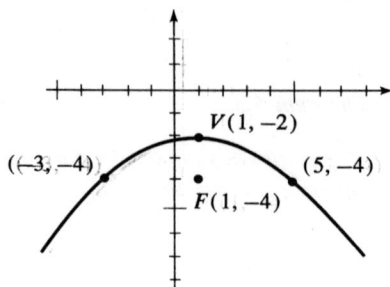

Figure 5.6

Example 2

Express $x^2 + 8x - 4y + 24 = 0$ in standard form.

We proceed in much the same way as we did with equations of circles—by completing the square.

$$x^2 + 8x - 4y + 24 = 0,$$
$$x^2 + 8x = 4y - 24,$$
$$x^2 + 8x + 16 = 4y - 8,$$
$$(x+4)^2 = 4(y-2).$$

Once the equation is in standard form, we can determine vertex, focus, and so forth.

Example 3

Express $9y^2 + 36x - 6y + 25 = 0$ in standard form.

$$9y^2 + 36x - 6y + 25 = 0,$$
$$y^2 + 4x - \frac{2}{3}y + \frac{25}{9} = 0,$$
$$y^2 - \frac{2}{3}y \phantom{+\tfrac{1}{9}} = -4x - \frac{25}{9},$$
$$y^2 - \frac{2}{3}y + \frac{1}{9} = -4x - \frac{8}{3},$$
$$\left(y - \frac{1}{3}\right)^2 = -4\left(x + \frac{2}{3}\right).$$

Unfortunately the converse of Theorem 5.5 is not true; that is, it is not true that every equation in one of the two forms

$$Ax^2 + Dx + Ey + F = 0, \quad A \neq 0,$$

or

5.3 Parabola with Vertex at (h, k)

$$Cy^2 + Dx + Ey + F = 0, \quad C \neq 0$$

represents a parabola. If $E = 0$ in the first form or $D = 0$ in the second, the resulting equation represents either a line (a pair of coincident lines), a pair of parallel lines or has no graph. These are sometimes referred to as degenerate cases of a parabola. We might note here that two of these cases—a pair of parallel lines and no graph—cannot be represented as the intersection of a plane and a cone.

Example 4

Sketch $y^2 + 2y - 3 = 0$.

We see that the equation cannot represent a parabola, since it does not involve both x and y. But, by factoring, we see that it represents a pair of parallel lines (see Figure 5.7).

$$y^2 + 2y - 3 = 0,$$
$$(y+3)(y-1) = 0;$$
$$y = -3, \quad y = 1.$$

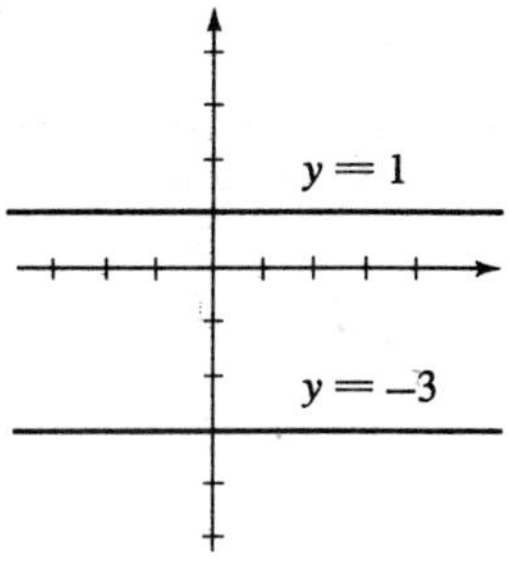

Figure 5.7

Problems

In Problems 1–16, sketch and discuss.

1. $(x-3)^2 = 8(y-2)$.
2. $(y+1)^2 = -4(x-1)$.
3. $(x+5)^2 = 6(y+2)$.
4. $(y-4)^2 = 2x$.
5. $x^2 + 4x - 4y + 8 = 0$.
6. $x^2 + 6x - 4y - 3 = 0$.
7. $y^2 - x + 2y + 4 = 0$.
8. $y^2 - 5y + 6 = 0$.
9. $y^2 + 2x + 4 = 0$.
10. $4x^2 - 4x - 8y - 19 = 0$.
11. $2x^2 - 6x + 5 = 0$.
12. $16y^2 - 64x + 8y + 33 = 0$.
13. $25y^2 - 200x - 20y - 116 = 0$.
14. $4x^2 - 4x - 4y - 3 = 0$.
15. $12x^2 - 12x - 24y + 11 = 0$.
16. $9y^2 - 36x - 12y + 22 = 0$.

In Problems 17–24, find an equation(s) of the parabola(s) described.

17. Focus: $(3, 5)$; directrix: $x = -1$.
18. Focus: $(-2, 1)$; directrix: $y = 0$.
19. Vertex: $(4, 1)$; focus: $(4, 4)$.
20. Vertex: $(3, 5)$; directrix: $x = 1$.
21. Vertex: $(-2, 4)$; axis $y = 4$; length of latus rectum: 8.
22. Axis parallel to the x axis, contains $(5, 3)$, $(2, -3)$, and $(10, 5)$.
23. Axis parallel to the y axis, contains $(0, 6)$, $(3, -6)$, and $(8, 14)$.
24. Vertex: $(1, -2)$; contains: $(5, 2)$.

A line is tangent to a parabola if it has one and only one point in common with the parabola and is not on or parallel to its axis. Use this fact to work Problems 25–28.

25. Find an equation of the line tangent to $y^2 + x + y - 1 = 0$ at $(-1, 1)$.
26. Find an equation of the line tangent to $x^2 - 2x + 4y - 3 = 0$ at $(1, 1)$.
27. Find an equation(s) of the line(s) tangent to $x^2 + 4x + y - 3 = 0$ and containing $(1, 2)$.
28. Find an equation(s) of the line(s) tangent to $y^2 - x + 2y + 3 = 0$ and containing $(2, -2)$.
29. If a parabola with a vertical axis contains the points (x_0, y_0), (x_1, y_1), and (x_2, y_2), show that its equation can be put into the form

$$\begin{vmatrix} x^2 & x & y & 1 \\ x_0^2 & x_0 & y_0 & 1 \\ x_1^2 & x_1 & y_1 & 1 \\ x_2^2 & x_2 & y_2 & 1 \end{vmatrix} = 0.$$

30. What happens to the determinant of Problem 29 if the three given points are collinear? (*Hint*: See Problem 25, page 82.)

5.4

The Ellipse

Definition

An *ellipse* is the set of all points (x, y) such that the sum of the distances from (x, y) to a pair of distinct fixed points (*foci*) is a fixed constant.

Let us choose the foci to be $(c, 0)$ and $(-c, 0)$ (see Figure 5.8) and let the fixed constant be $2a$. If (x, y) represents a point on the ellipse, we have

$$\sqrt{(x-c)^2 + y^2} + \sqrt{(x+c)^2 + y^2} = 2a,$$

$$\sqrt{(x-c)^2 + y^2} = 2a - \sqrt{(x+c)^2 + y^2},$$

$$x^2 - 2cx + c^2 + y^2 = 4a^2 - 4a\sqrt{(x+c)^2 + y^2} + x^2 + 2cx + c^2 + y^2,$$

$$4a\sqrt{(x+c)^2 + y^2} = 4a^2 + 4cx,$$

$$\sqrt{(x+c)^2 + y^2} = a + \frac{cx}{a},$$

5.4 The Ellipse

$$x^2 + 2cx + c^2 + y^2 = a^2 + 2cx + \frac{c^2 x^2}{a^2},$$

$$\frac{a^2 - c^2}{a^2} x^2 + y^2 = a^2 - c^2,$$

$$\frac{x^2}{a^2} + \frac{y^2}{a^2 - c^2} = 1.$$

The triangle of Figure 5.8, with vertices $(c, 0)$, $(-c, 0)$, and (x, y), has one side of length $2c$. The sum of the lengths of the other two sides is $2a$. Thus

$$2a > 2c,$$
$$a > c,$$
$$a^2 > c^2,$$
$$a^2 - c^2 > 0.$$

Since $a^2 - c^2$ is positive, we may replace it by another positive number, b^2. Thus

$$\frac{x^2}{a^2} + \frac{y^2}{b^2} = 1, \quad \text{where } b^2 = a^2 - c^2.$$

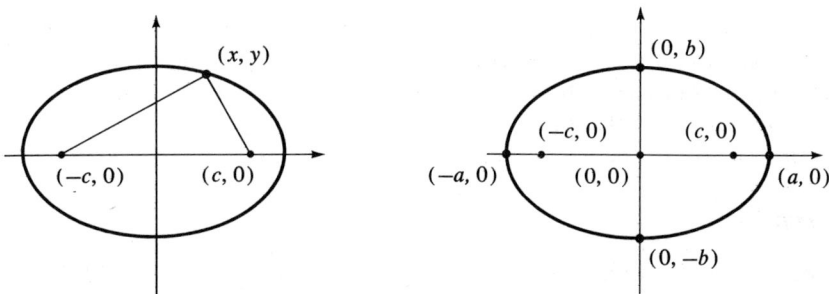

Figure 5.8 **Figure 5.9**

Note that we squared both sides of the equation at two of the steps. In both cases, both sides of the equation are nonnegative. Thus we have introduced no extraneous roots, and the steps may be reversed. (See Note 2 on page 89.)

Note that there are two axes of symmetry: the x axis and the y axis. Furthermore, $(\pm a, 0)$ are the x intercepts and $(0, \pm b)$ are the y intercepts where $a > b$ (since $b^2 = a^2 - c^2$). Thus the x axis is called the *major axis* and the y axis is the *minor axis*. The points $(\pm a, 0)$ on the major axis are called the *vertices*, the points $(0, \pm b)$ on the minor axis are the *covertices*, and the point of intersection of the axes, $(0, 0)$, is called the *center* (see Figure 5.9). The *foci* $(\pm c, 0)$ are on the major axis.

Theorem 5.6

A point (x, y) is on the ellipse with vertices $(\pm a, 0)$ and foci $(\pm c, 0)$ if and only if it satisfies the equation

$$\frac{x^2}{a^2} + \frac{y^2}{b^2} = 1,$$

where $b^2 = a^2 - c^2$.

An ellipse has two *latera recta* (plural of latus rectum), which are chords of the ellipse perpendicular to the major axis and containing the foci. If $x = \pm c$, then

$$\frac{c^2}{a^2} + \frac{y^2}{b^2} = 1,$$

$$\frac{y^2}{b^2} = \frac{a^2 - c^2}{a^2} = \frac{b^2}{a^2},$$

$$y^2 = \frac{b^4}{a^2},$$

$$y = \pm \frac{b^2}{a}.$$

Thus, one latus rectum has end points $(c, \pm b^2/a)$, while the other has end points $(-c, \pm b^2/a)$. In both cases the length is $2b^2/a$. This length may be used as an aid in sketching, as was done with the parabola; however, the vertices and covertices allow one to make a reasonable sketch.

Again, the role of the x and y may be reversed.

Theorem 5.7

A point (x, y) is on the ellipse with vertices $(0, \pm a)$ and foci $(0, \pm c)$ if and only if it satisfies the equation

$$\frac{y^2}{a^2} + \frac{x^2}{b^2} = 1,$$

where $b^2 = a^2 - c^2$.

One question that immediately arises is, How can we tell whether we have

$$\frac{x^2}{a^2} + \frac{y^2}{b^2} = 1 \quad \text{or} \quad \frac{y^2}{a^2} + \frac{x^2}{b^2} = 1?$$

The numbers in the denominator are not labeled a and b, so how do we know which is a and which is b? The answer is "size." In both cases $a > b$. Thus the larger denominator is a^2, and the smaller is b^2.

5.4 The Ellipse

Example 1
Sketch and discuss $9x^2 + 25y^2 = 225$.

First, we put the equation into standard form by dividing through by 225:
$$\frac{x^2}{25} + \frac{y^2}{9} = 1.$$

Now
$$a^2 = 25, \quad b^2 = 9,$$
and
$$c^2 = a^2 - b^2 = 16.$$

This ellipse has center $(0, 0)$, vertices $(\pm 5, 0)$, covertices $(0, \pm 3)$, and foci $(\pm 4, 0)$. The latera recta have length $2b^2/a = 2 \cdot 9/5 = 3.6$ (see Figure 5.10).

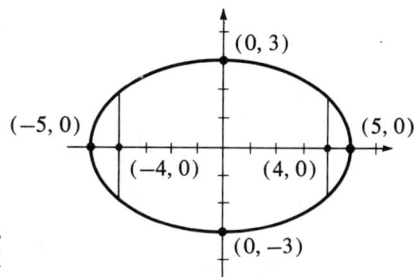

Figure 5.10

Example 2
Sketch and discuss $25x^2 + 16y^2 = 400$.

Putting the equation into standard form gives
$$\frac{x^2}{16} + \frac{y^2}{25} = 1.$$

Now
$$a^2 = 25, \quad b^2 = 16$$
and
$$c^2 = a^2 - b^2 = 9.$$

This ellipse has center $(0, 0)$, vertices $(0, \pm 5)$, covertices $(\pm 4, 0)$, and foci $(0, \pm 3)$. The latera recta have length
$$2b^2/a = 2 \cdot 16/5 = 6.4$$
(see Figure 5.11).

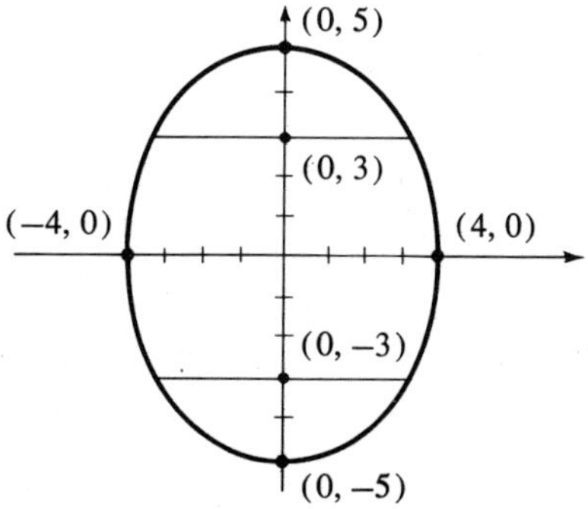

Figure 5.11

Example 3
Find an equation of the ellipse with vertices $(0, \pm 8)$ and foci $(0, \pm 5)$.

Since the vertices are on the y axis, we have the form
$$\frac{y^2}{a^2} + \frac{x^2}{b^2} = 1.$$

Furthermore, $a = 8$ and $c = 5$; thus $b^2 = a^2 - c^2 = 64 - 25 = 39$. The final result is
$$\frac{y^2}{64} + \frac{x^2}{39} = 1.$$

In addition to the quantities named, each ellipse is associated with a number, called the eccentricity. For any ellipse the *eccentricity* is

$$e = \frac{c}{a}.$$

The eccentricity of an ellipse satisfies the inequalities $0 < e < 1$. It gives a measure of the shape of the ellipse: the closer the eccentricity is to 0, the more nearly circular is the ellipse. For instance, in Example 1, $e = 4/5$, while in Example 2 $e = 3/5$. The ellipse of Example 2 is more nearly circular than the ellipse of Example 1, as can be easily seen by the sketches.

There is also a directrix associated with each focus of an ellipse. Associated with the focus $(c, 0)$ of the ellipse

$$\frac{x^2}{a^2} + \frac{y^2}{b^2} = 1$$

is the *directrix*

$$x = \frac{a}{e} = \frac{a^2}{c}.$$

If P is any point of the ellipse, the distance from P to the focus divided by the distance from P to the directrix is equal to the eccentricity. This is sometimes used as the definition of an ellipse.

Suppose we start with focus $(c, 0)$, directrix $x = a^2/c$, and eccentricity $e = c/a$. Now let us find the set of all points $P: (x, y)$ such that the distance from P to the focus divided by the distance from P to the directrix equals the eccentricity.

$$\frac{\sqrt{(x-c)^2 + y^2}}{\frac{a^2}{c} - x} = \frac{c}{a},$$

$$\sqrt{(x-c)^2 + y^2} = a - \frac{cx}{a},$$

$$x^2 - 2cx + c^2 + y^2 = a^2 - 2cx + \frac{c^2 x^2}{a^2},$$

$$\frac{a^2 - c^2}{a^2} x^2 + y^2 = a^2 - c^2,$$

$$\frac{x^2}{a^2} + \frac{y^2}{a^2 - c^2} = 1.$$

5.4 The Ellipse

With $b^2 = a^2 - c^2$, this becomes

$$\frac{x^2}{a^2} + \frac{y^2}{b^2} = 1.$$

The same result can be obtained using focus $(-c, 0)$, directrix $x = -a^2/c$, and eccentricity $e = c/a$.

For a parabola, the distance from a point on the parabola to the focus divided by the distance of the point from the directrix is always 1. Thus we define $e = 1$ for every parabola.

Problems

In Problems 1–10, sketch and discuss the given ellipse.

1. $\dfrac{x^2}{169} + \dfrac{y^2}{25} = 1.$
2. $\dfrac{x^2}{144} + \dfrac{y^2}{169} = 1.$
3. $\dfrac{x^2}{25} + \dfrac{y^2}{4} = 1.$
4. $\dfrac{x^2}{36} + \dfrac{y^2}{16} = 1.$
5. $\dfrac{x^2}{25} + \dfrac{y^2}{49} = 1.$
6. $x^2 + 4y^2 = 4.$
7. $9x^2 + 4y^2 = 36.$
8. $9x^2 + y^2 = 9.$
9. $16x^2 + 9y^2 = 144.$
10. $4x^2 + 25y^2 = 100.$

In Problems 11–18, find an equation(s) of the ellipse(s) described.

11. Center: $(0, 0)$; vertex: $(0, 13)$; focus: $(0, -5)$.
12. Center: $(0, 0)$; covertex: $(0, 5)$; focus: $(-12, 0)$.
13. Center: $(0, 0)$; vertex: $(5, 0)$; contains $(\sqrt{15}, 2)$.
14. Center: $(0, 0)$; axes on the coordinate axis; contains $(2, 2)$ and $(-4, 1)$.
15. Vertices: $(\pm 6, 0)$; length of latus rectum: 3.
16. Covertices: $(\pm 2, 0)$; length of latus rectum: 2.
17. Foci: $(\pm 6, 0)$; $e = 3/5$.
18. Foci: $(\pm 2, 0)$; directrices: $x = \pm 8$.
19. Prove Theorem 9.7.

A line is tangent to an ellipse if it has one and only one point in common with the ellipse. Use this fact to work Problems 20–23.

20. Find an equation of the line tangent to $x^2 + 4y^2 = 20$ at $(2, 2)$.
21. Find an equation of the line tangent to $2x^2 + 3y^2 = 11$ at $(2, 1)$.
22. Find an equation of the line containing $(3, -2)$ and tangent to $4x^2 + y^2 = 8$.
23. Find an equation of the line containing $(2, 4)$ and tangent to $3x^2 + 8y^2 = 84$.

24. Suppose, in Figure 5.12, that A and B are fixed pins on the arm ABP and that AP and BP have lengths a and b, respectively. Show that if A is free to slide in channel XX' and B in channel YY', the point P traces an ellipse.

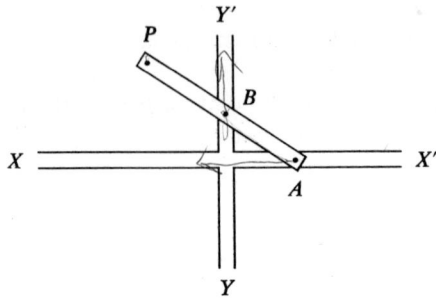

Figure 5.12

25. Given the focus $(-c, 0)$, directrix $x = -a^2/c$, and eccentricity $e = c/a$, show that these define the ellipse

$$\frac{x^2}{a^2} + \frac{y^2}{a^2 - c^2} = 1.$$

26. The earth moves in an elliptical orbit about the sun, with the sun at one focus. The least and greatest distances of the earth from the sun are 91,446,000 miles and 94,560,000 miles, respectively. What is the eccentricity of the ellipse?

5.5

Ellipse with Center (h, k)

Theorem 5.8

A point (x, y) is on the ellipse with center (h, k), vertices $(h \pm a, k)$, and covertices $(h, k \pm b)$ if and only if it satisfies the equation

$$\frac{(x - h)^2}{a^2} + \frac{(y - k)^2}{b^2} = 1.$$

The foci are $(h \pm c, k)$, where $c^2 = a^2 - b^2$.

5.5 Ellipse with Center (h, k)

Theorem 5.9

A point (x, y) is on the ellipse with center (h, k), vertices $(h, k \pm a)$, and covertices $(h \pm b, k)$ if and only if it satisfies the equation

$$\frac{(y-k)^2}{a^2} + \frac{(x-h)^2}{b^2} = 1.$$

The foci are $(h, k \pm c)$, where $c^2 = a^2 - b^2$.

These two theorems may be derived from Theorems 5.6 and 5.7, which are special cases of these, and they can be established by an argument similar to the one used for a point on the parabola with vertex (h, k).

Example 1

Sketch and discuss

$$\frac{(x-1)^2}{9} + \frac{(y+2)^2}{4} = 1.$$

The center of the ellipse is $(1, -2)$; the vertices are $(1 \pm 3, -2)$, and the covertices $(1, -2 \pm 2)$. Since $c^2 = a^2 - b^2 = 5$, the foci are $(1 \pm \sqrt{5}, -2)$. The length of the latera recta is

$$\frac{2b^2}{a} = 2 \cdot \frac{4}{3} = \frac{8}{3}$$

(see Figure 5.13).

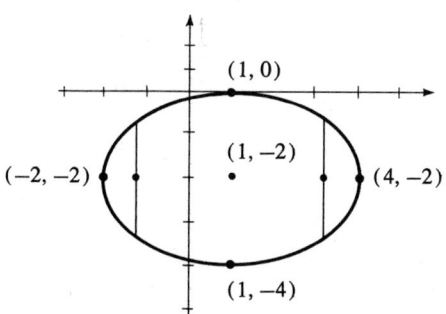

Figure 5.13

If the equations of Theorems 5.8 and 5.9 are cleared of fractions and multiplied out, we have

$$b^2x^2 + a^2y^2 - 2b^2hx - 2a^2ky + (b^2h^2 + a^2k^2 - a^2b^2) = 0$$

and

$$a^2x^2 + b^2y^2 - 2a^2hx - 2b^2ky + (a^2h^2 + b^2k^2 - a^2b^2) = 0.$$

Both of these are of the form

$$Ax^2 + Cy^2 + Dx + Ey + F = 0,$$

where A and C are both positive or both negative (if the equation were multiplied through by a negative number).

Theorem 5.10

An ellipse with axes parallel to (or on) the coordinate axes can be represented by an equation of the form:

$$Ax^2 + Cy^2 + Dx + Ey + F = 0,$$

where AC is positive.

Again, the form of the equation of Theorem 5.10 is called the *general form* of an ellipse, while those of Theorem 5.8 and 5.9 are called *standard forms*. It is a simple matter to change from a standard form to the general form, and the reverse is accomplished by completing squares.

Example 2

Express $25x^2 + 4y^2 + 50x - 24y - 39 = 0$ in standard form.

$$25(x^2 + 2x \quad) + 4(y^2 - 6y \quad) = 39,$$
$$25(x^2 + 2x + 1) + 4(y^2 - 6y + 9) = 39 + 25 + 36,$$
$$25(x + 1)^2 + 4(y - 3)^2 = 100,$$
$$\frac{(x + 1)^2}{4} + \frac{(y - 3)^2}{25} = 1.$$

The converse of Theorem 5.10 is not true: that is, it is not true that every equation of the form

$$Ax^2 + Cy^2 + Dx + Ey + F = 0,$$

with AC positive, represents an ellipse. If $A = C$, the equation may represent a circle, a point, or no graph; if $A \neq C$, the equation may represent an ellipse, a point, or no graph. The circle, point, and no graph are called degenerate cases of an ellipse.

Example 3

Determine whether $9x^2 + 4y^2 + 36x + 8y + 40 = 0$ represents an ellipse or a degenerate form of an ellipse.

$$9(x^2 + 4x \quad) + 4(y^2 + 2y \quad) = -40,$$
$$9(x^2 + 4x + 4) + 4(y^2 + 2y + 1) = -40 + 36 + 4,$$
$$9(x + 2)^2 + 4(y + 1)^2 = 0.$$

5.5 Ellipse with Center (h, k)

We can now see that, since neither of the two expressions on the left can be negative, the sum is zero only if both terms are zero. Thus $(-2, -1)$ is the only point satisfying the equation.

If the constant 40 of this example is replaced by a larger number, then the right-hand side of the last equation is negative and there is no point in the plane satisfying the equation.

Problems

In Problems 1–14, sketch and discuss.

1. $x^2 + 4y^2 - 24y + 35 = 0$.
2. $x^2 + 4y^2 - 2x - 3 = 0$.
3. $9x^2 + 25y^2 + 72x - 50y - 56 = 0$.
4. $4x^2 + 9y^2 - 24x + 36y + 36 = 0$.
5. $4x^2 + y^2 + 8x + 10y + 13 = 0$.
6. $16x^2 + 25y^2 - 160x + 200y + 400 = 0$.
7. $8x^2 + 9y^2 + 64x - 54y + 209 = 0$.
8. $9x^2 + 4y^2 + 54x - 16y + 133 = 0$.
9. $4x^2 + 9y^2 + 8x - 36y + 4 = 0$.
10. $25x^2 + 16y^2 - 160y = 0$.
11. $25x^2 + 4y^2 - 150x + 40y + 350 = 0$.
12. $4x^2 + 4y^2 - 32x - 24y + 99 = 0$.
13. $4x^2 + 9y^2 + 48x - 144y + 684 = 0$.
14. $25x^2 + 4y^2 - 250x + 56y + 821 = 0$.

In Problems 15–26, find an equation(s) of the ellipse(s) described.

15. Vertices: $(1, 0)$ and $(1, -8)$; covertices: $(2, -4)$ and $(0, -4)$.
16. Vertices: $(3, 0)$ and $(3, 10)$; focus: $(3, 2)$.
17. Vertices: $(-1, 8)$ and $(-1, -2)$; contains $(1, 0)$.
18. Vertex: $(3, 5)$; covertex: $(1, 0)$.
19. Covertices: $(-5, 0)$ and $(1, 0)$; length of latera recta: $9/2$.
20. Foci: $(-1, 0)$ and $(-1, -6)$; length of latera recta: $32/5$.
21. Contains $(6, -1)$, $(-4, -5)$, $(6, -5)$, and $(-12, -3)$.
22. Contains $(1, 1)$, $(7, -3)$, $(-1, -1)$, and $(4, 5/2)$.
23. Vertex: $(8, -1)$; focus: $(6, -1)$; $e = 3/5$.
24. Focus: $(5, 2)$; corresponding directrix: $x = 11$; $e = 1/2$.
25. Center: $(-3, 1)$; focus: $(-3, 4)$; corresponding directrix: $y = 28$.
26. Focus: $(5, 2)$; vertex: $(7, 2)$; length of latera recta: $32/5$.

A line is tangent to an ellipse if it has one and only one point in common with the ellipse. Use this fact to work Problems 27–30.

27. Find an equation of the line tangent to $x^2 + 4y^2 + 3x - 4y - 4 = 0$ at $(1, 1)$.
28. Find an equation of the line tangent to $2x^2 + 3y^2 - 4x + y - 4 = 0$ at $(2, 1)$.
29. Find an equation(s) of the line(s) tangent to $4x^2 + y^2 - 3x + 2y - 15 = 0$ and containing $(-32/3, -1)$.
30. Find an equation(s) of the line(s) tangent to $2x^2 + y^2 - 4x - 3y - 16 = 0$ and containing $(-7/2, 6)$.

5.6

The Hyperbola

Definition

A **hyperbola** is the set of all points (x, y) in a plane such that the positive difference between the distances from (x, y) to a pair of distinct fixed points (*foci*) is a fixed constant.

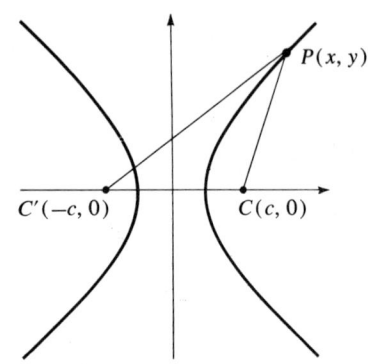

Figure 5.14

Again, let us choose the foci to be $(c, 0)$ and $(-c, 0)$ (see Figure 5.14) and choose the fixed constant to be $2a$. If (x, y) represents a point on the ellipse, we have

$$\sqrt{(x-c)^2 + y^2} - \sqrt{(x+c)^2 + y^2} = \pm 2a,$$

$$\sqrt{(x-c)^2 + y^2} = \sqrt{(x+c)^2 + y^2} \pm 2a,$$

$$x^2 - 2cx + c^2 + y^2 = x^2 + 2cx + c^2 + y^2 \pm 4a\sqrt{(x+c)^2 + y^2} + 4a^2,$$

$$\mp 4a\sqrt{(x+c)^2 + y^2} = 4a^2 + 4cx,$$

$$\mp \sqrt{(x+c)^2 + y^2} = a + \frac{cx}{a},$$

$$x^2 + 2cx + c^2 + y^2 = a^2 + 2cx + \frac{c^2 x^2}{a^2},$$

$$\frac{c^2 - a^2}{a^2} x^2 - y^2 = c^2 - a^2,$$

$$\frac{x^2}{a^2} - \frac{y^2}{c^2 - a^2} = 1.$$

In the triangle PCC' of Figure 5.14,

$$PC' < PC + CC',$$
$$PC' - PC < CC',$$
$$2a < 2c,$$
$$a < c,$$

5.6 The Hyperbola

$$c^2 - a^2 > 0.$$

Since $c^2 - a^2$ is positive, we may replace it by another positive number, b^2. Thus

$$\frac{x^2}{a^2} - \frac{y^2}{b^2} = 1,$$

where $b^2 = c^2 - a^2$.

Again we squared both sides of the equation at two of the steps. The first time, both sides of the equation were positive; the second time, they were either both positive or both negative. Thus we have introduced no extraneous roots, and the steps may be reversed. (See Note 2 on page 89.)

Again, both the x axis and the y axis are axes of symmetry and again $(\pm a, 0)$ are the x intercepts. However, there are no y intercepts; when $x = 0$, we have

$$-\frac{y^2}{b^2} = 1,$$

which is not satisfied by any real number y. The x axis (containing two points of the hyperbola) is called the *transverse axis*; the y axis is called the *conjugate axis*. The points $(\pm a, 0)$ on the transverse axis are called the *vertices*, and the point of intersection of the axes, $(0, 0)$, is called the *center* (see Figure 5.15).

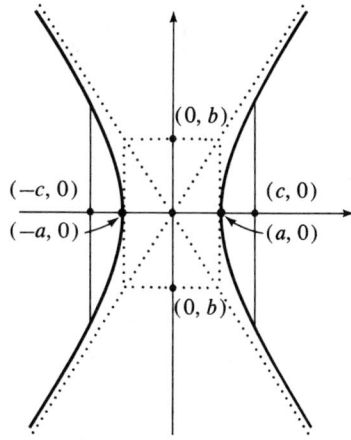

Figure 5.15

Theorem 5.11

A point (x, y) is on the hyperbola with vertices $(\pm a, 0)$ and foci $(\pm c, 0)$ if and only if it satisfies the equation

$$\frac{x^2}{a^2} - \frac{y^2}{b^2} = 1,$$

where $b^2 = c^2 - a^2$.

For every hyperbola there are two lines that the curve approaches more and more closely at its extremities. These two lines are called asymptotes (see Figure 5.15 in which the slanting dotted lines are the asymptotes). It might be noted that parabolas do not have asymptotes. Thus a hyperbola is not—as might appear from inaccurate diagrams—a pair of parabolas. The hyperbola

$$\frac{x^2}{a^2} - \frac{y^2}{b^2} = 1, \quad \text{or} \quad y = \pm \frac{b}{a}\sqrt{x^2 - a^2}$$

has asymptotes

$$y = \pm \frac{b}{a} x.$$

Let us verify this, at least for the portion in the first quadrant. Here we are dealing with

$$y = \frac{b}{a}\sqrt{x^2 - a^2} \quad \text{and} \quad y = \frac{b}{a} x$$

for positive values of x. For a given value of x, let us consider the difference d between the y coordinates of the points on the hyperbola and the line.

$$d = \frac{b}{a} x - \frac{b}{a}\sqrt{x^2 - a^2} = \frac{b}{a}(x - \sqrt{x^2 - a^2}).$$

Multiplying numerator and denominator by $x + \sqrt{x^2 + a^2}$, we have

$$d = \frac{b}{a} \frac{x^2 - (x^2 - a^2)}{x + \sqrt{x^2 - a^2}} = \frac{ab}{x + \sqrt{x^2 - a^2}}.$$

Now the numerator is a constant; but, for large positive values of x, both terms of the denominator are large and positive. In fact, the larger the value of x, the larger the denominator and, therefore, the smaller d is. Thus d approaches zero as x gets larger, which shows that the line is an asymptote of the hyperbola. Of course, similar arguments can be used to show the same thing in the other three quadrants.

A convenient way of sketching the asymptotes is to plot both $(\pm a, 0)$ and $(0, \pm b)$ (even though the second pair of points is not on the hyperbola) and sketch the rectangle determined by them (see Figure 5.15). The diagonals of this rectangle are the asymptotes.

Again, two *latera recta* contain the foci and are perpendicular to the transverse axis. By using the same method as in the case of the parabola and ellipse, we can show their length to be

$$\frac{2b^2}{a}.$$

As with the parabola and the ellipse, the roles of x and y can be reversed.

Theorem 5.12

A point (x, y) is on the hyperbola with vertices $(0, \pm a)$ and foci $(0, \pm c)$ if and only if it satisfies the equation

$$\frac{y^2}{a^2} - \frac{x^2}{b^2} = 1,$$

where $b^2 = c^2 - a^2$.

5.6 The Hyperbola

It might be noted that a and b are determined by the sign of the term in which they appear; a^2 is always the denominator of the positive term and b^2 the denominator of the negative term. There is no requirement that a be greater than b, as there was for an ellipse.

The asymptotes of the hyperbola

$$\frac{y^2}{a^2} - \frac{x^2}{b^2} = 1$$

are

$$y = \pm \frac{a}{b} x.$$

Since the formulas for the asymptotes for the two cases are rather easy to confuse, a method that always works is to replace the 1 by 0 in the standard form and solve for y.

Example 1

Sketch and discuss.

$$\frac{x^2}{9} - \frac{y^2}{16} = 1.$$

We see that $a^2 = 9$, $b^2 = 16$, and $c^2 = a^2 + b^2 = 25$. This hyperbola has center $(0, 0)$, vertices $(\pm 3, 0)$, and foci $(\pm 5, 0)$. Its asymptotes are $y = \pm 4x/3$, and the length of the latera recta is $2b^2/a = 32/3$ (see Figure 5.16).

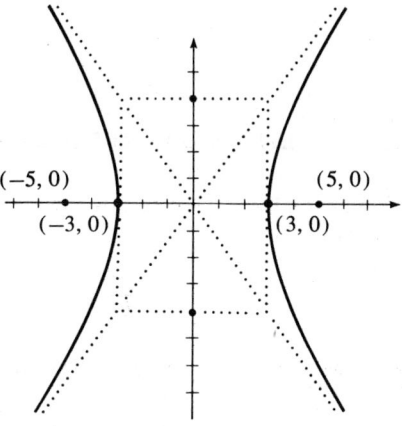

Figure 5.16

Example 2

Sketch and discuss $16x^2 - 9y^2 + 144 = 0$.

Putting this equation into standard form, we have

$$\frac{y^2}{16} - \frac{x^2}{9} = 1.$$

We see that $a^2 = 16$, $b^2 = 9$, and $c^2 = a^2 + b^2 = 25$. This hyperbola has center $(0, 0)$, vertices $(0, \pm 4)$ and foci $(0, \pm 5)$. Its asymptotes are $y = \pm 4x/3$ and the length of the latera recta is $2b^2/a = 9/2$ (see Figure 5.17).

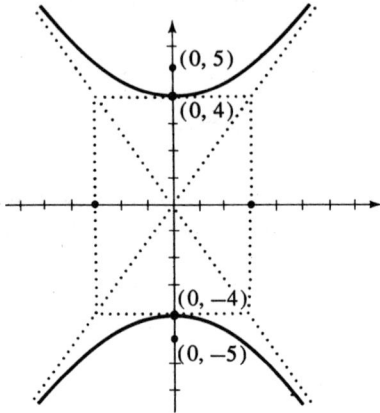

Figure 5.17

Note the relationship between the equations of these two examples when in the standard forms; the left-hand sides are simply opposite in sign. Such hyperbolas are called *conjugate hyperbolas*.

Example 3

Find an equation of the hyperbola with foci $(\pm 4, 0)$ and vertex $(2, 0)$.

Since the foci are on the transverse axis and we are given that they are on the x axis, we must have the form

$$\frac{x^2}{a^2} - \frac{y^2}{b^2} = 1.$$

The foci tell us that $c = 4$, and the vertex gives $a = 2$; thus $b^2 = c^2 - a^2 = 12$. The resulting equation is

$$\frac{x^2}{4} - \frac{y^2}{12} = 1,$$

or

$$3x^2 - y^2 = 12.$$

5.6 The Hyperbola

Hyperbolas, as well as the other conic sections, can be determined by a single focus, a directrix, and an eccentricity. For the hyperbola

$$\frac{x^2}{a^2} - \frac{y^2}{b^2} = 1,$$

we have *eccentricity*

$$e = \frac{c}{a}$$

and *directrices*

$$x = \pm \frac{a}{e} = \pm \frac{a^2}{c},$$

where $x = a^2/c$ is used in conjunction with the focus $(c, 0)$ and $x = -a^2/c$ with the focus $(-c, 0)$. Since $c > a$, $e = c/a > 1$. Furthermore a single focus and directrix gives the entire hyperbola—not merely one branch. Either focus with its corresponding directrix generates a hyperbola.

Problems

In Problems 1–14, sketch and discuss.

1. $\dfrac{x^2}{16} - \dfrac{y^2}{9} = 1.$
2. $\dfrac{x^2}{4} - \dfrac{y^2}{1} = 1.$
3. $\dfrac{y^2}{9} - \dfrac{x^2}{4} = 1.$
4. $\dfrac{y^2}{1} - \dfrac{x^2}{9} = 1.$
5. $\dfrac{x^2}{144} - \dfrac{y^2}{25} = 1.$
6. $\dfrac{y^2}{25} - \dfrac{x^2}{144} = 1.$
7. $\dfrac{y^2}{25} - \dfrac{x^2}{9} = 1.$
8. $4x^2 - 9y^2 = 36.$
9. $4x^2 - y^2 = 4.$
10. $4x^2 - y^2 + 16 = 0.$
11. $x^2 - y^2 = 9.$
12. $16x^2 - 9y^2 = -36.$
13. $36y^2 - 100x^2 = 225.$
14. $9x^2 - 4y^2 - 9 = 0.$

In Problems 15–26, find an equation(s) of the hyperbola(s) described.

15. Vertices: $(\pm 2, 0)$; focus: $(-4, 0)$.
16. Foci: $(0, \pm 5)$; vertex: $(0, 2)$.
17. Asymptotes: $y = \pm 2x/3$; vertex: $(6, 0)$.
18. Asymptotes: $y = \pm 3x/4$; focus: $(0, -10)$.
19. Asymptotes: $y = \pm 4x/3$; contains $(3\sqrt{2}, 4)$.
20. Asymptotes: $y = \pm 3x/4$; length of latera recta: $9/2$.
21. Vertices: $(\pm 5, 0)$; contains $(9/5, -4)$.

22. Foci: $(\pm 2\sqrt{61}, 0)$; contains $(65/6, 5)$.
23. Vertices: $(0, \pm 3)$; $e = 5/3$.
24. Foci: $(\pm 10, 0)$; $e = 5/2$.
25. Directices: $x = \pm 9/5$; $e = 5/3$.
26. Directrices: $y = \pm 25/13$; focus: $(0, -13)$.

A line is tangent to a hyperbola if it has one and only one point in common with the hyperbola. Use this fact to work Problems 27–30.

27. Find an equation of the line tangent to $16x^2 - 9y^2 = 144$ at $(13/4, 5/3)$.
28. Find an equation of the line tangent to $x^2 - y^2 = 16$ at $(-5, 3)$.
29. Find an equation(s) of the line(s) tangent to $x^2 - y^2 = 9$ and containing $(9, 9)$.
30. Find an equation(s) of the line(s) tangent to $4x^2 - 9y^2 = 7$ and containing $(-7, 7)$.
31. Show that there is a number k such that, if P is any point of a hyperbola, the product of the distances of P from the asymptotes of the hyperbola is k.

5.7

Hyperbola with Center (h, k)

Theorem 5.13

A point (x, y) is on the hyperbola with center (h, k), vertices $(h \pm a, k)$, and foci $(h \pm c, k)$ if and only if it satisfies the equation

$$\frac{(x-h)^2}{a^2} - \frac{(y-k)^2}{b^2} = 1,$$

where $b^2 = c^2 - a^2$.

Theorem 5.14

A point (x, y) is on the hyperbola with center (h, k), vertices $(h, k \pm a)$, and foci $(h, k \pm c)$, if and only if it satisfies the equation

$$\frac{(y-k)^2}{a^2} - \frac{(x-h)^2}{b^2} = 1,$$

where $b^2 = c^2 - a^2$.

5.7 Hyperbola with Center (h, k)

Example 1

Sketch and discuss

$$\frac{(x-1)^2}{4} - \frac{(y+3)^2}{9} = 1.$$

The center of the hyperbola is $(1, -3)$. $a^2 = 4$, $b^2 = 9$, and $c^2 = a^2 + b^2 = 13$. Thus the vertices are $(1 \pm 2, -3)$ and the foci are $(1 \pm \sqrt{13}, -3)$. The length of the latera recta is $2b^2/a = 2 \cdot 9/2 = 9$, and the equations of the asymptotes are

$$y + 3 = \pm \frac{3}{2}(x - 1),$$

or

$$3x - 2y - 9 = 0 \quad \text{and} \quad 3x + 2y + 3 = 0.$$

The method of replacing the 1 by 0 in the original equation is especially useful here (see Figure 5.18).

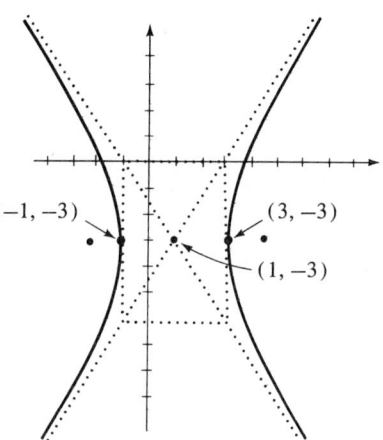

Figure 5.18

If we clear fractions and carry out the multiplication in the equations of Theorems 5.13 and 5.14, we have

$$b^2x^2 - a^2y^2 - 2b^2hx + 2a^2ky + (b^2h^2 - a^2k^2 - a^2b^2) = 0$$

and

$$-a^2x^2 + b^2y^2 + 2a^2hx - 2b^2ky + (-a^2h^2 + b^2k^2 - a^2b^2) = 0.$$

Both of these are in the form

$$Ax^2 + Cy^2 + Dx + Ey + F = 0,$$

where A and C have opposite signs.

Theorem 5.15

A hyperbola with axes parallel to (or on) the coordinate axes can be represented by an equation of the form:

$$Ax^2 + Cy^2 + Dx + Ey + F = 0,$$

where AC is negative.

Again, the equations of Theorems 5.13 and 5.14 are called *standard forms* while that of equation 5.15 is called the *general form* of a hyperbola.

Example 2

Express $9x^2 - 4y^2 + 36x + 32y + 8 = 0$ in standard form.

$$9(x^2 + 4x \quad) - 4(y^2 - 8y \quad) = -8$$
$$9(x^2 + 4x + 4) - 4(y^2 - 8y + 16) = -8 + 36 - 64,$$
$$9(x + 2)^2 - 4(y - 4)^2 = -36,$$
$$\frac{(y-4)^2}{9} - \frac{(x+2)^2}{4} = 1.$$

Again, the converse of Theorem 5.15 is not true. Equations of the form
$$Ax^2 + Cy^2 + Dx + Ey + F = 0,$$
where AC is negative, may represent either a hyperbola or a pair of intersecting lines.

Example 3

Determine whether $25x^2 - 4y^2 - 150x - 16y + 209 = 0$ represents a hyperbola or a pair of intersecting lines.

$$25(x^2 - 6x \quad) - 4(y^2 + 4y \quad) = -209,$$
$$25(x^2 - 6x + 9) - 4(y^2 + 4y + 4) = -209 + 225 - 16,$$
$$25(x - 3)^2 - 4(y + 2)^2 = 0.$$

At this point we know that the equation represents a pair of intersecting lines. Continuing, we have
$$4(y + 2)^2 = 25(x - 3)^2,$$
$$2(y + 2) = \pm 5(x - 3),$$
$$5x - 2y - 19 = 0 \quad \text{and} \quad 5x + 2y - 11 = 0.$$

Problems

In Problems 1–12, sketch and discuss.

1. $16x^2 - 9y^2 + 54y - 225 = 0.$
2. $4x^2 - y^2 + 8x + 8 = 0.$
3. $x^2 - y^2 - 10x - 2y - 40 = 0.$
4. $4x^2 - 9y^2 - 4x + 36y - 71 = 0.$
5. $4x^2 - 16y^2 + 12x + 16y + 69 = 0.$
6. $16x^2 - y^2 - 16x + 6y - 1 = 0.$
7. $9x^2 - 4y^2 - 36x - 8y + 32 = 0.$
8. $9x^2 - 16y^2 + 18x - 16y - 139 = 0.$
9. $9x^2 - 9y^2 - 6x - 12y - 39 = 0.$
10. $25x^2 - 16y^2 + 200x + 160y = 0.$
11. $36x^2 - 36y^2 + 24x + 36y + 31 = 0.$
12. $4x^2 - y^2 - 2x - y - 16 = 0.$

In Problems 13–24, find an equation(s) of the hyperbola(s) described.

13. Vertices: (4, 1) and (0, 1); focus: (6, 1).
14. Foci: (−2, 5) and (−2, −3); vertex: (−2, 2).
15. Vertex: (6, −1); asymptotes: $3x − 2y − 6 = 0$ and $3x + 2y − 2 = 0$.
16. Asymptotes: $4x − 3y + 13 = 0$ and $4x + 3y − 5 = 0$; focus: (−1, −2).
17. Center: (2, 5); vertex: (2, 7); focus: (2, 0).
18. Asymptotes: $x − y − 7 = 0$ and $x + y − 3 = 0$; contains (0, 2).
19. Contains (2, −2), (−3, 8), (−1, −1), and (2, 8).
20. Contains (−1/4, 0), (9/4, 6), (0, 3), and (−1/4, 6).
21. Foci: (4, 0) and (−6, 0); $e = 5/2$.
22. Focus: (26/5, −2); directrix: $x = 2$; $e = 5/3$.
23. Vertices: (8, 1) and (2, 1); length of latera recta: 24.
24. Foci: (1, 14) and (1, −12); length of latera recta: 25/6.

A line is tangent to a hyperbola if it has one and only one point in common with the hyperbola. Use this fact to work Problems 25–28.

25. Find an equation of the line tangent to $x^2 − 4y^2 + 2x + y = 0$ at (1, 1).
26. Find an equation(s) of the line tangent to $9x^2 − 4y^2 + x + 8y − 42 = 0$ at (2, 1).
27. Find an equation(s) of the line(s) tangent to $x^2 − 5y^2 + 2x + 10y − 24 = 0$ and containing the point (−1, −3).
28. Find an equation of the line(s) tangent to $x^2 − 5y^2 + x + 5 = 0$ and containing the point (−1/2, 19/20).

5.8

Conics and a Right Circular Cone

As we indicated in Section 5.1, the conic sections are curves formed by the intersection of a plane with a right circular cone. We shall prove here that this is really the case. Let us first recall that a cone consists of two portions, or nappes, separated from each other by the vertex.

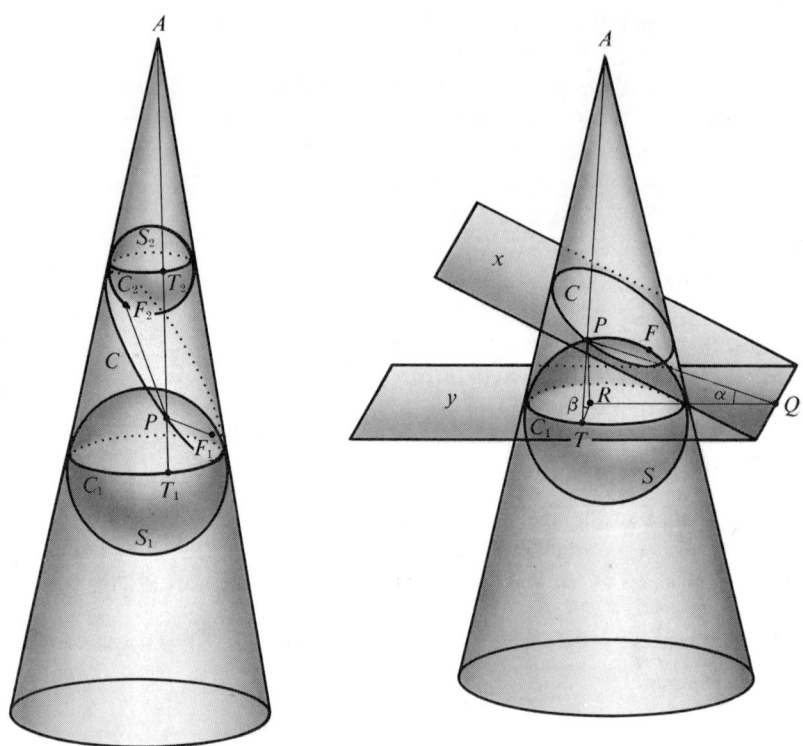

Figure 5.19   **Figure 5.20**

Suppose a plane intersects a right circular cone to give the curve C of Figure 5.19. Let S_1 and S_2 be spheres which are inscribed in the cone and tangent to the plane at the points F_1 and F_2, respectively. (We are considering the case in which there are two distinct tangent spheres which lie in the same nappe of the cone and are tangent at two distinct points. See Problems 1–3 for other cases.) The spheres S_1 and S_2 have the circles C_1 and C_2, respectively, in common with the cone. Of course C_1 and C_2 lie in planes which are perpendicular to the axis of the cone and, therefore, parallel to each other. Let P be any point of the curve C. Consider the line determined by P and the apex A of the cone. This line intersects the circles C_1 and C_2 at points T_1 and T_2, respectively. Since PT_1 and PF_1 are two tangents to the sphere S_1 from P,

$$\overline{PT_1} = \overline{PF_1}.$$

Similarly

$$\overline{PT_2} = \overline{PF_2}.$$

Thus
$$\overline{PF_1} + \overline{PF_2} = \overline{PT_1} + \overline{PT_2} = \overline{T_1 T_2}.$$

Since the circles C_1 and C_2 lie in parallel planes, the length of $T_1 T_2$ is independent of the choice of the point P. Thus $\overline{PF_1} + \overline{PF_2}$ is constant, proving that C is an ellipse with foci F_1 and F_2.

The conic can also be related to the cone by the focus-directrix property. Suppose a plane x intersects a right circular cone to give the curve C of Figure 5.20. Let S be a sphere inscribed in the cone and tangent to the given plane at F. The sphere S has a circle C_1 in common with the cone and lies in a plane y, perpendicular to the axis of the cone. Planes x and y intersect at a line d and form the angle α. Let P be any point of C. The line determined by P and the apex A of the cone intersects the circle C_1 at T. Now we drop a perpendicular from P to the plane y, intersecting it at R. Finally RQ is taken perpendicular to d. Let us consider the triangles PQR and PRT. Angle PQR is α and we shall designate the angle PTR by β. Thus

$$\sin \alpha = \frac{\overline{PR}}{\overline{PQ}}$$

and

$$\sin \beta = \frac{\overline{PR}}{\overline{PT}},$$

giving

$$\frac{\sin \alpha}{\sin \beta} = \frac{\overline{PT}}{\overline{PQ}}.$$

Again, since PF and PT are tangent to the sphere S from P,

$$\overline{PT} = \overline{PF},$$

giving

$$\frac{\sin \alpha}{\sin \beta} = \frac{\overline{PF}}{\overline{PQ}}.$$

Of course α and β are angles determined by the cone and the intersecting plane x; they are independent of the choice of the point P on the curve. Thus C is a conic with focus F, directrix d and eccentricity

$$e = \frac{\sin \alpha}{\sin \beta}.$$

Furthermore, if $0 < \alpha < \beta$, then $e < 1$ and the curve C is an ellipse; if $\alpha = \beta$, $e = 1$ and the curve is a parabola; and if $\alpha > \beta$, then $e > 1$ and the curve is a hyperbola.

Problems

1. Show that if there is only one sphere inscribed in a cone and tangent to the intersecting plane, then the conic section is a parabola.
2. Show that if there are two spheres inscribed in a cone and tangent to the intersecting plane, and if these two spheres are in different nappes of the cone, then the conic section is a hyperbola.
3. Show that if there are two spheres inscribed in a cone and tangent to the intersecting plane at the same point, then the conic section is a circle.
4. Show that the intersection of a plane with a right circular cylinder is either a circle, an ellipse, or a pair of parallel lines.

6

Transformation of Coordinates

6.1

Translation

The *general second-degree equation* is

$$Ax^2 + Bxy + Cy^2 + Dx + Ey + F = 0.$$

We have seen that any conic section with axes parallel to the coordinate axes can be put into this form with $B = 0$ and A and C not both 0. Furthermore, any second-degree equation with $B = 0$ and A and C not both 0 is a conic section (or degenerate conic) with axes parallel to the coordinate axes. This is summarized in the following table.

Conic	AC	Degenerate cases
Parabola	0	One line (two coincident lines) Two parallel lines No graph
Ellipse	+	Circle Point No graph
Hyperbola	−	Two intersecting lines

The coordinate axes are something of an artificiality, which we introduced on the plane in order to represent points and curves algebraically. Since the axes are of this nature, their placement is quite arbitrary. Thus we might prefer to move them in order to simplify some equation. Any change in the position of the axes may be represented by a combination of a translation and a rotation. A translation of the axes gives a new set of axes parallel to the old ones (see Figure 6.1a), while in a rotation, the axes are rotated about the origin (see Figure 6.1b).

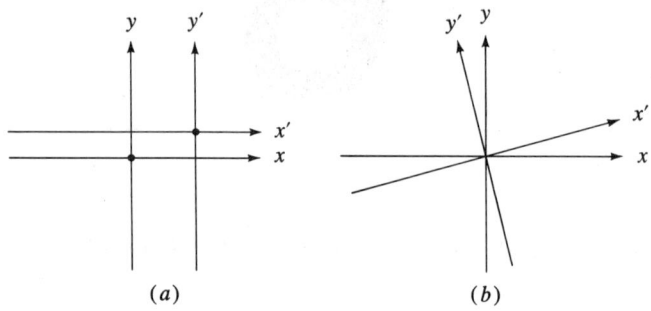

Figure 6.1

Let us consider translation first. If the axes are translated in such a way that the origin of the new coordinate system is the point (h, k) of the old system (see Figure 6.2), then every point has two representations: (x, y) in the old coordinate system,

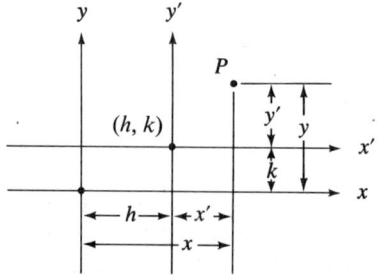

Figure 6.2

and (x', y') in the new. The relationship between the old and new coordinate system is easily seen from Figure 6.2 to be

$$x = x' + h \quad \text{or} \quad x' = x - h,$$
$$y = y' + k \quad \text{or} \quad y' = y - k.$$

These equations (either set) are called equations of translation. This is exactly the situation we had in the discussion of parabolas in Section 5.3.

6.1 Translation

We can also consider translation from a vector point of view. The vector $x\mathbf{i} + y\mathbf{j}$ can be used to represent the point (x, y). A graphical representation of this vector is a directed line segment with its tail at the origin and its head at (x, y). It is easily seen from Figure 6.3 that $\mathbf{u} = \mathbf{v} + \mathbf{w}$ or

$$x\mathbf{i} + y\mathbf{j} = (x' + h)\mathbf{i} + (y' + k)\mathbf{j}.$$

Thus

$$x = x' + h,$$
$$y = y' + k.$$

Thus, translation of the axes for a conic section consists of simply putting its equation into standard form and replacing $x - h$ by x' and $y - k$ by y'.

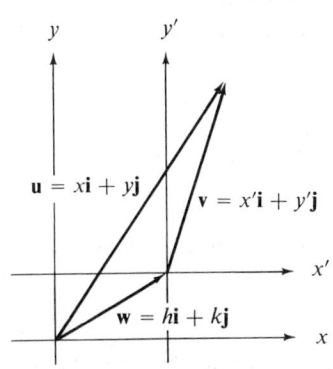

Figure 6.3

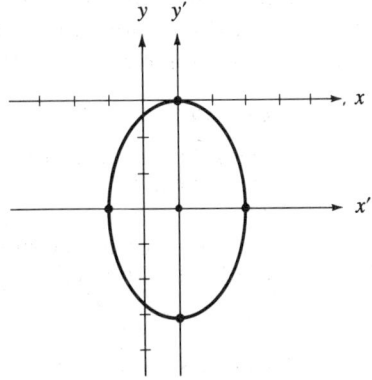

Figure 6.4

Example 1

Translate axes to eliminate the first-degree terms of $9x^2 + 4y^2 - 18x + 24y + 9 = 0$. Sketch, showing both the old and new coordinate systems.

Putting the equation into standard form, we have

$$\frac{(x-1)^2}{4} + \frac{(y+3)^2}{9} = 1.$$

The equations of translation,

$$x' = x - 1 \quad \text{and} \quad y' = y + 3,$$

transform this into the equation in the new coordinate system

$$\frac{x'^2}{4} + \frac{y'^2}{9} = 1.$$

The origin of the new coordinate system is the point $(1, -3)$ of the old system. Note that we have the same ellipse—it simply has a new representation in the new system (see Figure 6.4).

Note that any of the conic sections with axes parallel to the coordinate axes can be represented by a translation of the axes that puts the center or vertex at the origin. Thus we have the cases shown in the following table.

Conic	Before translation	After translation
Parabola	$Ax^2 + Dx + Ey + F = 0$ $Cy^2 + Dx + Ey + F = 0$	$Ax'^2 + Ey' = 0$ $Cy'^2 + Dx' = 0$
Ellipse Hyperbola	$Ax^2 + Cy^2 + Dx + Ey + F = 0$	$Ax'^2 + Cy'^2 + F' = 0$

The method of completing the square is simple to use, but it is rather limited in scope. It can be used only on second-degree equations with no xy term. If there is an xy term or if the equation is not of the second degree, another method, illustrated by the following example, can be used.

Example 2

Translate axes so that the constant and the x term of $y = x^3 - 5x^2 + 7x - 5$ are eliminated.

Since we do not know what values of h and k to choose, we simply use the equations of translation,

$$x = x' + h \quad \text{and} \quad y = y' + k,$$

and see what values of h and k are needed to eliminate the terms specified.

$$y' + k = (x' + h)^3 - 5(x' + h)^2 + 7(x' + h) - 5,$$

$$y' = x'^3 + (3h - 5)x'^2 + (3h^2 - 10h + 7)x' + (h^3 - 5h^2 + 7h - 5 - k).$$

Now we must choose h and k so that

$$3h^2 - 10h + 7 = 0,$$

$$h^3 - 5h^2 + 7h - 5 - k = 0.$$

The first of these two equations gives

$$h = 1 \quad \text{or} \quad h = 7/3.$$

Substituting these values into the second, we have

$$k = -2 \quad \text{or} \quad k = -86/27.$$

Using $h = 1$ and $k = -2$, we get

$$y' = x'^3 - 2x'^2.$$

Using $h = 7/3$ and $k = -86/27$, we get

$$y' = x'^3 + 2x'^2.$$

6.1 Translation

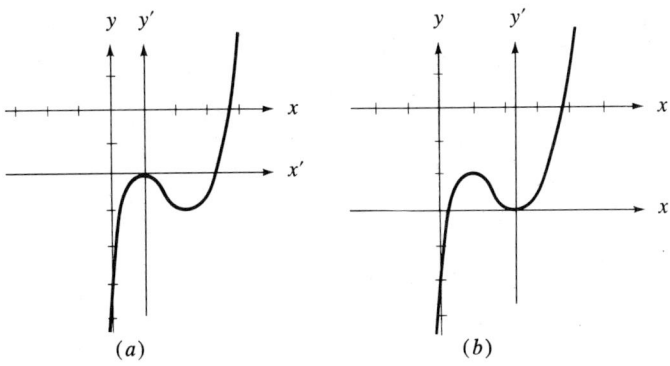

Figure 6.5

The graphs of both cases are given in Figure 6.5. While there are two different translations giving two different equations, both graphs are the same when referred to the original xy system. As an added bonus, this method has located the local maximum, $(1, -2)$, and minimum, $(7/3, -86/27)$. This method can also be used on second-degree equations with no xy term, but completing the square is so much simpler that most would prefer to use it.

Problems

In Problems 1–10, translate so as to have the center or vertex of the conic section at the origin of the new coordinate system. Sketch the curve showing both the old and new axes.

1. $y^2 - 4x - 2y + 9 = 0$.
2. $x^2 - 8x - 8y + 8 = 0$.
3. $4x^2 + y^2 + 24x - 2y + 21 = 0$.
4. $x^2 + 9y^2 - 10x + 36y + 52 = 0$.
5. $9x^2 - 4y^2 + 90x + 32y + 125 = 0$.
6. $9x^2 - 16y^2 + 72x + 96y + 144 = 0$.
7. $9x^2 + 4y^2 - 72x + 16y + 160 = 0$.
8. $4x^2 - y^2 - 40x + 6y + 91 = 0$.
9. $4x^2 - 4x - 4y - 5 = 0$.
10. $16x^2 + 36y^2 + 48x - 180y + 257 = 0$.

In Problems 11–24, translate so as to eliminate the terms indicated.

11. $x^2 - 2xy + 4y^2 + 8x - 26y + 38 = 0$; first degree terms.
12. $2x^2 - xy - y^2 + 5x - 8y - 3 = 0$; first-degree terms.
13. $x^2 + 4xy - y^2 - 2x - 14y - 3 = 0$; first-degree terms.
14. $3x^2 + xy + y^2 - 16x - 10y + 30 = 0$; first-degree terms.
15. $xy - 5x + 4y - 4 = 0$; first-degree terms.
16. $x^2 + xy + 9x + 5y + 20 = 0$; first-degree terms.
17. $y = x^3 - 6x^2 + 11x - 8$; constant, x^2 term.
18. $y = x^3 - 3x + 6$; constant, x term.
19. $y = x^4 - 8x^3 + 24x^2 - 28x + 7$; constant, x term.
20. $y = x^4 - 10x^3 + 37x^2 - 120x + 138$; constant, x term.
21. $y = x^5 - 5x^4 + 8x^3 - 4x^2 + x - 3$; constant, x^2 term.
22. $y = x^5 + 7x^4 + 19x^3 + 25x^2 + 16x + 7$; constant, x term.

23. $x^2y - 2x^2 + 2xy + y - 4x - 6 = 0$; first-degree terms.
24. $x^2y + x^2 + 2xy + x + y - 1 = 0$; second-degree terms.
25. Suppose that a translation changes the equation
$$Ax^2 + Bxy + Cy^2 + Dx + Ey + F = 0$$
into
$$A'x'^2 + B'x'y' + C'y'^2 + D'x' + E'y' + F' = 0.$$
Show that $A' = A$, $B' = B$, and $C' = C$ for any translation. A, B, and C are said to be invariant under translation.

6.2

Rotation

The second transformation of the axes that we wish to consider is a rotation of the axes about the origin (see Figure 6.1(b)). If the axes are rotated through an angle θ, then every point of the plane has two representations: (x, y) in the original coordinate system and (x', y') in the new coordinate system. Alternatively, every vector **v** in the plane has two representations: $\mathbf{v} = x\mathbf{i} + y\mathbf{j}$ in the original coordinate system and $\mathbf{v} = x'\mathbf{i'} + y'\mathbf{j'}$ in the new coordinate system (see Figure 6.6). In order to find the relationships between the x and y of one coordinate system and the x' and y' of the other, let us consider the relationships of **i** and **j** with **i'** and **j'**. Remembering that **i**, **j**, **i'** and **j'** are all unit vectors, we see from Figure 6.6 that

$$\mathbf{i'} = \cos\theta \mathbf{i} + \sin\theta \mathbf{j},$$

$$\mathbf{j'} = -\sin\theta \mathbf{i} + \cos\theta \mathbf{j}.$$

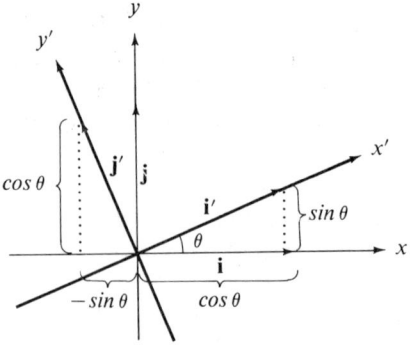

Figure 6.6

6.2 Rotation

Thus
$$\begin{aligned}\mathbf{v} &= x'\mathbf{i}' + y'\mathbf{j}' \\ &= x'(\cos\theta\,\mathbf{i} + \sin\theta\,\mathbf{j}) + y'(-\sin\theta\,\mathbf{i} + \cos\theta\,\mathbf{j}) \\ &= (x'\cos\theta - y'\sin\theta)\mathbf{i} + (x'\sin\theta + y'\cos\theta)\mathbf{j} \\ &= x\mathbf{i} + y\mathbf{j},\end{aligned}$$

which gives the equations of rotation
$$x = x'\cos\theta - y'\sin\theta,$$
$$y = x'\sin\theta + y'\cos\theta.$$

Example 1

Find the new representation of
$$x^2 - xy + y^2 - 2 = 0$$
after rotating through an angle of 45°. Sketch the curve, showing both the old and new coordinate systems.

Since $\sin 45° = \cos 45° = 1/\sqrt{2}$, the equations of rotation are
$$x = \frac{x' - y'}{\sqrt{2}} \quad \text{and} \quad y = \frac{x' + y'}{\sqrt{2}}.$$

Substituting into the original equation, we have

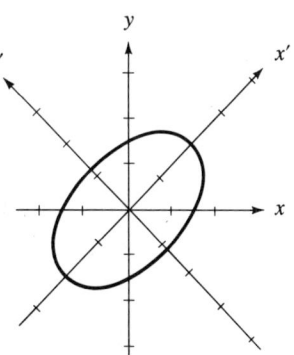

Figure 6.7

$$\frac{(x'-y')^2}{2} - \frac{x'-y'}{\sqrt{2}} \cdot \frac{x'+y'}{\sqrt{2}} + \frac{(x'+y')^2}{2} - 2 = 0,$$

$$\frac{x'^2 - 2x'y' + y'^2 - x'^2 + y'^2 + x'^2 + 2x'y' + y'^2}{2} = 2,$$

$$x'^2 + 3y'^2 = 4.$$

Figure 6.7 shows the final result.

Example 2

Find a new representation of $x^2 + 4xy - 2y^2 - 6 = 0$ after rotating through an angle $\theta =$ Arctan 1/2. Sketch the curve, showing both the old and new coordinate systems.

Figure 6.8 shows that
$$\sin\theta = \frac{1}{\sqrt{5}} \quad \text{and} \quad \cos\theta = \frac{2}{\sqrt{5}},$$

giving equations of rotation

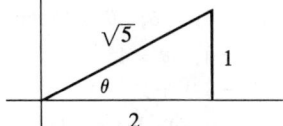

Figure 6.8

$$x = \frac{2x' - y'}{\sqrt{5}} \quad \text{and} \quad y = \frac{x' + 2y'}{\sqrt{5}}.$$

Substituting into the original equation, we have

$$\frac{(2x' - y')^2}{5} + 4 \frac{2x' - y'}{\sqrt{5}} \cdot \frac{x' + 2y'}{\sqrt{5}} - 2 \frac{(x' + 2y')^2}{5} - 6 = 0,$$

$$\frac{4x'^2 - 4x'y' + y'^2 + 8x'^2 + 12x'y' - 8y'^2 - 2x'^2 - 8x'y' - 8y'^2}{5} = 6,$$

$$2x'^2 - 3y'^2 = 6.$$

Figure 6.9 shows the final result. Note that Figure 6.8 can be used to determine the position of the new coordinate axes.

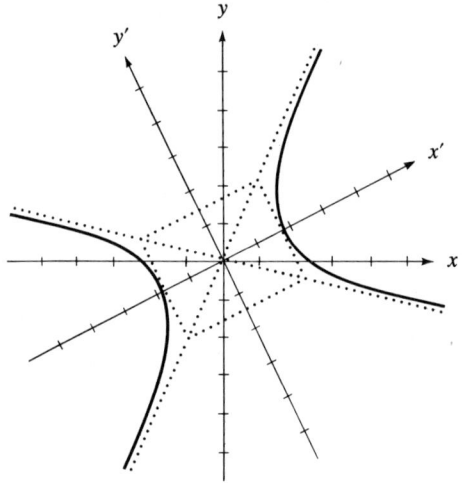

Figure 6.9

In both of these examples, we have seen that the given rotation has eliminated the xy term. Of course, not every rotation will do so—it must be specially chosen. We shall see in the next section how to choose θ to eliminate the xy term.

Problems

In Problems 1–10, find a new representation of the given equation after rotating through the given angle. Sketch the curve, showing both the old and new coordinate systems.

1. $2x + 3y = 6$; $\theta = \text{Arctan } 3/2$.
2. $3x - y = 5$; $\theta = \text{Arctan } 3$.
3. $xy = 4$; $\theta = 45°$.
4. $2x^2 - xy + 2y^2 - 15 = 0$; $\theta = 45°$.
5. $x^2 - 2xy + y^2 + x + y = 0$; $\theta = 45°$.
6. $31x^2 + 10\sqrt{3}xy + 21y^2 - 144 = 0$; $\theta = 30°$.
7. $x^2 + 2\sqrt{3}xy + 3y^2 + 8\sqrt{3}x - 8y = 0$; $\theta = 60°$.

8. $11x^2 - 50\sqrt{3}xy - 39y^2 + 576 = 0;\quad \theta = 60°$.
9. $8x^2 + 5xy - 4y^2 - 4 = 0;\quad \theta = \text{Arctan } 1/5$.
10. $6x^2 - 5xy - 6y^2 + 26 = 0;\quad \theta = \text{Arctan } (-1/5)$.

In Problems 11–16, find a new representation of the given equation after rotating through the given angle.

11. $3x^2 - 3xy - y^2 + 4 = 0;\quad \theta = \text{Arctan } (-1/2)$.
12. $4x^2 + 3xy - 5 = 0;\quad \theta = \text{Arctan } 1/2$.
13. $4x^2 + 3xy - 5 = 0;\quad \theta = 45°$.
14. $x^2 - 3xy + y^2 + 5 = 0;\quad \theta = 30°$.
15. $3x^2 - 3xy - y^2 + 4 = 0;\quad \theta = 60°$.
16. $x^2 - 5xy + 2 = 0;\quad \theta = \text{Arctan } 1/5$.
17. Show that $x^2 + y^2 = 25$ is invariant under rotation through any angle.
18. Given the equation

$$Ax^2 + Bxy + Cy^2 + Dx + Ey + F = 0,$$

which yields

$$A'x'^2 + B'x'y' + C'y'^2 + D'x' + E'y' + F' = 0$$

after rotation through the angle θ, show that $A' + C' = A + C$ for any value of θ: that is, $A + C$ is invariant under rotation.

19. Show that, in the general equation of second degree, $B^2 - 4AC$ is invariant under rotation. (See Problem 18 for definition of invariant.)
20. Show that a second form for the equations of rotation is

$$x' = x \cos \theta + y \sin \theta,$$
$$y' = -x \sin \theta + y \cos \theta.$$

6.3

The General Equation of Second Degree

We have seen that any conic section with axes parallel to the coordinate axes can be represented by a second-degree equation with $B = 0$; furthermore, any second-degree equation with $B = 0$ represents a conic or degenerate conic with axes parallel to the coordinate axes. We now extend this concept to conic sections in any position. It is an easy matter to see that any conic can be represented by a second-degree equation, starting from our standard forms and translating and rotating.

Suppose, given a second-degree equation with $B \neq 0$, we rotate axes through an angle θ. If our assumption that this equation represents a conic or degenerate conic is correct, then a rotation of axes through some positive angle less than 90° should

eliminate the xy term. Thus we shall assume throughout this discussion that $0° < \theta < 90°$ and
$$Ax^2 + Bxy + Cy^2 + Dx + Ey + F = 0.$$
Substituting the equations of rotation we have
$$A(x' \cos \theta - y' \sin \theta)^2 + B(x' \cos \theta - y' \sin \theta)(x' \sin \theta + y' \cos \theta)$$
$$+ C(x' \sin \theta + y' \cos \theta)^2 + D(x' \cos \theta - y' \sin \theta)$$
$$+ E(x' \sin \theta + y' \cos \theta) + F = 0.$$

After carrying out the multiplication and combining similar terms, we find that the coefficient of $x'y'$ is
$$(C - A)2 \sin \theta \cos \theta + B(\cos^2 \theta - \sin^2 \theta) = (C - A) \sin 2\theta + B \cos 2\theta.$$

We want this coefficient to be zero for the proper choice of θ. Let us set it equal to zero and see what θ should be.
$$(C - A) \sin 2\theta + B \cos 2\theta = 0,$$
$$(A - C) \sin 2\theta = B \cos 2\theta,$$
$$\frac{\sin 2\theta}{\cos 2\theta} = \frac{B}{A - C}, \quad A \neq C,$$
$$\tan 2\theta = \frac{B}{A - C}, \quad A \neq C.$$

We can easily solve this equation for θ, but it would involve us in inverse trigonometric functions—let us try to get around them. To do this we shall try to find expressions for $\sin \theta$ and $\cos \theta$ that we can use in the equations of rotation.

First, we note that if $A \neq C$, then $\tan 2\theta$ exists, $2\theta \neq 90°$, and $\theta \neq 45°$. We start with the identity
$$\sin^2 2\theta + \cos^2 2\theta = 1.$$
Dividing through by $\cos^2 2\theta$ (which is not zero since $\theta \neq 45°$), we get
$$\tan^2 2\theta + 1 = \frac{1}{\cos^2 2\theta}, \quad \cos^2 2\theta = \frac{1}{1 + \tan^2 2\theta}, \quad \cos 2\theta = \frac{\pm 1}{\sqrt{1 + \tan^2 2\theta}}.$$

The $\pm$ presents the question of which one to use. Since $0° < \theta < 90°$, $0° < 2\theta < 180°$. Both the tangent and cosine are positive for a first-quadrant angle and both are negative for a second-quadrant angle. Thus we choose the sign to agree with the sign of $\tan 2\theta$.

Now let us recall the half-angle identities
$$\sin \frac{A}{2} = \pm \sqrt{\frac{1 - \cos A}{2}} \quad \text{and} \quad \cos \frac{A}{2} = \pm \sqrt{\frac{1 + \cos A}{2}}.$$

Replacing A by 2θ and noting that both $\sin \theta$ and $\cos \theta$ must be positive since $0° < \theta < 90°$, we have

6.3 The General Equation of Second Degree

$$\sin\theta = \sqrt{\frac{1-\cos 2\theta}{2}}, \quad \cos\theta = \sqrt{\frac{1+\cos 2\theta}{2}}.$$

Finally, if $A = C$, then

$$B\cos 2\theta = 0,$$
$$\cos 2\theta = 0,$$
$$2\theta = 90°,$$
$$\theta = 45°.$$

Thus, in either case, we are able to rotate axes to eliminate the xy term. The resulting equation must then represent a conic or degenerate conic.

Theorem 6.1

Any conic section can be represented by the second-degree equation

$$Ax^2 + Bxy + Cy^2 + Dx + Ey + F = 0$$

where A, B, and C are not all zero. Any second-degree equation represents either a conic or a degenerate conic.

If $B \neq 0$, then the axes may be rotated to eliminate the xy term in the following way:

$A = C$ $\qquad\qquad A \neq C$

$\theta = 45°$ $\qquad \tan 2\theta = \dfrac{B}{A-C}$

$\qquad\qquad\qquad \cos 2\theta = \dfrac{\pm 1}{\sqrt{1+\tan^2 2\theta}}$ (sign agrees with the sign of $\tan 2\theta$)

$\qquad\qquad\qquad \sin\theta = \sqrt{\dfrac{1-\cos 2\theta}{2}}$

$\qquad\qquad\qquad \cos\theta = \sqrt{\dfrac{1+\cos 2\theta}{2}}$

Example 1

Rotate axes to eliminate the xy term of $x^2 + 4xy - 2y^2 - 6 = 0$. Sketch, showing both sets of axes.

$$\tan 2\theta = \frac{B}{A-C} = \frac{4}{1-(-2)} = \frac{4}{3},$$

$$\cos 2\theta = \frac{1}{\sqrt{1+\tan^2 2\theta}} \quad \text{(since } \tan 2\theta \text{ is positive,} \\ \cos 2\theta \text{ is also positive)}$$

$$= \frac{1}{\sqrt{1+\left(\frac{4}{3}\right)^2}} = \frac{3}{5},$$

$$\sin \theta = \sqrt{\frac{1-\cos 2\theta}{2}} = \sqrt{\frac{1-\frac{3}{5}}{2}} = \frac{1}{\sqrt{5}},$$

$$\cos \theta = \sqrt{\frac{1+\cos 2\theta}{2}} = \sqrt{\frac{1+\frac{3}{5}}{2}} = \frac{2}{\sqrt{5}},$$

$$\tan \theta = \frac{\sin \theta}{\cos \theta} = \frac{1}{2} \quad \text{(we shall use this for sketching);}$$

$$x = \frac{2x'-y'}{\sqrt{5}}, \quad y = \frac{x'+2y'}{\sqrt{5}}.$$

Substituting these equations of rotation into the original equation (see Example 2 of the previous section), we have

$$2x'^2 - 3y'^2 = 6.$$

The sketch is given in Figure 6.9.

Example 2

Rotate axes to eliminate the xy term of

$$2x^2 - xy + 2y^2 - 2 = 0.$$

Sketch, showing both sets of axes.

Since $A = C$, $\theta = 45°$ and the equations of rotation are

$$x = \frac{x'-y'}{\sqrt{2}}, \quad y = \frac{x'+y'}{\sqrt{2}}.$$

Substituting these into the original equation, we have

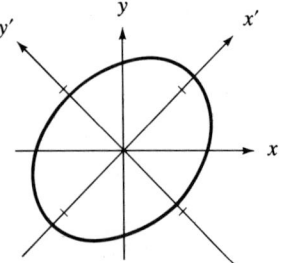

Figure 6.10

6.3 The General Equation of Second Degree

$$2\frac{(x'-y')^2}{2} - \frac{x'-y'}{\sqrt{2}}\frac{x'+y'}{\sqrt{2}} + 2\frac{(x'+y')^2}{2} - 2 = 0,$$

$$\frac{2x'^2 - 4x'y' + 2y'^2 - x'^2 + y'^2 + 2x'^2 + 4x'y' + 2y'^2}{2} = 2,$$

$$3x'^2 + 5y'^2 = 4,$$

$$\frac{x'^2}{4/3} + \frac{y'^2}{4/5} = 1.$$

The sketch is given in Figure 6.10.

Problems

In Problems 1–14, rotate axes to eliminate the xy term. Sketch, showing both sets of axes.

1. $x^2 + xy + y^2 + 4\sqrt{2}x - 4\sqrt{2}y = 0.$
2. $5x^2 + 6xy + 5y^2 - 8 = 0.$
3. $7x^2 + 6xy - y^2 - 32 = 0.$
4. $4x^2 + 4xy + y^2 + 8\sqrt{5}x - 16\sqrt{5}y = 0.$
5. $241x^2 + 252xy + 136y^2 - 1300 = 0.$
6. $12x^2 + 10xy - 12y^2 + 13 = 0.$
7. $5x^2 - 4xy + 8y^2 - 36 = 0.$
8. $27x^2 - 78xy - 77y^2 + 360 = 0.$
9. $4x^2 + 12xy + 9y^2 + 8\sqrt{13}x + 12\sqrt{13}y - 65 = 0.$
10. $8x^2 - 12xy + 17y^2 + 20 = 0.$
11. $9x^2 - 6xy + y^2 - 12\sqrt{10}x - 36\sqrt{10}y = 0.$
12. $x^2 + 8xy + 7y^2 - 36 = 0.$
13. $8x^2 + 12xy - 8y^2 + 2\sqrt{10}x + 14\sqrt{10}y - 40 = 0.$
14. $5x^2 - 6xy + 5y^2 + 20\sqrt{2}x - 28\sqrt{2}y + 72 = 0.$
15. It can easily be seen graphically that two conic sections have at most four points in common. But

$$2x^2 + xy - y^2 + 3y - 2 = 0,$$
$$2x^2 + 3xy + y^2 - 6x - 5y + 4 = 0$$

have in common the five points $(1, 0)$, $(-2, 3)$, $(5, -4)$, $(-6, 7)$, and $(10, -9)$. Why?

Curve Sketching

7.1

Intercepts and Asymptotes

In the first chapter we sketched the graph of an equation by the tedious process of point-by-point plotting—a method that sometimes causes one to overlook some "interesting" portions of the graph or to sketch certain portions incorrectly. Suppose for example, you are asked to sketch the graph of

$$y = \frac{10x(x+8)}{(x+10)^2}.$$

The methods of Chapter 1 might lead you to the graph of Figure 7.1. A better sketch of the graph is given in Figure 7.2. While the earlier method produced correct results for the portion we were sketching, it provided no means for determining which portions of the curve are most "interesting."

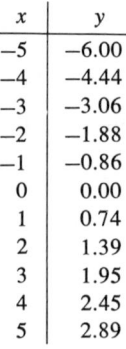

x	y
−5	−6.00
−4	−4.44
−3	−3.06
−2	−1.88
−1	−0.86
0	0.00
1	0.74
2	1.39
3	1.95
4	2.45
5	2.89

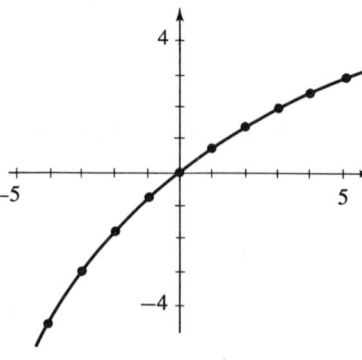

Figure 7.1

7.1 Intercepts and Asymptotes

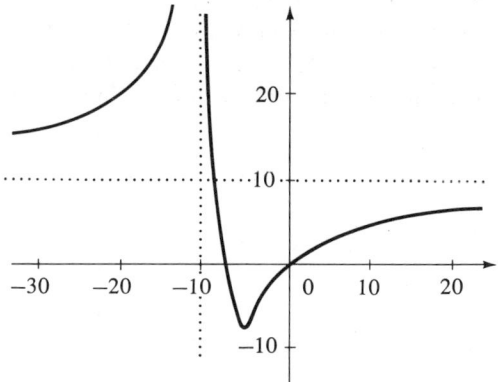

Figure 7.2

Let us consider one more example. Suppose we want to graph

$$y = \frac{2x(2x-1)}{4x-1}.$$

The methods of Chapter 1 lead us to the set of points shown in Figure 7.3. Now, what does the graph look like? How would you join the points? Many would join them as indicated in (a) of Figure 7.4. The correct graph is shown in (b). These examples demonstrate the need for better methods of sketching curves. We begin by considering intercepts and asymptotes.

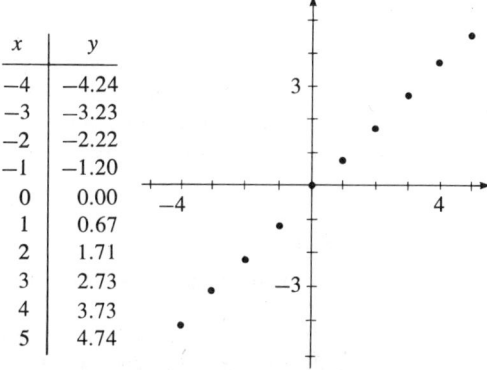

x	y
−4	−4.24
−3	−3.23
−2	−2.22
−1	−1.20
0	0.00
1	0.67
2	1.71
3	2.73
4	3.73
5	4.74

Figure 7.3

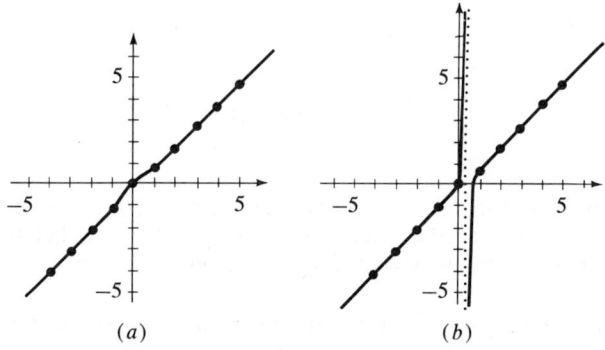

Figure 7.4

The intercepts of a curve are simply the points of the curve that lie on the coordinate axes; those on the x axis are the x intercepts, while those on the y axis are the y intercepts (the origin is both an x intercept and a y intercept). They are very simply determined by setting x and y equal to zero in turn and solving for y and x, respectively.

Example 1

Find the intercepts of $\dfrac{x^2}{4} + \dfrac{y^2}{9} = 1$.

When $y = 0$, $x^2 = 4$ and $x = \pm 2$. When $x = 0$, $y^2 = 9$ and $y = \pm 3$. Thus, the intercepts are $(2, 0)$, $(-2, 0)$, $(0, 3)$, and $(0, -3)$.

An equation frequently encountered is one of the type $y = P(x)$ or $y = P(x)/Q(x)$, where $P(x)$ and $Q(x)$ are polynomials having no common factor. If $P(x)$ can be factored in the form

$$P(x) = c(x - a_1)^{n_1}(x - a_2)^{n_2} \cdots (x - a_k)^{n_k},$$

where $c, a_1, \ldots, a_k$ are real numbers, then the x intercepts are $(a_1, 0), (a_2, 0), \ldots, (a_k, 0)$. The y intercept (there can only be one when the equation is in this form) is still found by setting x equal to zero.

Example 2

Find the intercepts of $y = (x + 1)^2(x - 3)$.

The x intercepts can be taken from the two factors: $(-1, 0)$ from $(x + 1)^2$ and $(3, 0)$ from $(x - 3)$. When $x = 0$,

$$y = 1^2(-3) = -3.$$

Thus the y intercept is $(0, -3)$.

Example 3

Find the intercepts of

$$y = \frac{(x - 2)^2(x + 1)}{(x - 3)(x - 1)^2}.$$

From the factors $(x - 2)^2$ and $(x + 1)$, we get $(2, 0)$ and $(-1, 0)$. When $x = 0$, $y = -4/3$, and so the y intercept is $(0, -4/3)$. Note that the factors of the denominator have no part in determining the x intercepts.

Let us now turn to *asymptotes* (you are encouraged to study the spelling of that word). We encountered asymptotes earlier when we studied the hyperbola (see Sections 5.6 and 5.7). Let us consider some other examples. In Figure 7.2, the lines $x = -10$

7.1 Intercepts and Asymptotes

and $y = 10$ are asymptotes. We see that portions of the curve approach $x = -10$ and $y = 10$. This is the main feature to be considered in determining asymptotes. Note that the curve contains the point $(-25/3, 10)$ of the line $y = 10$. This does not prevent $y = 10$ from being an asymptote. A curve *can* have one or more (even infinitely many) points in common with its asymptote; however, a line is not an asymptote of itself, nor does $y = |x|$ have an asymptote. In Figure 7.4(b), the line $x = 1/4$ is an asymptote. Again we see that portions of the curve approach this line.

Although it is possible for any line to be an asymptote, we shall consider only horizontal and vertical asymptotes here (slant asymptotes are considered later in this chapter). First we take up vertical asymptotes. In Figure 7.2 you can see that, as x approaches -10 from either side, y approaches no definite number but gets larger and larger. As x approaches $1/4$ from the right in Figure 7.4(b), y gets large and negative; and, as x approaches $1/4$ from the left, y gets large and positive. To find vertical asymptotes, we are not concerned with whether y gets large and positive or large and negative. We are interested only in determining values of x for which y gets large in absolute value. If the equation is in the form

$$y = \frac{P(x)}{Q(x)},$$

then, as x approaches a, y gets large in absolute value if $Q(x)$ approaches zero and $P(x)$ does not. Thus we need determine only the values of x which make $Q(x) = 0$ and $P(x) \neq 0$.

Example 4

Determine the vertical asymptotes of

$$y = \frac{(x+1)(x-3)}{(2x-1)(x+2)^2}.$$

The denominator is zero when either one of the two factors is zero.

$$2x - 1 = 0 \quad \text{gives} \quad x = 1/2.$$
$$x + 2 = 0 \quad \text{gives} \quad x = -2.$$

Since neither value of x gives zero for the numerator, $x = 1/2$ and $x = -2$ are the vertical asymptotes.

Let us now consider horizontal asymptotes. Of course, if the given equation is in the form

$$x = \frac{P(y)}{Q(y)},$$

or can easily be put into that form, we can simply use the methods given for vertical asymptotes. We merely reverse the role of the x and y here. Unfortunately, it is often difficult or impossible to solve for x as a function of y (consider the equation of Example 4), so another method must be found.

If $y = k$ is a horizontal asymptote for $y = f(x)$, then the distance between a point of the graph of $y = f(x)$ and the line $y = k$ must approach zero as x gets large in absolute value. Thus we must investigate the behavior of y as x gets large in one direction or the other; if y approaches the number k ($y \to k$) as $|x|$ gets large, then $y = k$ is a horizontal asymptote.

Example 5

Determine the horizontal asymptote of

$$y = \frac{x^2 - 4}{x^2 + 3x}.$$

As x gets large and positive, both the numerator and denominator are also getting large and positive. Suppose we alter the equation by dividing both numerator and denominator by x^2. Then

$$y = \frac{x^2 - 4}{x^2 + 3x} = \frac{1 - 4/x^2}{1 + 3/x}.$$

Now as x gets large,

$$\frac{4}{x^2} \to 0 \quad \text{and} \quad \frac{3}{x} \to 0.$$

Thus

$$y = \frac{1 - 4/x^2}{1 + 3/x} \to 1,$$

and $y = 1$ is the horizontal asymptote. Similarly, as x gets large and negative, $y \to 1$. Thus the asymptote $y = 1$ is approached by the curve in both directions.

In finding the asymptote in the preceding example, we first divided both numerator and denominator by the highest power of x (x^2 in this case) in the given expression. This trick often helps in finding asymptotes.

Example 6

Determine the horizontal asymptote of

$$y = \frac{(x+1)(x-3)}{(2x-1)(x+2)^2}.$$

If we multiplied out the numerator, the highest power of x would be x^2; in the denominator it would be x^3. Thus we shall divide the numerator and denominator by x^3 (do *not* divide the numerator by x^2 and the denominator by x^3; the result would *not* equal y).

$$y = \frac{(x+1)(x-3)}{(2x-1)(x+2)^2} = \frac{\left(\dfrac{x+1}{x}\right)\left(\dfrac{x-3}{x}\right)\left(\dfrac{1}{x}\right)}{\left(\dfrac{2x-1}{x}\right)\left(\dfrac{x+2}{x}\right)^2} = \frac{\left(1+\dfrac{1}{x}\right)\left(1-\dfrac{3}{x}\right)\left(\dfrac{1}{x}\right)}{\left(2-\dfrac{1}{x}\right)\left(1+\dfrac{2}{x}\right)^2}.$$

7.1 Intercepts and Asymptotes

As x gets large, all of the expressions with x in the denominator approach zero and

$$y \to \frac{(1+0)(1-0)(0)}{(2-0)(1+0)^2} = 0.$$

Thus $y = 0$ is the only horizontal asymptote.

Example 7

Find the horizontal asymptote of

$$y = \frac{x(x+1)(x-2)}{(x-4)(x+2)}.$$

The highest power of x in this expression is x^3. Dividing numerator and denominator by x^3, we have

$$y = \frac{x(x+1)(x-2)}{(x-4)(x+2)} = \frac{\left(\frac{x}{x}\right)\left(\frac{x+1}{x}\right)\left(\frac{x-2}{x}\right)}{\left(\frac{x-4}{x}\right)\left(\frac{x+2}{x}\right)\left(\frac{1}{x}\right)} = \frac{(1)\left(1+\frac{1}{x}\right)\left(1-\frac{2}{x}\right)}{\left(1-\frac{4}{x}\right)\left(1+\frac{2}{x}\right)\left(\frac{1}{x}\right)}.$$

As x gets large in absolute value, the numerator approaches 1 and the denominator, 0. Thus the fraction becomes arbitrarily large, rather than leveling off to some number k. There is, therefore, no horizontal asymptote.

Problems

In Problems 1–16, find the intercepts.

1. $y = x^2 + 3x$.
2. $y = x^2 - x - 2$.
3. $y = (x+1)(x^2-1)$.
4. $y = (2x-1)^2(3x+2)^3$.
5. $y = (4x+1)(x-2)(2x+3)^2$.
6. $y = (2x-1)^3(3x+2)^2(x-3)$.
7. $y = (x-1)(x^2+1)$.
8. $y = (2x+3)(x^2+x+1)$.
9. $y = (3x-1)^2(x^2+2)^3(2x+1)^4$.
10. $y = (2x-5)(x^2-x+1)^3(x^2+2)$.
11. $y = \dfrac{x}{x+1}$.
12. $y = \dfrac{(x+1)(x-2)}{(3x+1)^2}$.
13. $y = \dfrac{(x-3)^2}{2x+1}$.
14. $y = \dfrac{1}{3x+2}$.
15. $y = \dfrac{2}{x^2}$.
16. $y = \dfrac{(x+1)^2(x^2+1)}{x^2}$.

In Problems 17–34, find all horizontal and vertical asymptotes.

17. $y = (x+1)(x-2)$.
18. $y = (4x+3)(x-2)$.
19. $y = \dfrac{1}{x-1}$.
20. $y = \dfrac{4x-2}{x+1}$.
21. $y = \dfrac{x}{x+3}$.
22. $y = \dfrac{(x+1)^2}{2x(x-2)}$.

23. $y = \dfrac{2x(x-2)}{(x+1)^2}$.

24. $y = \dfrac{(x+1)(x-3)^2}{(x+2)^2}$.

25. $y = \dfrac{(2x-3)(x-2)}{(x+1)(x-3)^2}$.

26. $y = \dfrac{(4x-7)(x-1)^2}{(x+1)(x+2)(x+3)}$.

27. $y = \dfrac{(x+1)^2}{x^2+1}$.

28. $y = \dfrac{(2x+1)^2(x-2)^2}{x(4x-3)}$.

29. $y = \dfrac{(3x+2)^3(x-4)}{(2x+3)^2(x+1)^3}$.

30. $y = \dfrac{(2x+1)^2(x-3)^3}{x(2x-3)^2}$.

31. $y = x - \sqrt{x^2+1}$.
    (Hint: Rationalize the numerator.)

32. $y = x + \sqrt{x^2+1}$.

33. $y = 2x - \sqrt{4x^2+3}$.

34. $y = 3x + \sqrt{9x^2-1}$.

35. If $y = \dfrac{a_n x^n + a_{n-1} x^{n-1} + \cdots + a_1 x + a_0}{b_m x^m + b_{m-1} x^{m-1} + \cdots + b_1 x + b_0}$, where $a_n \neq 0$ and $b_m \neq 0$, what can be said about horizontal asymptotes in case (a) $n < m$? (b) $n = m$? (c) $n > m$?

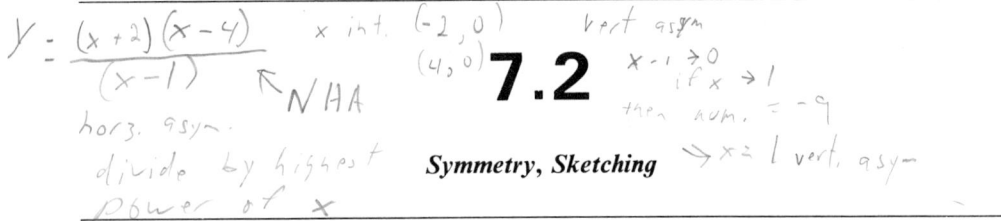

7.2 Symmetry, Sketching

Another characteristic that helps in sketching a curve is symmetry. There are two types: symmetry about a line and symmetry about a point. If a curve is symmetric about a line, then one-half of it is the mirror image of the other half, with the mirror as the line of symmetry. More precisely, for every point P of the curve, on one side of the line there is another point P' of the curve such that PP' is perpendicular to the line of symmetry and is bisected by it. An example of this type of symmetry occurs with the graph of $y = 1/x^2$, in which the y axis is the line of symmetry (see Figure 7.5).

A curve is symmetric about a point O if for every point $P \neq O$ of the curve, there corresponds a point P' such that PP' is bisected by the point O. The origin is the point of symmetry of the graph of $y = 1/x$ (see Figure 7.6).

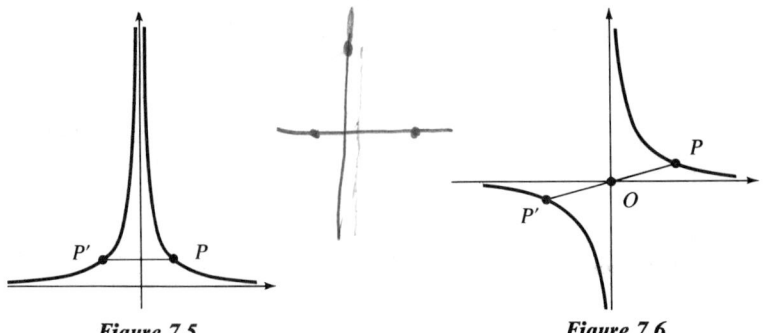

Figure 7.5   Figure 7.6

7.2 Symmetry, Sketching

While there is no restriction on the lines or points that may be lines or points of symmetry, we shall consider here only symmetry about the axes and about the origin. We begin with symmetry about the y axis. If a curve is symmetric about the y axis, then, corresponding to every point $P(x, y)$ on the curve, there is a point $P'(-x, y)$ (see Figure 7.5) with the same y coordinate and an x coordinate that is the negative of the x coordinate of P. In this situation, we get the same value for y whether we substitute a positive number x into the equation or its negative, $-x$.

Theorem 7.1

If every x in an equation is replaced by $-x$ and the resulting equation is equivalent to the original (has the same graph), then its graph is symmetric about the y axis.

Example 1

We have already noted that $y = 1/x^2$ is symmetric about the y axis. We now test for symmetry. Replacing x by $-x$, we have

$$y = \frac{1}{(-x)^2}.$$

Since $(-x)^2 = x^2$, we see that the substitution has produced an equation equivalent to the original equation, proving symmetry about the y axis. A similar argument can be used to prove the next theorem.

Theorem 7.2

If every y in an equation is replaced by $-y$ and the resulting equation is equivalent to the original, then its graph is symmetric about the x axis.

If a curve is symmetric about the origin, then, for every point $P(x, y)$ on the curve, there is a point $P'(-x, -y)$ (see Figure 7.6). This is the statement of the next theorem.

Theorem 7.3

If every x in an equation is replaced by $-x$ and every y by $-y$ and the resulting equation is equivalent to the original, then its graph is symmetric about the origin.

Example 2

We test the theorem on $y = 1/x$, which we have already noted is symmetric about the origin. Replacing x by $-x$ and y by $-y$ gives

$$-y = \frac{1}{-x}.$$

This is equivalent to the original equation, since we get the original equation if we multiply both sides by -1.

Before using what we have observed about intercepts, asymptotes, and symmetry to sketch the curve, we shall look at one other characteristic that will help greatly in sketching curves represented by equations of the form

$$y = \frac{P(x)}{Q(x)}.$$

The factors of $P(x)$ determine the x intercepts, and the factors of $Q(x)$ determine the vertical asymptotes of the curve. It might be noted that these are the only two places at which y can change from positive to negative or from negative to positive. This is not to say that the value of y *must* change there—only that it cannot do so elsewhere. We can easily determine whether or not the change occurs at a given intercept or asymptote by considering the exponent on the factor that produces it.

Theorem 7.4

Given an equation of the form

$$y = \frac{P(x)}{Q(x)}$$

in reduced form, if $(x - a)^n$ (where n is a positive integer) is a factor of either $P(x)$ or $Q(x)$ and if $(x - a)^{n+1}$ is a factor of neither, then

(a) *the graph crosses the x axis at $x = a$ if and only if n is odd, and*
(b) *the graph stays on the same side of the x axis at $x = a$ if and only if n is even.*

The expression "the graph crosses the x axis" is not intended to mean that the graph has a point in common with the x axis. It means that the graph is above (or below) the x axis for $c < x < a$ and below (or above) for $a < x < d$ for some c and d.

Although the following discussion does not constitute a proof of this theorem, it serves to show why the theorem works. Let us consider the case in which $(x - a)^n$ is a factor of $P(x)$ (a similar argument can be used for the other case). Then

$$y = \frac{P(x)}{Q(x)} = \frac{R(x)}{Q(x)}(x - a)^n.$$

For all values of x at and "near" $x = a$, $R(x)/Q(x)$ is either positive throughout or negative throughout, not making any sign change. But

$$\begin{aligned} x - a &< 0 \quad \text{for} \quad x < a, \\ x - a &= 0 \quad \text{for} \quad x = a, \\ x - a &> 0 \quad \text{for} \quad x > a. \end{aligned}$$

7.2 Symmetry, Sketching

In other words, $x - a$ changes sign at $x = a$. If n is odd, then $(x - a)^n$ also changes sign; thus y changes sign at $x = a$. If n is even, then $(x - a)^n$ is positive whether $x < a$ or $x > a$; that is, $(x - a)^n$ does not change sign and y does not change sign. Let us now use all of this information to sketch the graph of an equation.

Example 3

Sketch $y = (x - 3)(x + 1)^2$.

From the "numerator," we get x intercepts $(3, 0)$ with an odd exponent and $(-1, 0)$ with an even exponent. If $x = 0$, then $y = -3$, which gives $(0, -3)$. Since the denominator is 1, there are no vertical asymptotes. As x gets large and positive, y gets large and positive; as x gets large and negative, y gets large and negative, which means that there are no horizontal asymptotes. It is easy to see that no symmetry exists about either axis or the origin. Summing up, we have:

Intercepts: $(3, 0)$ odd, $(-1, 0)$ even, $(0, -3)$;
No asymptotes: as $x \to +\infty$, $y \to +\infty$;
as $x \to -\infty$, $y \to -\infty$.
No symmetry.

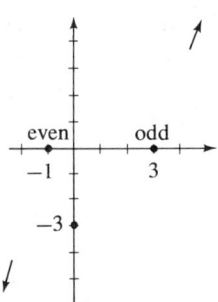

Figure 7.7

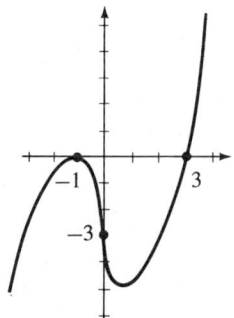

Figure 7.8

All of this is indicated in Figure 7.7. Let us sketch the graph, starting at the far left and working to the right (this choice is quite arbitrary; we might just as well go from right to left or start in the middle and work outward). We keep in mind that the curve must go through all intercepts and that y is a *function* of x; that is, it is single-valued. Since $y \to -\infty$ as $x \to -\infty$, we start in the lower left-hand corner. Going to the right, we first reach the intercept $(-1, 0)$. Since it is an even intercept, the graph merely touches the x axis but stays below it. Next, the graph goes through $(0, -3)$ and then turns back up in order to go through $(3, 0)$. Since $(3, 0)$ is an odd intercept, the graph crosses the x axis there and proceeds upward. The result is given in Figure 7.8.

Note that we put the lowest point of the "dip" at approximately $x = 1$. How did we know to put it there? We didn't. We made no attempt to locate it—we simply guessed. Without further work, the best we can say is that it is between $x = -1$ and $x = 3$. Furthermore, how do we know that the graph does not have some extra "turns" and "wiggles" and perhaps look like Figure 7.9? Again, we don't. As a general rule, unless there is some special reason to put in some extra "turn" or "wiggle," we shall leave it out. This rule will not necessarily give us the correct graph every time but there is no point in needlessly complicating the situation. These methods give only a general idea of the graph.

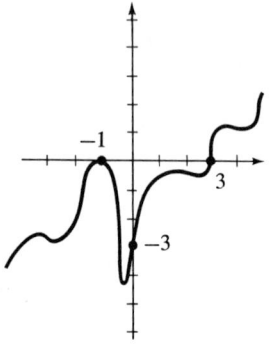

Figure 7.9

Note that, with the exception of the three intercepts, we have not plotted a single point! Yet we have some idea (within the restrictions noted above) of the main features of the curve. With a little practice, you should be able to sketch such curves quite quickly and thus achieve the principal aim here.

Example 4

Sketch $y = \dfrac{(2x-1)(x+2)^2}{(x+1)^2(x-3)}$.

Intercepts: $(1/2, 0)$, odd; $(-2, 0)$, even; $(0, 4/3)$.

Asymptotes:
From the denominator: $x = -1$, even; $x = 3$, odd,

$$y = \frac{(2x-1)(x+2)^2}{(x+1)^2(x-3)} = \frac{\left(\dfrac{2x-1}{x}\right)\left(\dfrac{x+2}{x}\right)^2}{\left(\dfrac{x+1}{x}\right)^2\left(\dfrac{x-3}{x}\right)}$$

$$= \frac{\left(2-\dfrac{1}{x}\right)\left(1+\dfrac{2}{x}\right)^2}{\left(1+\dfrac{1}{x}\right)^2\left(1-\dfrac{3}{x}\right)}.$$

As $x \to \pm\infty$, $y \to \dfrac{2 \cdot 1}{1 \cdot 1} = 2$.

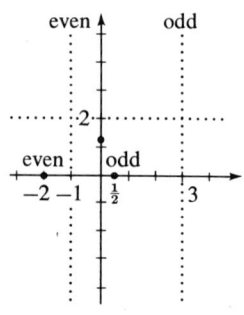

Figure 7.10

7.2 Symmetry, Sketching

Thus $y = 2$ is the horizontal asymptote.
No symmetry.

All of this information is summarized in Figure 7.10. If we begin sketching at one end or the other, we have the problem of not knowing whether the curve is approaching the asymptote from above or below. Similar problems exist at the vertical asymptotes and x intercepts. Suppose, then, we start at $(0, 4/3)$. Going to the right, we first come to $(1/2, 0)$. Since it is an odd intercept, the graph crosses the x axis there and then goes down to the vertical asymptote $x = 3$ (it cannot go up, since it cannot cross the x axis anywhere between $x = 1/2$ and $x = 3$). Since this asymptote is also odd, the graph now jumps above the x axis. Finally it comes down to the horizontal asymptote $y = 2$.

Going back to $(0, 4/3)$ and proceeding to the left, we see that the graph must go up to the vertical asymptote $x = -1$ (remember there is nothing to prevent the graph from crossing a horizontal asymptote). Since $x = -1$ is an even asymptote, the curve stays above the x axis. It must then proceed down to the intercept $(-2, 0)$. This is also even, so the graph again remains above the x axis, finally going up to the horizontal asymptote. Thus, we have the graph indicated in Figure 7.11.

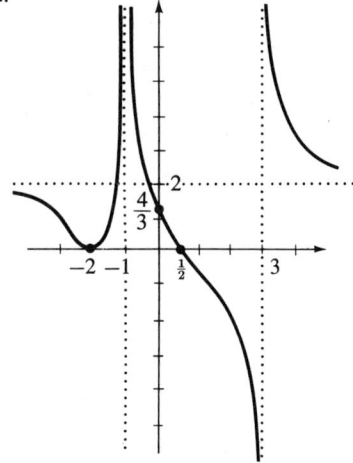

Figure 7.11

Problems

In Problems 1–10, check for symmetry about both axes and the origin.

1. $y = x^4 - x^2$.
2. $y = x^3 - x$.
3. $y = x^3 - x^2$.
4. $\dfrac{x^2}{4} + \dfrac{y^2}{9} = 1$.
5. $y^2 = \dfrac{x+1}{x}$.
6. $y^3 = \dfrac{x+1}{x}$.
7. $xy = 1$.
8. $x^2 y^2 = 1$.
9. $y = \dfrac{x}{x^2+1}$.
10. $y = \dfrac{(x+1)(x-1)}{x^2}$.

In Problems 11–30, use the methods of this and the preceding section to sketch. Do not plot the graph point by point.

11. $y = (x+1)(x-3)$.
12. $y = (x+2)(x-1)^2$.
13. $y = x^2 - 5x - 6$.
14. $y = x^3 + x^2 - 2x$.
15. $y = x^4 - x^2$.
16. $y = x^3 - x$.

17. $y = \dfrac{x+1}{x}$.

18. $y = \dfrac{x-2}{x+2}$.

19. $y = \dfrac{(2x+1)(x-1)^2}{(x-2)(x+1)^2}$.

20. $y = \dfrac{x-3}{(x+1)(x-2)}$.

21. $y = \dfrac{(x+2)(x-4)}{x-1}$.

22. $y = \dfrac{x-1}{(x+2)(x-4)}$.

23. $y = \dfrac{(x+2)^2(x-4)}{(x-1)^2}$.

24. $y = \dfrac{x}{x^2+1}$.

25. $y = \dfrac{x^2+1}{x}$.

26. $xy = 2x + 1$.

27. $x^2 y = 2x + 1$.

28. $x^2 y - y = x^2$.

29. $x^2 y - y = x^3$.

30. $x^2 y - y = x$.

31. Show that if a graph has any two of the three types of symmetry—about the x axis, about the y axis, about the origin—then it must have the third.

32. Give an example of a curve with exactly two lines of symmetry.

33. Can a graph have two points of symmetry?

34. Give an example of a curve with infinitely many lines of symmetry.

35. Show that if two perpendicular lines are lines of symmetry of a given curve, then their point of intersection is a point of symmetry.

7.3

Radicals and the Domain of the Equation

There are two things that keep us from getting a value for y when we substitute a value of x into an equation: a zero in the denominator and an even root of a negative number. A zero in the denominator gives a vertical asymptote. Even roots of negative numbers simply cause gaps in the domain of the equation.

Example 1

$$\text{Sketch } y = \dfrac{2x}{\sqrt{x^2-4}} = \dfrac{2x}{\sqrt{(x+2)(x-2)}}.$$

Using the previous methods, we have:

Intercepts: (0, 0) odd.

Asymptotes: $x = 2$, $x = -2$.

7.3 Radicals and the Domain of the Equation

The radical is equivalent to the one-half power, which is neither odd nor even. We have a special problem in finding the horizontal asymptotes. The highest power of x in the numerator is clearly x. The highest power in the denominator appears to be x^2. But it is under the radical; so the highest power is really $(x^2)^{1/2} = x$. Thus we shall want to divide the numerator and denominator by x. But we shall want to put the x under the radical in the denominator, which leads to further complications. The symbol $\sqrt{\,}$ means the *non-negative* square root. Thus $x = \sqrt{x^2}$ is true only when $x \geq 0$; when $x < 0$, $\sqrt{x^2} = -x$ (note that, since x itself is negative, $-x$ is positive), and we have two cases to consider:

$$\frac{2x}{\sqrt{x^2-4}} = \frac{\frac{2x}{x}}{\sqrt{\frac{x^2-4}{x^2}}} = \frac{2}{\sqrt{1-\frac{4}{x^2}}} \quad \text{when } x > 0,$$

$$\frac{2x}{\sqrt{x^2-4}} = \frac{\frac{2x}{-x}}{\sqrt{\frac{x^2-4}{x^2}}} = \frac{-2}{\sqrt{1-\frac{4}{x^2}}} \quad \text{when } x < 0.$$

Thus,

$$\text{as } x \to +\infty, \quad y = \frac{2x}{\sqrt{x^2-4}} = \frac{2}{\sqrt{1-\frac{4}{x^2}}} = 2;$$

$$\text{as } x \to -\infty, \quad y = \frac{2x}{\sqrt{x^2-4}} = \frac{-2}{\sqrt{1-\frac{4}{x^2}}} = -2;$$

giving two horizontal asymptotes: $y = 2$, which is approached on the right, and $y = -2$, which is approached on the left. Replacing x by $-x$ and y by $-y$ gives

$$-y = \frac{2(-x)}{\sqrt{(-x)^2-4}} = \frac{-2x}{\sqrt{x^2-4}},$$

which is equivalent to the original equation. Thus we have symmetry about the origin.

Finally, $\sqrt{x^2-4}$ represents a real number only when $x^2 - 4 \geq 0$, which gives

$$x^2 \geq 4 \quad \text{or} \quad \begin{cases} x \geq 2 \\ x \leq -2. \end{cases}$$

But y is real for one additional value of x, namely, $x = 0$. If $x = 0$, y equals zero divided by a complex number, which is still zero. Thus the domain is

$$\{x \mid x \geq 2 \text{ or } x \leq -2 \text{ or } x = 0\}.$$

We see here that $(0, 0)$ is an isolated point of the graph.

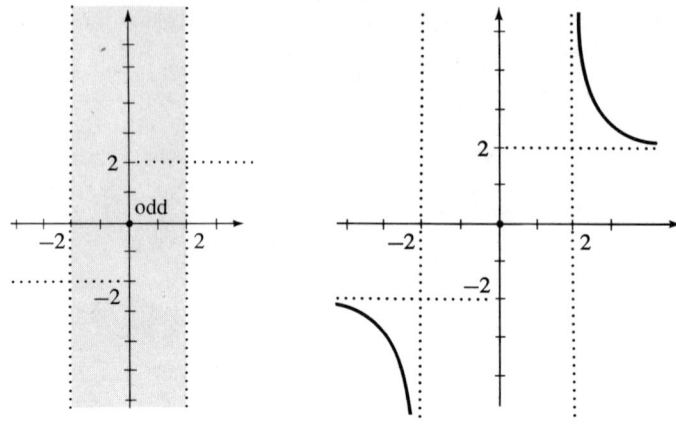

Figure 7.12   **Figure 7.13**

All of this information is represented graphically in Figure 7.12. We can now see that the fact that the intercept is odd is of no use, since it is an isolated point. Note one thing more: Since $\sqrt{x^2 - 4}$ is never negative, y is positive whenever x is positive and negative whenever x is negative. This additional information makes it easy for us to sketch the curve (see Figure 7.13).

Example 2

Sketch $y^2 = x^4 - x^2$.

To graph this equation, we use the following device: Since $y = \pm\sqrt{x^4 - x^2}$, we first graph $z = x^4 - x^2$ and then, from the values of z, get $y = \pm\sqrt{z}$. Graphing $z = x^4 - x^2 = x^2(x^2 - 1) = x^2(x + 1)(x - 1)$, we have

Intercepts:

(0, 0), even,   (1, 0), odd,   (−1, 0), odd.

No asymptotes.
Symmetry about the z axis (see Figure 7.14).

We see on the dotted graph that, for each value of x, we have a value of $z = x^4 - x^2$. Now let us find the corresponding values for $y = \pm\sqrt{z}$. But first, we note the following points to keep in mind.

(1) $\sqrt{z} = z$ if $z = 0$ or $z = 1$,
(2) $\sqrt{z} > z$ if $0 < z < 1$,
(3) $\sqrt{z} < z$ if $z > 1$,
(4) $\sqrt{z}$ is not real if $z < 0$.

The final result is given by the solid graph of Figure 7.14. The origin is again an isolated point of this graph.

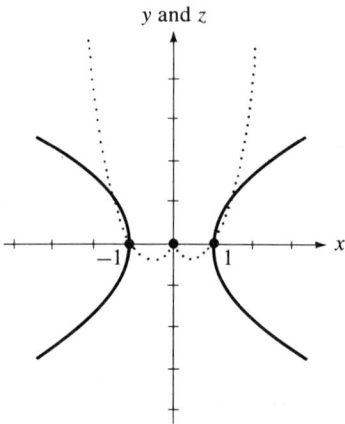

Figure 7.14

7.3 Radicals and the Domain of the Equation

The same method could be used to sketch $y = \sqrt{x^4 - x^2}$. The only difference would be that we would have only the top half of the result in Figure 7.14. We might also have used this method in Example 1, starting with

$$y^2 = \frac{4x^2}{x^2 - 4}.$$

In that case, we would have to be careful which branch we chose; we would have to choose the top portion when x is positive and the bottom portion when x is negative.

One final point. Let us recall that when we had an equation of the form

$$y = \frac{P(x)}{Q(x)},$$

we noted that x intercepts come from factors in the numerator and vertical asymptotes from factors in the denominator, *provided there is no value of x for which both numerator and denominator are zero.* In the examples we have been considering, this is equivalent to the provision that there is no factor common to both numerator and denominator. What happens if there *are* common factors? The answer is simple. You simply cancel the common factors and sketch the resulting equation. But remember that if you cancel the factor $x - a$, the original equation is not defined at $x = a$ (it gives 0/0) and there is no point on the graph with x coordinate a.

Example 3

Sketch $y = \dfrac{x^2 - 1}{x - 1}$.

Since the numerator and denominator have the common factor $x - 1$, we cancel them to get

$$y = x + 1,$$

which gives a straight line. But recall that the original equation gives no value of y when $x = 1$. Thus the point $(1, 2)$ should be deleted from the graph, as in Figure 7.15.

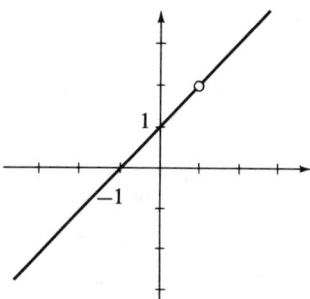

Figure 7.15

Problems

Sketch the graphs of the following equations.

1. $y = x\sqrt{x^2 - 1}$.

2. $y = \dfrac{x}{\sqrt{x - 1}}$.

3. $y = \dfrac{-x}{\sqrt{x^2 - 1}}$.

4. $y = \dfrac{x - 1}{\sqrt{x(x + 1)}}$.

5. $y = x + \sqrt{x^2 - 1}$.
6. $y = x - \sqrt{x^2 - 1}$.
7. $y^2 = \dfrac{x}{x+1}$.
8. $y^2 = \dfrac{x^2}{(x+1)(x-2)}$.
9. $y^2 = \dfrac{2x}{(x-1)^2}$.
10. $y^2 = \dfrac{(x-1)^2}{x}$.
11. $y^2 = \dfrac{x(x+1)}{(x-2)^2}$.
12. $y^2 = \dfrac{x(x+1)^2}{x-2}$.
13. $y^2 = \dfrac{x(x-1)}{(x+1)^2}$.
14. $y^2 = \dfrac{x(1-x)}{(x+1)^2}$.
15. $y^2 = \dfrac{(x^2-1)^2}{x-2}$.
16. $y^2 = (1-x)(3-x)^2$.
17. $y^2 = (x-1)(x-3)^2$.
18. $y^2 = -(x-1)(x-3)^2$.
19. $y = \dfrac{x^2-4}{x-2}$.
20. $y = \dfrac{x^2+x}{x}$.
21. $y = \dfrac{x^2+x}{x^2}$.
22. $y = \dfrac{x^3+x^2}{x}$.
23. $y = \dfrac{x(x+1)^2}{(x-1)(x+1)^3}$.
24. $y = \dfrac{2x(x-1)}{x(x+1)}$.
25. $y = \dfrac{1-(1+h)^2}{h}$.
26. $y = \dfrac{-1-[(1+h)^2 - 2(1+h)]}{h}$.

(Hint: Simplify the numerator.)

27. $y = \dfrac{2 - \dfrac{2+h}{1+h}}{h}$.
28. $y = \dfrac{1 - \sqrt{1+h}}{h}$.

7.4

Direct Sketching of Conics

We have been able to sketch conics by putting them into a standard form in Chapters 5 and 6. But this was often quite tedious, especially when we had to rotate the axes. Let us see if we can determine some methods of sketching without going through the process of rotating axes.

First of all, there is a method of determining which conic we have without rotating axes. It is based on the fact that certain expressions are invariant under rotation; that is, they have the same value before and after any rotation. Although there are several such expressions (see Problems 18 and 19 of Section 6.2), the one in which we

7.4 Direct Sketching of Conics

are interested is $B^2 - 4AC$ for the equation

$$Ax^2 + Bxy + Cy^2 + Dx + Ey + F = 0.$$

Although a proof of the following theorem is not difficult, it is long and tedious and is omitted here.

Theorem 7.5

If the equation

$$Ax^2 + Bxy + Cy^2 + Dx + Ey + F = 0$$

is transformed into the equation

$$A'x'^2 + B'x'y' + C'y'^2 + D'x' + E'y' + F' = 0$$

by rotating the axes, then

$$B^2 - 4AC = B'^2 - 4A'C'.$$

If we choose the angle of rotation properly, $B' = 0$ and the type of conic can be determined by looking at A' and C' (see the table on page 119). Thus we have the following results.

Theorem 7.6

The equation

$$Ax^2 + Bxy + Cy^2 + Dx + Ey + F = 0$$

represents a hyperbola, ellipse, or parabola (or a degenerate case of one of these) according to whether $B^2 - 4AC$ is positive, negative, or zero, respectively.

This theorem gives us a general idea of the graph of an equation before we start. Remember that this test does not distinguish between the conics and their degenerate cases. Thus, for instance, if $B^2 - 4AC$ is positive, we may have either a hyperbola or two intersecting lines.

If we are dealing with a hyperbola, the greatest single aid in sketching the graph is determination of the asymptotes. If they are horizontal or vertical, the determination is relatively easy, so let us go to slant asymptotes. We shall consider two cases: the equation is linear in y ($C = 0$) and the equation is quadratic in y ($C \neq 0$). In either case, we first solve for y. Examples of each follow.

Example 1

Sketch $x^2 - xy - 3y - 1 = 0$ without rotating axes.

First of all, $B^2 - 4AC = (-1)^2 - 4(1)(0) = 1$, indicating that the conic is a hyperbola or a degenerate case of one. Solving for y, we have

$$y = \frac{x^2 - 1}{x + 3}.$$

The methods of this chapter give intercepts $(\pm 1, 0)$, $(0, -1/3)$ and vertical asymptote, $x = -3$. There is no horizontal asymptote, but we know that there must be a second asymptote. To find it, we carry out the division.

$$y = \frac{x^2 - 1}{x + 3} = x - 3 + \frac{8}{x + 3}.$$

We now see that, for numerically large values of x, $8/(x+3)$ is almost zero and y is very near $x - 3$. Thus the slant asymptote is

$$y = x - 3.$$

With this we can easily sketch the hyperbola (see Figure 7.16).

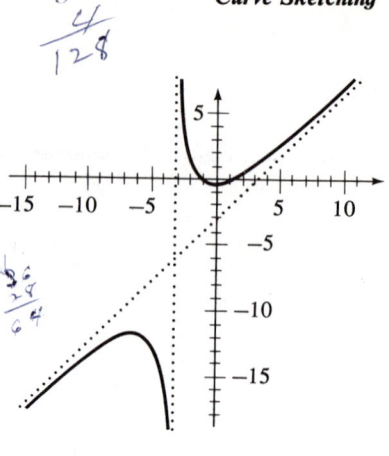

Figure 7.16

Example 2
Sketch $7x^2 + 6xy - y^2 - 32 = 0$ without rotating axes.

This equation is quadratic in y.

$$y^2 - 6xy + (32 - 7x^2) = 0.$$

Using the quadratic formula, we have

$$y = 3x \pm 4\sqrt{x^2 - 2}.$$

Again, for large values of x, $\sqrt{x^2 - 2}$ is almost $\sqrt{x^2}$, and y is very near $3x \pm 4x$. Thus, the slant asymptotes are (see Figure 7.17)

$$y = 7x \quad \text{and} \quad y = -x.$$

This and the intercepts give us a good idea of the curve.

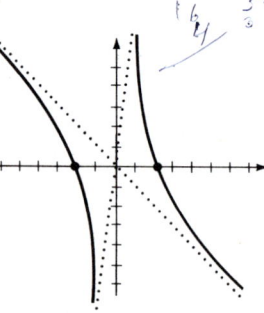

Figure 7.17

Another useful procedure in sketching conics (as well as other curves) is the method of addition of ordinates. Let us consider an example.

Example 3

Sketch $2x^2 - 2xy + y^2 - 9 = 0$ without rotating axes.

Since $B^2 - 4AC = -4$, the curve is an ellipse. Again, the equation is quadratic in y,

7.4 Direct Sketching of Conics

$$y^2 - 2xy + (2x^2 - 9) = 0.$$

By the quadratic formula, we have

$$y = x \pm \sqrt{9 - x^2}.$$

Instead of trying to sketch this curve directly let us sketch

$$y = x \quad \text{and} \quad y = \pm\sqrt{9 - x^2}.$$

By squaring both sides, we can put the second equation into the form

$$x^2 + y^2 = 9.$$

These two are easily sketched (see Figure 7.18). For each value of x in the interval $[-3, 3]$, there is an ordinate on the line and one (or two) on the circle. Adding them, we have the ellipse of Figure 7.18.

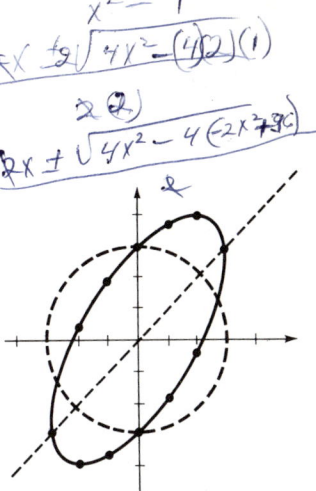

Figure 7.18

Since values of x outside the interval $[-3, 3]$ give complex values of y in the equation $y = x \pm \sqrt{9 - x^2}$, there is no graph to the right of $x = 3$ or to the left of $x = -3$.

Example 4

Sketch $4x^2 - 4xy + y^2 - 5x + 2y + 1 = 0$ without rotating axes.

Since $B^2 - 4AC = 0$, the curve is a parabola. Again solving for y, we have

$$y = 2x - 1 \pm \sqrt{x}.$$

This gives the two equations

$$y = 2x - 1 \quad \text{and} \quad y = \pm\sqrt{x},$$

where the latter can be written $y^2 = x$. Sketching these two and adding ordinates, we have the result given in Figure 7.19. The line $y = 2x - 1$ is *not* the axis of the parabola.

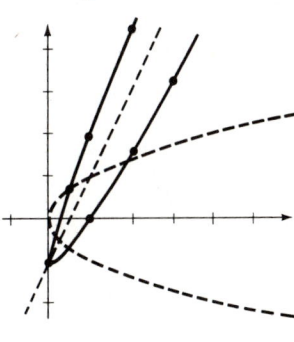

Figure 7.19

Addition of ordinates can be used for hyperbolas as well as for ellipses and parabolas. One disadvantage of this method is that the two curves must be graphed relatively accurately, or the final result is likely to be extremely inaccurate.

One final word; it is *not* maintained that the methods of this section will *always* provide the simplest method of sketching conics. They are alternate methods that are useful in many cases.

Problems

In Problems 1–20, sketch without rotating axes.

1. $xy - x + y + 3 = 0$.
2. $2xy - x - y - 2 = 0$.
3. $x^2 - xy - y - 4 = 0$.
4. $x^2 - xy + x + 2y = 0$.
5. $2x^2 - 2xy + y^2 - 1 = 0$.
6. $5x^2 - 4xy + y^2 - 4 = 0$.
7. $x^2 - 2xy + y^2 - x = 0$.
8. $x^2 - 2xy + y^2 + x - 2y + 2 = 0$.
9. $2xy - y^2 - 4 = 0$.
10. $2x^2 - 2xy + y^2 + 4x - 4y - 5 = 0$.
11. $2xy - y^2 + 6x - 6y - 18 = 0$.
12. $3x^2 - 4xy + y^2 - 4x + 2y + 5 = 0$.
13. $4x^2 + 4xy + y^2 - 3x + 2y + 1 = 0$.
14. $x^2 - 2xy + y^2 - 12x + 8y + 24 = 0$.
15. $3x^2 + 2xy - y^2 + 10x + 2y + 8 = 0$.
16. $x^2 - 2xy + y^2 - 2x + 2y - 3 = 0$.
17. $x^2 - xy - x - 2 = 0$.
18. $xy - y^2 - y + 2 = 0$.
19. $10x^2 - 6xy + y^2 + 12x - 4y + 4 = 0$.
20. $2xy + y^2 - 4 = 0$.
21. Show that $\sqrt{x} + \sqrt{y} = \sqrt{a}$ is a portion of a parabola.

8

Transcendental Curves

8.1

Trigonometric Functions

We now turn to some of the nonalgebraic (or transcendental) functions, beginning with the trigonometric functions and their inverses. Although the measurement of angles in degrees is convenient for many purposes, in advanced mathematics radian measure is the more natural way of measuring angles. We shall assume that all angles are measured in radians unless otherwise stated.

The graphs of the six common trigonometric functions are given in Figure 8.1. Note that all of them are periodic (repeating). The period of $y = \tan x$ and $y = \cot x$ is π; the other four functions have period 2π. Note also that four of them have vertical asymptotes, although there is no denominator to be zero. This is reasonable when we consider that all of them are defined as *ratios* of certain lengths. Let us see how these curves are altered by changing certain constants.

154   *Transcendental Curves*

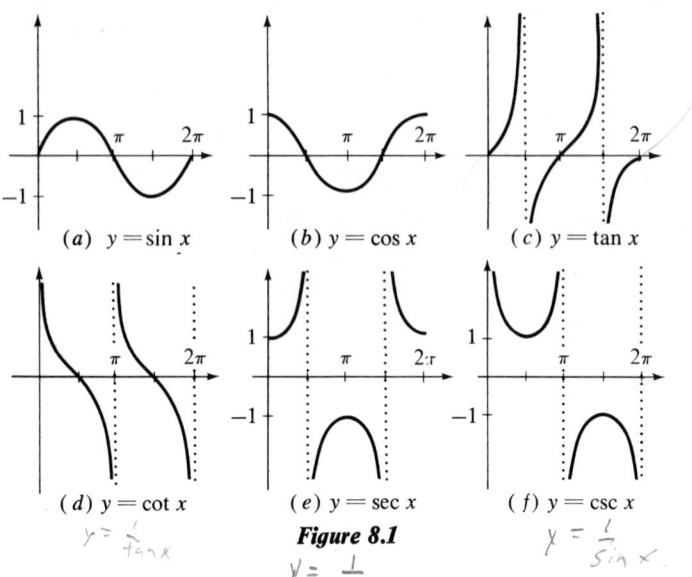

(a) $y = \sin x$ (b) $y = \cos x$ (c) $y = \tan x$

(d) $y = \cot x$ (e) $y = \sec x$ (f) $y = \csc x$

Figure 8.1

Example 1

Sketch $y = 3 \sin 2x$.

First of all, note that $-1 \leq \sin x \leq 1$. Thus the factor 3 in $3 \sin 2x$ changes this range by a factor of 3. The fact that we have $\sin 2x$ instead of $\sin x$ does not alter the range. Now it takes one complete cycle for whatever angle we are taking the sine of to go from 0 to 2π; that is, we have one complete cycle for

$$0 \leq 2x \leq 2\pi \quad \text{or} \quad 0 \leq x \leq \pi.$$

Thus, the 3 in $3 \sin 2x$ triples the amplitude (or height) of the wave, while the 2 halves the period (or gives two complete cycles in the normal period of 2π). The result is shown in Figure 8.2.

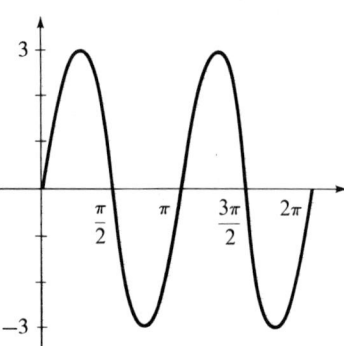

Figure 8.2

Example 2

Sketch $y = 4 \cos\left(2x + \dfrac{\pi}{2}\right)$.

First let us write the equation in the form

$$y = 4 \cos 2\left(x + \dfrac{\pi}{4}\right).$$

8.1 Trigonometric Functions

Now we see that the amplitude is 4 and the period is $2\pi/2 = \pi$. When $x = -\pi/4$, $\pi/4$, $3\pi/4$, and so on, $x + \pi/4 = 0$, $\pi/2$, π, and so forth. Therefore $\pi/4$ has the effect of shifting the curve a distance $\pi/4$ to the left, as shown in Figure 8.3.

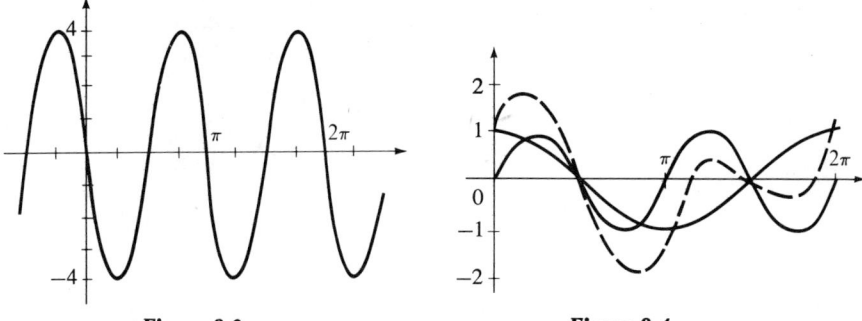

Figure 8.3   Figure 8.4

Example 3

Sketch $y = \cos x + \sin 2x$.

The method of addition of ordinates (see Section 7.4) is quite useful for equations of this type. Sketching $y = \cos x$ and $y = \sin 2x$ and adding the ordinates, we have the result given in Figure 8.4.

Example 4

Sketch $y = x + \sin x$.

Perhaps you wonder how we can add x and $\sin x$, if x is an angle and $\sin x$ is a number. Actually both x and $\sin x$ are numbers. We take trigonometric functions not of angles, but of numbers. The numbers are simply the *measures* of angles. It is quite possible to consider trigonometric functions of numbers quite independently of any angular interpretations; but if we do want to impose such an interpretation, the value of x is the measure of an angle in radians. Again, addition of ordinates works very well and Figure 8.5 is self-explanatory.

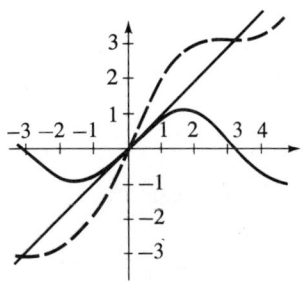

Figure 8.5

Problems

1. Express the following degree measures in radian measure: $45°$, $-210°$, $270°$, $30°$, $-180°$, $-60°$, $135°$, $150°$.
2. Express the following radian measures in degree measure: $\pi/3$, π, $3\pi/4$, $-\pi/2$, $5\pi/6$, $-2\pi/3$, $3\pi/2$, $10\pi/6$.

In Problems 3–26, sketch one complete cycle.

3. $y = 3 \cos x$.
4. $y = 2 \sec x$.
5. $y = -2 \sin x$.
6. $y = 2 \tan 3x$.
7. $y = 4 \sin \pi x$.
8. $y = -2 \csc(\pi x/2)$.
9. $y = 3 \cos 4x$.
10. $y = 2 \sin(2x + \pi)$.
11. $y = 3 \cos(2\pi x + \pi/2)$.
12. $y = \tan(3x - \pi)$.
13. $y = 2 \sec(4x - 2\pi)$.
14. $y = -\cos(x - \pi/3)$.
15. $y = -2 \sin(\pi/4 - x)$.
16. $y = \sin x - \cos x$.
17. $y = 2 \sin x + \sin 2x$.
18. $y = \cos x - \sin 2x$.
19. $y = 3 \cos x + \sin x$.
20. $y = 4 \sin x + 2 \sin 2x - \sin 4x$.
21. $y = 2 \sin x - \sin 2x + \frac{2}{3} \sin 3x$.
22. $y = 1 - \cos x$.
23. $y = 1 + 2 \sin x$.
24. $y = 2 + \cos x$.
25. $y = \sin^2 x$.
26. $y = 2 - 3 \sin x$.

In Problems 27–30, sketch

27. $y = x - \sin x$.
28. $y = x^2 + \sin x$.
29. $y = x \sin x$.
30. $y = \dfrac{\sin x}{x}$.

31. Sketch

$$y = \frac{\pi}{2} + 2 \sin x, \qquad y = \frac{\pi}{2} + 2 \sin x + \frac{2}{3} \sin 3x,$$

and

$$y = \frac{\pi}{2} + 2 \sin x + \frac{2}{3} \sin 3x + \frac{2}{5} \sin 5x$$

on the same coordinates. What do you think the graph of

$$y = \frac{\pi}{2} + 2\left(\sin x + \frac{1}{3} \sin 3x + \frac{1}{5} \sin 5x + \cdots\right)$$

looks like?

8.2

Inverse Trigonometric Functions

Suppose we have the equation $y = \sin x$ and want to express x in terms of y. To do so, we introduce a new notation for the solution,

$$x = \arcsin y \quad \text{or} \quad x = \sin^{-1} y.$$

8.2 Inverse Trigonometric Functions

This is read: x is an inverse sine of y. Thus, $x = \arcsin y$ is equivalent to $y = \sin x$, or $y = \arcsin x$ is equivalent to $x = \sin y$. To graph $y = \arcsin x$, we merely graph $x = \sin y$ (see Figure 8.6). This looks exactly like the graph of $y = \sin x$ with the x and y axes reversed. Note that arcsin x is *not* single valued; one value of x gives many values of arcsin x. The remaining five trigonometric functions have inverses that are defined analogously.

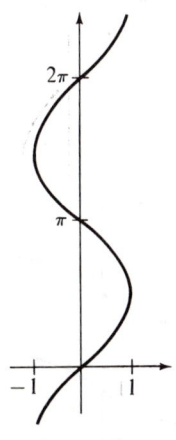

Figure 8.6

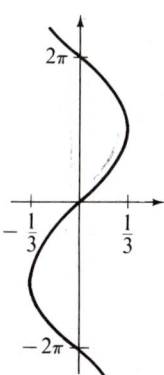

Figure 8.7

Example 1

Sketch $y = 2 \arcsin 3x$.

We first convert this to the equivalent equation involving the sine.

$$\frac{y}{2} = \arcsin 3x,$$

$$3x = \sin \frac{y}{2},$$

$$x = \frac{1}{3} \sin \frac{y}{2}.$$

Graphing this by the methods of the previous section, we have Figure 8.7.

Note that the graph above is a sine wave on the y axis. Furthermore, the 3 in the equation gives an amplitude of $1/3$, while the 2 gives a period of $2 \cdot 2\pi = 4\pi$.

Example 2

Sketch $y = \frac{1}{3} \arccos 2x$.

Again we can convert this to the equivalent equation involving the cosine,

$$x = \frac{1}{2} \cos 3y,$$

and graph by the methods of the previous section. The result is given in Figure 8.8. This could also have been sketched by noting that we have a cosine wave on the y axis with amplitude $1/2$ and period $1/3 \cdot 2\pi = 2\pi/3$.

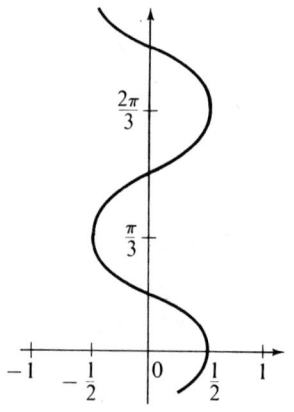

Figure 8.8

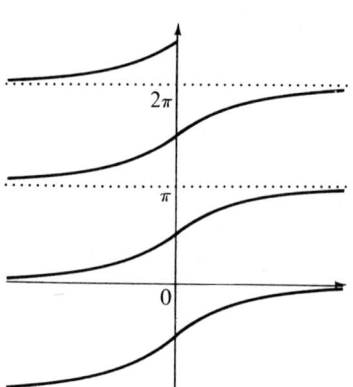

Figure 8.9

Example 3

Sketch $y = \pi/2 + \arctan x$.

In the tangent form, the equation is

$$x = \tan\left(y - \frac{\pi}{2}\right);$$

giving the graph of Figure 8.9. The $\pi/2$ has the effect of raising the curve a distance $\pi/2$.

Example 4

Sketch $y = \text{arcsec}\,(x+1)$.

8.2 Inverse Trigonometric Functions

In secant form we have

$$x = \sec y - 1,$$

which gives the graph of Figure 8.10. The 1 has the effect of shifting the graph one unit to the left.

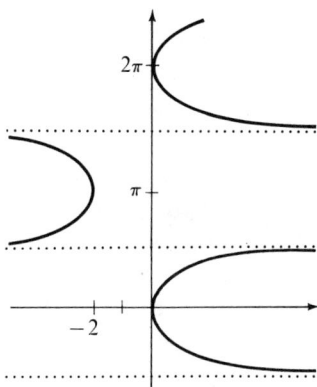

Figure 8.10

Problems

Sketch the graph of each of the following equations.

1. $y = \arccos x$.
2. $y = \arctan x$.
3. $y = \text{arccot } x$.
4. $y = \text{arcsec } x$.
5. $y = 2 \arcsin x$.
6. $y = 3 \arccos x$.
7. $y = 3 \text{ arcsec } x$.
8. $y = 4 \text{ arccsc } x$.
9. $y = \arcsin 3x$.
10. $y = \arctan 2x$.
11. $y = \arccos 4x$.
12. $y = \text{arcsec } 2x$.
13. $y = 3 \arcsin 2x$.
14. $y = 2 \arccos 4x$.
15. $y = 2 \arctan 3x$.
16. $y = 4 \text{ arcsec } 2x$.
17. $y = -2 \arcsin \dfrac{x}{2}$.
18. $y = -2 \arccos \dfrac{x}{3}$.
19. $y = \dfrac{\pi}{2} + \arcsin x$.
20. $y = \dfrac{\pi}{4} + \arccos x$.
21. $y = -\dfrac{\pi}{4} + 2 \arcsin x$.
22. $y = \dfrac{\pi}{4} - 2 \arctan x$.
23. $y = \arcsin (x + 2)$.
24. $y = \arccos (x - 1)$.
25. $y = 3 \arcsin 2(x - 1)$.
26. $y = 2 \arctan (3x + 2)$.
27. $y = \dfrac{\pi}{4} + 2 \arcsin 3(x + 2)$.
28. $y = \dfrac{\pi}{3} - 2 \text{ arcsec } 4(x - 2)$.
29. $y = -\dfrac{2\pi}{3} + \dfrac{1}{2} \arccos \dfrac{x+1}{3}$.
30. $y = \dfrac{\pi}{4} + \dfrac{1}{3} \arctan \dfrac{2x+1}{3}$.

8.3

Exponential and Logarithmic Functions

Suppose we consider the graph of $y = 2^x$. By plotting a few points, we have the graph shown in Figure 8.11. The curve has the x axis as a horizontal asymptote, but it approaches this asymptote only at the left end. At the right end the graph increases very rapidly; there is, however, no vertical asymptote. The graph crosses the y axis at $(0, 1)$ and is always above the x axis. These are general characteristics of the graph of $y = a^x$, where $a > 1$. If $a < 1$, then the graph rises steeply on the left and approaches the x axis on the right. The graph of $y = 1^x$ is the horizontal line $y = 1$, since $1^x = 1$ for all x. The graphs of several such equations are given in Figure 8.12. Note that, for $a > 1$, the larger the base a, the more steeply the curve rises on the right and the more rapidly it approaches the x axis on the left; while, for $a < 1$, the smaller the base, the more steeply the curve rises on the left and the more rapidly it approaches the x axis on the right.

A base 0 gives the x axis without the origin (since 0^0 is meaningless), and negative bases are not considered at all, since powers of negative bases are not defined for any but integer exponents.

A number that is often used for a base is the number $e = 2.71828\ldots$. This base is especially useful in calculus and, for this reason, is frequently encountered. Of course, the graph of $y = e^x$ lies between the graphs of $y = 2^x$ and $y = 3^x$, as in Figure 8.12.

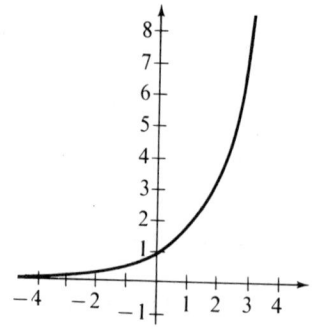

Figure 8.11

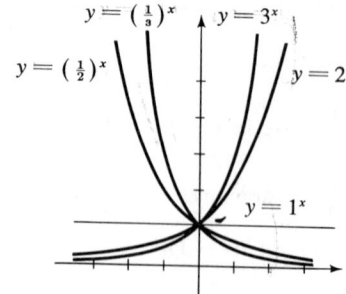

Figure 8.12

8.3 Exponential and Logarithmic Functions

Example 1

Sketch the graph of $y = 2^{x-2}$.

This can be considered from two different points of view. One way is to consider the desired graph to be the graph of $y = 2^x$ translated two units to the right. Another is to note that

$$y = 2^{x-2} = 2^x \cdot 2^{-2} = \frac{2^x}{4}.$$

Thus the y coordinate of $y = 2^{x-2}$ is one-fourth of the corresponding y coordinate of $y = 2^x$. The result is given in Figure 8.13.

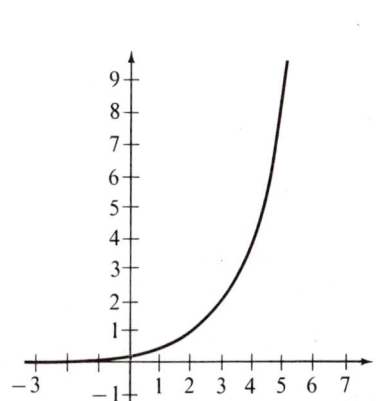

Figure 8.13

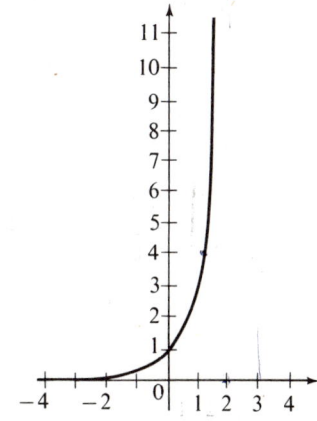

Figure 8.14

Example 2

Sketch the graph of $y = 2^{2x}$.

Since

$$y = 2^{2x} = (2^2)^x = 4^x,$$

we have the graph shown in Figure 8.14.

Note the distinction between $y = 2^x$ and $y = x^2$. The first, with a constant base and variable exponent, is an exponential function such as we have been discussing. The other, with a variable base and constant exponent, is a power function; its graph is a parabola.

Let us now consider the logarithm, which we define in the customary way.

Definition

The **logarithm**, base a $(a > 0, a \neq 1)$, of the number x $(x > 0)$ is the number y such that $a^y = x$. Thus,

$$y = \log_a x \quad \text{means} \quad x = a^y.$$

We see that the logarithm and exponential functions are inverses of each other; that is, $y = \log_a x$, which is equivalent to $x = a^y$, is simply the exponential function $y = a^x$ with the x and y reversed. Thus, we may use our knowledge of the exponential function in order to sketch the graph of a logarithm. The graph of $y = \log_a x$ for various values of a is given in Figure 8.15. Note that they are defined only for positive values of x—there is no graph to the left of the y axis. The y axis is a vertical asymptote and $y = \log_a x$ increases slowly for $a > 1$ and decreases slowly for $a < 1$. All contain the point $(1, 0)$.

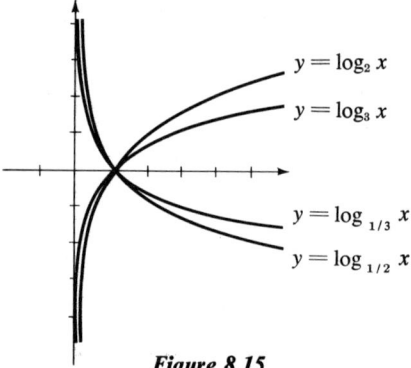

Figure 8.15

Example 3

Sketch the graph of $y = \log_2 (x + 2)$.

The expression $x + 2$ merely translates the graph of $y = \log_2 x$ two units to the left. That is, there is a vertical asymptote when $x + 2 = 0$ or $x = -2$, and the graph crosses the x axis when $x + 2 = 1$ or $x = -1$. Thus we have the graph of Figure 8.16.

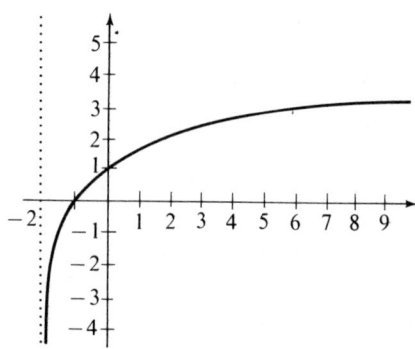

Figure 8.16

8.3 Exponential and Logarithmic Functions

Example 4

Sketch the graph of $y = \log_4(-x)$.

The logarithm is defined only when $-x$ is positive, or when x is negative. The result is the mirror image of $y = \log_4 x$ reflected in the y axis (see Figure 8.17).

Two bases that are frequently used are 10 and e. The following abbreviations are frequently encountered and are used throughout this book.

$$\log x = \log_{10} x,$$

$$\ln x = \log_e x.$$

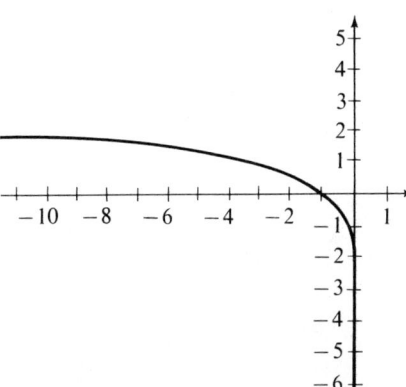

Figure 8.17

Problems

Sketch the graph of each of the following equations.

1. $y = 2^{x+1}$.
2. $y = 2^{x-1}$.
3. $y = 3^{-x}$.
4. $y = e^x$.
5. $y = 4^{x-3}$.
6. $y = 3^{1-x}$.
7. $y = 3^{2x}$.
8. $y = 4^{-2x}$.
9. $y = 2^{2x+1}$.
10. $y = 5^{2x-1}$.
11. $y = \left(\frac{1}{2}\right)^{x+1}$.
12. $y = \left(\frac{1}{3}\right)^{x-2}$.
13. $y = \left(\frac{1}{4}\right)^{2x+1}$.
14. $y = \left(\frac{1}{2}\right)^{2-x}$.
15. $y = 2^{|x|}$.
16. $y = 2^{x^2}$.
17. $y = \log_2(x+1)$.
18. $y = \log_2(x-1)$.
19. $y = \log_3(x+4)$.
20. $y = \log_4(x-3)$.
21. $y = \log_3(-x)$.
22. $y = \ln x$.
23. $y = \log_2(2x)$.
24. $y = \log_3(9x)$.
25. $y = \log_2(4x-1)$.
26. $y = \log_4(4x+1)$.
27. $y = \log_2|x|$.
28. $y = \log_2 x^2$.
29. $y = \log_{1/2}(x+1)$.
30. $y = \log_{1/3}(x-2)$.

8.4

Hyperbolic Functions

The hyperbolic functions, which occur relatively frequently, are defined in terms of exponential functions, but they are like the trigonometric functions in many ways. They are called the hyperbolic sine, hyperbolic cosine, and so on, and are abbreviated sinh, cosh, and so forth, respectively.

Definition

(1) $\sinh x = \dfrac{e^x - e^{-x}}{2}.$ (2) $\cosh x = \dfrac{e^x + e^{-x}}{2}.$

(3) $\tanh x = \dfrac{\sinh x}{\cosh x}.$ (4) $\coth x = \dfrac{\cosh x}{\sinh x}.$

(5) $\operatorname{sech} x = \dfrac{1}{\cosh x}.$ (6) $\operatorname{csch} x = \dfrac{1}{\sinh x}.$

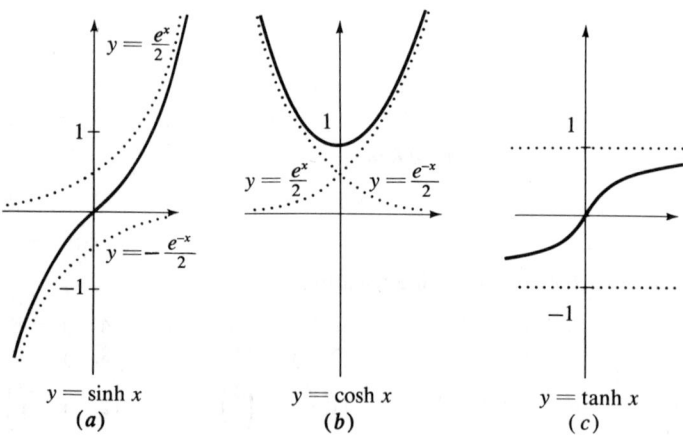

$y = \sinh x$
(a)

$y = \cosh x$
(b)

$y = \tanh x$
(c)

Figure 8.18

The first two hyperbolic functions are easily graphed by addition of ordinates, and the remaining four by division of ordinates. For example, since

8.4 Hyperbolic Functions

$$\sinh x = \frac{e^x}{2} - \frac{e^{-x}}{2},$$

we graph $y = e^x/2$ and $y = -e^{-x}/2$ and add the ordinates. This is given in (a) of Figure 8.18. The graph of $y = \tanh x$ is found by noting that

$$\tanh x = \frac{\sinh x}{\cosh x}.$$

For each value of x, we divide $y_1 = \sinh x$ by $y_2 = \cosh x$ to find $\tanh x$. The result is given in Figure 8.18(c).

Example 1

Sketch the graph of $y = \sinh \frac{x}{2}$.

Note that the substitution of $x = 1$ into the given equation is equivalent to the substitution of $x = 1/2$ into $y = \sinh x$. Thus the graph of $y = \sinh x/2$ is like that of $y = \sinh x$ but "stretched" horizontally so that each point is twice as far from the y axis. This is given in Figure 8.19.

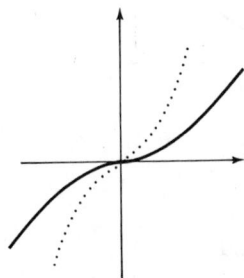

Figure 8.19

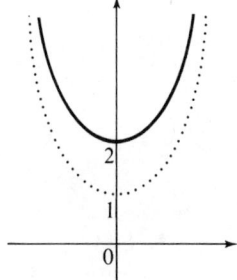

Figure 8.20

Example 2

Sketch the graph of $y = 2 \cosh x$.

The effect of the 2 here is to "stretch" the graph of $y = \cosh x$ vertically so that each point is twice as far from the x axis. This is given in Figure 8.20.

The inverses of hyperbolic functions can be considered in the same way as those of trigonometric functions. They are called the inverse hyperbolic sine, and so forth, and are represented by $\sinh^{-1}$, and so forth.

Definition

$y = \sinh^{-1} x$ *means* $x = \sinh y$. *The other five inverse hyperbolic functions are defined similarly.*

Again the graphs of the inverses can be determined from those of the corresponding hyperbolic functions.

Example 3

Sketch the graph of $y = \sinh^{-1} x$.

Since this is equivalent to $x = \sinh y$, its graph is the same as the graph of $y = \sinh x$ with the x and y reversed. The result is given in Figure 8.21.

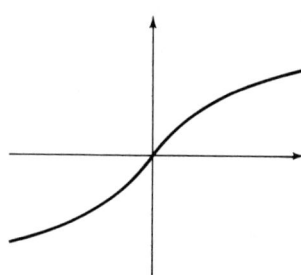

Figure 8.21

Problems

Sketch the graph of each of the following equations.

1. $y = \coth x$.
2. $y = \operatorname{sech} x$.
3. $y = \operatorname{csch} x$.
4. $y = \sinh 2x$.
5. $y = \frac{1}{2} \cosh x$.
6. $y = \tanh \frac{x}{3}$.
7. $y = 2 \sinh x$.
8. $y = 3 \cosh x$.

8.4 Hyperbolic Functions

9. $y = 2 \cosh 3x$.
10. $y = 2 \tanh 3x$.
11. $y = \sinh |x|$.
12. $y = \sinh x^2$.
13. $y = \sinh (x - 1)$.
14. $y = \cosh (x + 2)$.
15. $y = \tanh |x + 1|$.
16. $y = 2 \cosh (x - 3)$.
17. $y = \cosh^{-1} x$.
18. $y = \tanh^{-1} x$.
19. $y = 2 \sinh^{-1} x$.
20. $y = \sinh^{-1} 2x$.

9

Polar Coordinates

9.1

Polar Coordinates

Up to now, a point in the plane has been represented by a pair of numbers, (x, y), which represent (for perpendicular axes) the distances of the point from the y and x axes, respectively. Another way of representing points is by *polar coordinates*. In this case, we need only one axis (the *polar axis*) and a point on it (the *pole*). These correspond to the x axis and the origin of the rectangular coordinate system. Normally we shall include the y axis, even though it is not necessary to do so.

Before considering points in polar coordinates, let us recall that an angle in the standard position has its vertex at the origin (or pole) and its initial side on the positive end of the x axis (or polar axis). The terminal side is another ray (or half-line) with the origin as its end point. The ray with the same end point and on the same line as the terminal side is called the ray opposite the terminal side. For example, the terminal side of a 90° angle in standard position is the positive end of the y axis together with the origin; the ray opposite the terminal side is the negative end of the y axis together with the origin.

A point P is represented, in polar coordinates, by an ordered pair of numbers (r, θ). (See Figure 9.1.) It is determined in the following way: first find the terminal side of the angle θ in standard position; if $r \geq 0$, then P is on this terminal

9.2 Graphs in Polar Coordinates

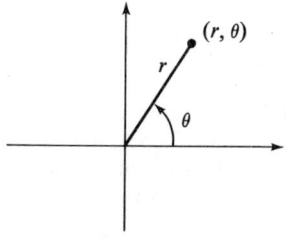

Figure 9.1

side and at a distance r from the pole; if $r < 0$, then P is on the ray opposite the terminal side and at a distance $|r|$ from the pole. A few points are given with their polar coordinates in Figure 9.2.

It might be noted that while the terminal side of the angle $-\pi/3$ is in the fourth quadrant, $(-1, -\pi/3)$ is in the second quadrant. The quadrant that a point is in is *not* determined by the signs of the two polar coordinates, as it is with rectangular coordinates. It is determined by the size of θ and the sign of r. If r is positive, the point is in whatever quadrant θ is in; if r is negative, the point is in the opposite quadrant.

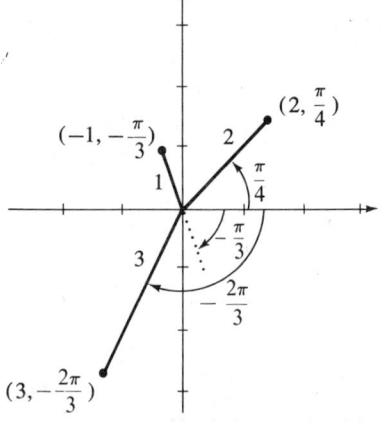

Figure 9.2

Polar coordinates present one problem we did not have with rectangular coordinates—a point has more than one representation. For example: $(2, \pi/2)$ and $(-2, -\pi/2)$ represent the same point. In fact, if (r, θ) is one representation of a point, then $(r, \theta + \pi n)$, where n is an even integer, and $(-r, \theta + \pi n)$, where n is an odd integer, are representations of the same point. Furthermore, $(0, \theta)$ is the pole for any choice of θ.

9.2

Graphs in Polar Coordinates

Equations in polar coordinates can be graphed by point-by-point plotting, as we graphed rectangular coordinates.

Example 1

Graph $r = \sin\theta$.

Note in Figure 9.3 that we have the entire graph for $0° \leq \theta < 180°$. The remaining values of θ simply repeat the graph a second time, since $(0, 0°) = (0, 180°)$, $(.5, 30°) = (-.5, 210°)$, and so forth. Of course, values of θ outside the range $0° \leq \theta \leq 360°$ would give no new points.

θ	r
0°	0.00
30°	0.50
60°	0.87
90°	1.00
120°	0.87
150°	0.50
180°	0.00
210°	−0.50
240°	−0.87
270°	−1.00
300°	−0.87
330°	−0.50
360°	0.00

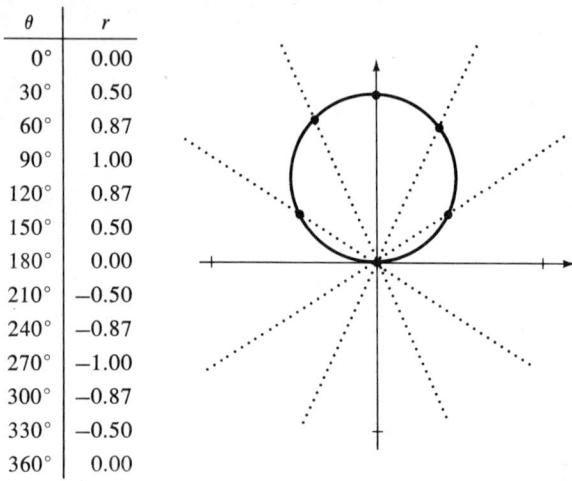

Figure 9.3

This method of point-by-point plotting is quite cumbersome here, as it was in the case of rectangular coordinates. One way to simplify the proceedings is to represent the table of values of r and θ by means of a graph. This may sound as if we are going in circles—we can get the graph from a table of values of r and θ that is represented by a graph. Actually, this is not so bad as it sounds. We shall represent the table by a graph in *rectangular coordinates*.

Example 2

Graph $r = 1 + \cos\theta$.

We can easily graph this equation in rectangular coordinates by using addition of ordinates. The result is given in Figure 9.4. Now we can read off values of r and θ just as we would from a table. As θ increases from 0° to 90°, r goes from 2 to 1. This gives the portion of the curve shown in (a) of Figure 9.5.

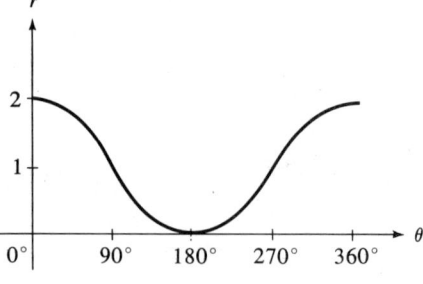

Figure 9.4

9.2 Graphs in Polar Coordinates

As θ goes from 90° to 180°, r goes from 1 down to 0 (shown in (b)). As θ goes from 180° to 270°, r goes from 0 back up to 1 (as in (c)); and finally, as θ goes from 270° to 360°, we see in (d) that θ goes from 1 to 2. The same path is traced for values of θ beyond 360° or less than 0°. Putting all of this together, we have the desired graph, shown in (e).

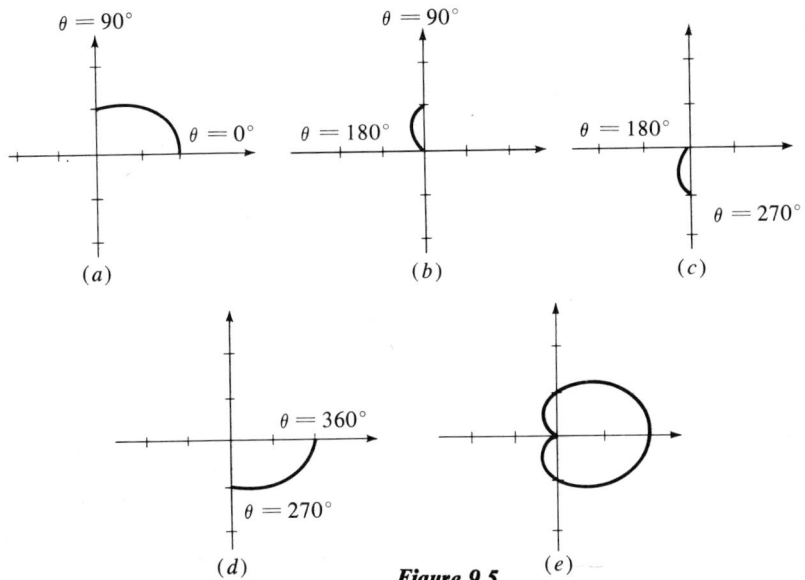

Figure 9.5

Example 3

Graph $r = \sin 2\theta$.

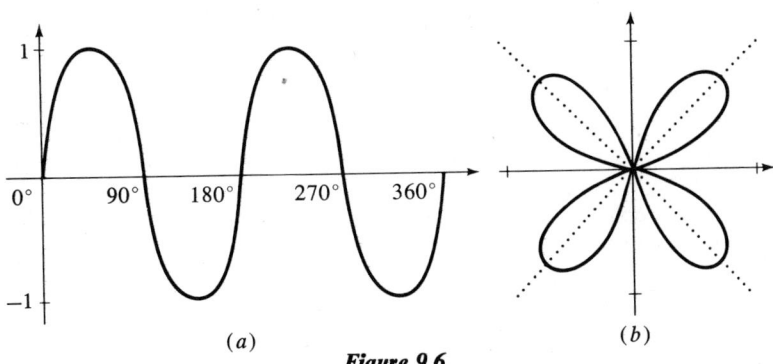

Figure 9.6

The graph is given in rectangular coordinates in (a) of Figure 9.6. This is then put on the polar graph shown in (b). Note that for θ in the range $90° < \theta < 180°$, r is negative. Thus instead of giving the loop in the second quadrant, it gives the one in the fourth quadrant. Similarly, r is negative for θ in the range $270° < \theta < 360°$. This gives the loop in the second quadrant.

Example 4

Graph $r^2 = \sin 2\theta$.

Graphing in rectangular coordinates by the methods of Section 7.3, we have the result given in Figure 9.7(a). There are a couple of things of interest here. First of all, $r^2 = \sin 2\theta$ has two values of r for each θ in the ranges $0° < \theta < 90°$ and $180° < \theta < 270°$, while it has no value at all for $90° < \theta < 180°$ and $270° < \theta < 360°$. Since it has two values in the range $0° < \theta < 90°$, we get both loops for $0° \le \theta \le 90°$, shown in (b). Similarly we get both loops a second time for $180° \le \theta \le 270°$. Because there is no value of r for $90° < \theta < 180°$ and $270° < \theta < 360°$, there are no points of the graph in the second or fourth quadrants.

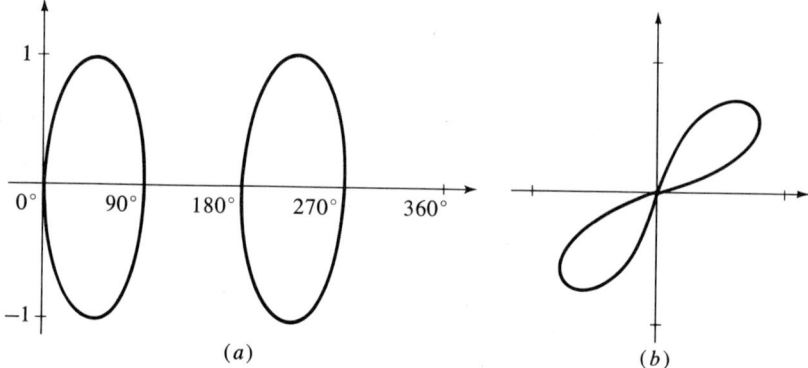

Figure 9.7

Problems

1. Plot the following points: $(1, \pi/3)$, $(2, 45°)$, $(0, 30°)$, $(-2, 90°)$, $(-1, 3\pi/4)$, $(2, 300°)$.
2. Give an alternate polar representation with $0° \le \theta < 180°$: $(4, 330°)$, $(-2, 420°)$, $(1, 210°)$, $(0, 283°)$, $(-3, 270°)$, $(2, 240°)$.
3. Give an alternate polar representation with $r \ge 0$ and $0° \le \theta < 360°$: $(-4, 120°)$, $(3, -60°)$, $(0, 530°)$, $(-1, 330°)$, $(-2, 390°)$, $(-2, 135°)$.
4. Give an alternate polar representation: $(1, 30°)$, $(-2, 180°)$, $(4, 210°)$, $(0, 60°)$, $(-1, 30°)$, $(2, 90°)$.

In Problems 5–34, sketch the graph of the given equation.

5. $r = \cos \theta$.
6. $r = 2 \sin \theta$.
7. $r = 1 - \cos \theta$.
8. $r = 1 + \sin \theta$.
9. $r = 1 - \sin \theta$.
10. $r = \sin \theta - 1$.
11. $r = \cos 2\theta$.
12. $r = \sin 4\theta$.

9.3 Points of Intersection

13. $r = \sin 3\theta$.
14. $r = \cos 3\theta$.
15. $r = \cos 5\theta$.
16. $r = \sin 6\theta$.
17. $r = 1 + 2 \sin \theta$.
18. $r = 1 - 2 \cos \theta$.
19. $r = 2 + \cos \theta$.
20. $r = 2 + 3 \sin \theta$.
21. $r = \tan \theta$.
22. $r = \sec \theta$.
23. $r^2 = \sin \theta$.
24. $r^2 = \cos 3\theta$.
25. $r^2 = \cos 4\theta$.
26. $r^2 = \sin^2 \theta$.
27. $r^2 = 1 + \cos \theta$.
28. $r^2 = 1 - \sin \theta$.
29. $r = \theta$.
30. $r = |\theta|$.
31. $r^2 = \theta^2$.
32. $r = \dfrac{2}{1 - \cos \theta}$.
33. $r = \dfrac{2}{1 - 2 \cos \theta}$.
34. $r = \dfrac{2}{2 - \cos \theta}$.
35. Show that if θ is replaced by $-\theta$ and the result is equivalent to the original equation, then the graph is symmetric about the x axis.
36. Show that if θ is replaced by $\pi - \theta$ and the result is equivalent to the original equation, then the graph is symmetric about the y axis.
37. Show that if r is replaced by $-r$ and the result is equivalent to the original equation, then the graph is symmetric about the pole.

9.3

Points of Intersection

Suppose we have a pair of equations in polar form that we solve simultaneously to obtain pairs of numbers satisfying both equations—that is, points of intersection of the two curves.

Example 1

Find the points of intersection of $r = 1$ and $r = 2 \sin \theta$.

Eliminating r from this pair of equations, we get

$$\sin \theta = \tfrac{1}{2}, \quad \text{or} \quad \theta = 30°, 150°,$$

giving the points $(1, 30°)$ and $(1, 150°)$. The graphs of these two curves, showing the two points of intersection are given in Figure 9.8.

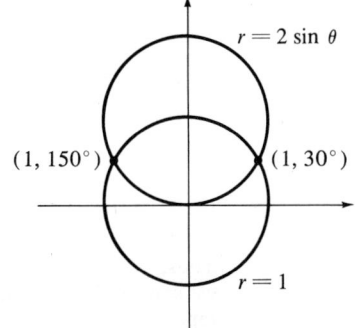

Figure 9.8

While the solutions of a pair of simultaneous equations must be points of intersection of the curves represented by the equations, some points of intersection cannot be found in this way. The reason is that they have different representations on the two curves. Thus *we must graph both curves* to be sure that we have found all points of intersection.

Example 2

Find the points of intersection of $r = \sin \theta$ and $r = \cos \theta$.

Eliminating r between the two equations, we have

$$\sin \theta = \cos \theta.$$

If we divide by $\cos \theta$, then

$$\tan \theta = 1 \quad \text{and} \quad \theta = 45° + 180° \cdot n.$$

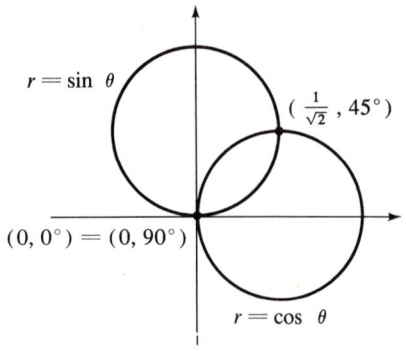

Figure 9.9

In the range $0° \leq \theta < 360°$, we have $(1/\sqrt{2}, 45°)$ and $(-1/\sqrt{2}, 225°)$. But these are different representations for the same point. Thus we have found only one point of intersection. As we can see from Figure 9.9, there are really two points of intersection—the one we found and the pole.

The pole has many different representations. On the curve $r = \sin \theta$ it is represented by $(0, 180° \cdot n)$; on $r = \cos \theta$ it is represented by $(0, 90° + 180° \cdot n)$. Thus, while the pole is common to both curves, it does not have a common representation that satisfies both equations. So we cannot find this point of intersection by finding simultaneous solutions of the two equations. We might represent this point by $(0, 0°) = (0, 90°)$.

Example 3

Find all points of intersection of $r = \cos 2\theta$ and $r = \sin \theta$.

$$\cos 2\theta = \sin \theta,$$
$$1 - 2 \sin^2 \theta = \sin \theta,$$
$$2 \sin^2 \theta + \sin \theta - 1 = 0,$$
$$(2 \sin \theta - 1)(\sin \theta + 1) = 0;$$
$$\sin \theta = \frac{1}{2}, \quad \sin \theta = -1;$$
$$\theta = 30°, 150°, 270°.$$

Thus, we have the points $(1/2, 30°)$, $(1/2, 150°)$, and $(-1, 270°)$. In addition we can see from Figure 9.10 that the pole is a point of intersection; it may be represented by $(0, 45°) = (0, 0°)$. It might also be noted that the point $(-1, 270°)$ can also be written $(1, 90°)$, but this form satisfies only $r = \sin \theta$.

9.4 Relationships between Rectangular and Polar Coordinates

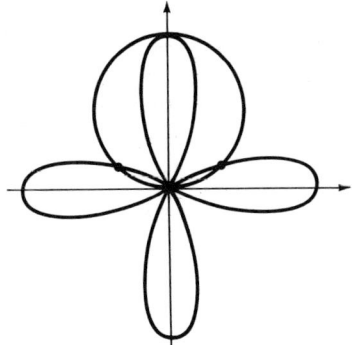

Figure 9.10

Problems

Find all points of intersection of the given curves.

1. $r = \sqrt{2}$, $r = 2\cos\theta$.
2. $r = \sqrt{3}$, $r = 2\sin\theta$.
3. $r = 2$, $r = \sin\theta + 2$.
4. $r = 1$, $r = 2\cos 2\theta$.
5. $r = \cos\theta$, $r = 1 - \cos\theta$.
6. $r = \cos\theta$, $r = 1 + \sin\theta$.
7. $r = \sin 2\theta$, $r = \sin\theta$.
8. $r = \sin 2\theta$, $r = \sqrt{2}\cos\theta$.
9. $r = \sec\theta$, $r = \csc\theta$.
10. $r = \sec\theta$, $r = \tan\theta$.
11. $r = 3\cos\theta + 4$, $r = 3$.
12. $r = \sin 2\theta$, $r = \cos 2\theta$.
13. $r = 2(1 + \cos\theta)$, $r(1 - \cos\theta) = 1$.
14. $r = 1 - \sin\theta$, $r(1 - \sin\theta) = 1$.
15. $r = 1 - \sin\theta$, $r = 1 - \cos\theta$.
16. $r^2 = \sin\theta$, $r^2 = \cos\theta$.
17. $r^2 = \cos\theta$, $r^2 = \sec\theta$.
18. $r = 2\cos\theta + 1$, $r = 2\cos\theta - 1$.
19. $r^2 = \sin\theta$, $r = \sin\theta$.
20. $r^2 = \sin\theta$, $r = \cos\theta$.

9.4

Relationships between Rectangular and Polar Coordinates

There are some simple relationships between rectangular and polar coordinates. These can be found easily by a consideration of Figure 9.11.

$$x = r \cos \theta,$$
$$y = r \sin \theta,$$
$$r^2 = x^2 + y^2$$
$$\tan \theta = \frac{y}{x}.$$

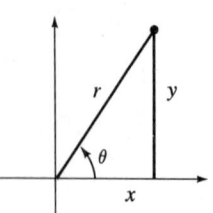

Figure 9.11

The last two, which may be solved for r and θ, would give us expressions involving $\pm$ and arctan. Thus, we prefer to leave them in their present form.

With these we can now change from one coordinate system to the other.

Example 1

Express $(2, 30°)$ in rectangular coordinates.

$$x = r \cos \theta \qquad y = r \sin \theta$$
$$= 2 \cos 30° \qquad = 2 \sin 30°$$
$$= 2 \cdot \frac{\sqrt{3}}{2} \qquad = 2 \cdot \frac{1}{2}$$
$$= \sqrt{3}; \qquad = 1.$$

Thus $(2, 30°) = (\sqrt{3}, 1)$.

Example 2

Express $(4, -4)$ in polar coordinates.

$$r^2 = x^2 + y^2 \qquad \tan \theta = \frac{y}{x}$$
$$= 16 + 16 \qquad = \frac{-4}{4}$$
$$= 32 \qquad = -1$$
$$r = \pm 4\sqrt{2}; \qquad \theta = 135° + 180° \cdot n.$$

We have a choice for both r and θ. The values of r and θ cannot be selected independently; the value we choose for one will limit the available choices for the other. In this case, the point $(4, -4)$ is in the fourth quadrant. Thus, we may choose either a fourth-quadrant angle and a positive r or a second-quadrant angle and a negative r. Thus

$$(4, -4) = (4\sqrt{2}, 315°) = (-4\sqrt{2}, 135°) = (4\sqrt{2}, -45°), \text{ etc.}$$

Of course we can use these equations to find a polar equation corresponding to one in rectangular coordinates, and vice versa.

Example 3

Express $y = x^2$ in polar coordinates.

9.4 Relationships between Rectangular and Polar Coordinates

$$y = x^2,$$
$$r \sin \theta = r^2 \cos^2 \theta,$$
$$\sin \theta = r \cos^2 \theta \quad \text{or} \quad r = 0,$$
$$r = \frac{\sin \theta}{\cos^2 \theta},$$
$$r = \sec \theta \tan \theta.$$

Since $r = 0$ represents only the pole and it is included in $r = \sec \theta \tan \theta$, we may drop $r = 0$. The result is

$$r = \sec \theta \tan \theta.$$

Example 4

Express $r = 1 - \cos \theta$ in rectangular coordinates.

First, multiply through by r.

$$r^2 = r - r \cos \theta.$$

At this point we could make the substitutions $r^2 = x^2 + y^2$, $r \cos \theta = x$, and $r = \pm\sqrt{x^2 + y^2}$. The last is rather bothersome, since it involves a $\pm$. In order to avoid this, let us isolate r on one side of the equation and square.

$$r = r^2 + r \cos \theta,$$
$$r^2 = (r^2 + r \cos \theta)^2,$$
$$x^2 + y^2 = (x^2 + y^2 + x)^2.$$

We have done two things that might introduce extraneous roots: (1) Multiplying by r may introduce only a single point, the pole, to the graph. Since the pole is already a point of the graph of $r = 1 - \cos \theta$, no new point is introduced here. (2) Squaring may introduce several new points. The equation

$$r^2 = (r^2 + r \cos \theta)^2$$

is equivalent to

$$r = \pm(r^2 + r \cos \theta).$$

Now $r = r^2 + r \cos \theta$ is equivalent to our original equation, $r = 1 - \cos \theta$, while $r = -(r^2 + r \cos \theta)$ is equivalent to $r = -1 - \cos \theta$. Thus

$$x^2 + y^2 = (x^2 + y^2 + x)^2$$

is equivalent to $r = 1 - \cos \theta$ together with $r = -1 - \cos \theta$. But $r = 1 - \cos \theta$ and $r = -1 - \cos \theta$ have the same graph. Thus we have introduced no new points by squaring.

Problems

1. The following points are given in polar coordinates. Give the rectangular coordinate representation of each. $(1, \pi)$, $(\sqrt{3}, \pi/3)$, $(-1, 3\pi)$, $(\sqrt{2}, 3\pi/4)$, $(2\sqrt{3}, 5\pi/3)$, $(-3, 7\pi/6)$, $(0, 5\pi/4)$, $(4, 0)$, $(-2, 7\pi/4)$.

2. The following points are given in rectangular coordinates. Give a polar coordinate representation of each. $(\sqrt{2}, -\sqrt{2})$, $(-1, \sqrt{3})$, $(4, 0)$, $(-1, -1)$, $(0, -2)$, $(0, 0)$, $(-2\sqrt{3}, 2)$, $(-3, 1)$, $(4, 3)$, $(-2, 4)$.

In Problems 3–18, express the given equation in polar coordinates.

3. $x = 2$.
4. $y = 5$.
5. $x^2 + y^2 = 1$.
6. $x^2 - y^2 = 4$.
7. $y = x^2$.
8. $y = x^3$.
9. $(x + y)^2 = x - y$.
10. $x = y$.
11. $y = 3x$.
12. $y^2 = x^3$.
13. $x + 2y - 4 = 0$.
14. $x^2 + y^2 - 2x = 0$.
15. $x^2 + y^2 - 2x - 2y + 1 = 0$.
16. $x^2 + 9y^2 = 9$.
17. $xy = 1$.
18. $y = \dfrac{x}{x + 1}$.

In Problems 19–34, express the given equation in rectangular coordinates.

19. $r = a$.
20. $\theta = \pi/4$.
21. $\theta = \pi/3$.
22. $r = 2 \sin \theta$.
23. $r = 4 \cos \theta$.
24. $r = \sin 2\theta$.
25. $r = \cos 2\theta$.
26. $r = 1 - \cos \theta$.
27. $r = 3 + 2 \sin \theta$.
28. $r^2 = \sin \theta$.
29. $r^2 = 1 + \sin \theta$.
30. $r^2 = \sin 2\theta$.
31. $r = \dfrac{1}{1 - \cos \theta}$.
32. $r = \dfrac{1}{1 + \sin \theta}$.
33. $r = 2 \sin \theta + 3 \cos \theta$.
34. $r = \sec \theta$.

9.5

Conics in Polar Coordinates

We found earlier that the equations of conic sections (in rectangular coordinates) have very simple forms if the center or vertex is at the origin and the axes are the coordinate axes. There are, however, three different forms corresponding to the three different types of conics. We find that conics can be easily represented in polar coordinates if a focus is at the origin and one axis is a coordinate axis. Furthermore, the same type of equation represents all three types of conics if we use the unifying concept of eccentricity.

Recall that any conic can be determined by a single focus, the corresponding directrix, and the eccentricity. If P is a point on the conic, then the distance from P to the focus divided by the distance from P to the directrix equals the eccentricity.

9.5 Conics in Polar Coordinates

The particular conic we get depends upon the eccentricity; the eccentricity is a positive number and

if $e < 1$, the conic is an ellipse,
if $e = 1$, the conic is a parabola,
if $e > 1$, the conic is a hyperbola.

If $P(r, \theta)$ is a point on a conic with focus 0, directrix $x = p$ (p positive), and eccentricity e, then

$$\frac{OP}{PD} = e, \quad \text{or} \quad \frac{|r|}{|p - r\cos\theta|} = e.$$

There are now two cases to consider:

$$\frac{r}{p - r\cos\theta} = e \quad \text{and} \quad \frac{r}{p - r\cos\theta} = -e.$$

Either of these yields an equation of the desired conic (see Problems 21 and 22); however, the first yields the commonly used form. Solving for r in this equation, we have

$$r = \frac{ep}{1 + e\cos\theta}.$$

If the directrix is $x = -p$ (p positive), then the equation is

$$r = \frac{ep}{1 - e\cos\theta}.$$

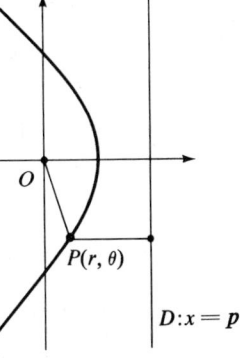

Figure 9.12

If the directrix is $y = \pm p$ (p positive), then the equation is

$$r = \frac{ep}{1 \pm e\sin\theta}.$$

Theorem 9.1

The conic section with focus at the origin, directrix $x = \pm p$ (p positive), and eccentricity e has polar equation

$$r = \frac{ep}{1 \pm e\cos\theta};$$

if the directrix is $y = \pm p$ (p positive), it has equation

$$r = \frac{ep}{1 \pm e\sin\theta}.$$

Example 1

Describe $r = \dfrac{6}{4 + 3\cos\theta}$.

Dividing numerator and denominator by 4, we have

$$r = \frac{\frac{3}{2}}{1+\frac{3}{4}\cos\theta} = \frac{\frac{3}{4}\cdot 2}{1+\frac{3}{4}\cos\theta}.$$

Thus the eccentricity is 3/4 and the directrix is $x = 2$. The conic is an ellipse with focus at the origin, directrix $x = 2$, and eccentricity 3/4.

Example 2

Sketch $r = \dfrac{15}{2 - 3\cos\theta}$.

Dividing by 2, we have

$$r = \frac{\frac{15}{2}}{1-\frac{3}{2}\cos\theta} = \frac{\frac{3}{2}\cdot 5}{1-\frac{3}{2}\cos\theta}.$$

Thus we have a hyperbola with focus at the origin, eccentricity 3/2, and directrix $x = -5$. The vertices are on the x axis, one between the focus and directrix and the other to the left of the directrix. When $\theta = 0°$, $r = -15$; when $\theta = 180°$, $r = 3$. Thus the vertices are $(-15, 0°)$ and $(3, 180°)$. When $\theta = 90°$ or $270°$, $r = 15/2$. Thus, the ends of one of the latera recta are $(15/2, 90°)$ and $(15/2, 270°)$. This information is enough to give a reasonably accurate picture of the hyperbola. If the asymptotes are desired, they can best be found by considering some of the above points in rectangular coordinates. Thus the vertices are $(-3, 0)$ and $(-15, 0)$, and the center is $(-9, 0)$, giving $a = 6$ and $c = 9$. We can now use the equation

$$b^2 = c^2 - a^2$$

to find $b^2 = 45$ or $b = 3\sqrt{5}$. Once we have this, the asymptotes are easily found (see Figure 9.13).

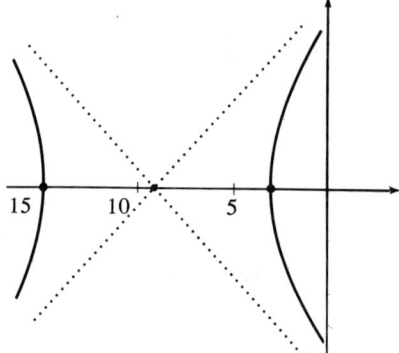

Figure 9.13

Example 3

Find a polar equation of the parabola with focus at the origin and directrix $y = -4$.

The equation is in the form

9.5 Conics in Polar Coordinates

$$r = \frac{ep}{1 - e \sin \theta},$$

since the directrix is a horizontal line below the focus. Furthermore, $e = 1$, since the conic is a parabola, and directrix $y = -4$ gives $p = 4$. Thus the equation is

$$r = \frac{4}{1 - \sin \theta}.$$

Problems

In Problems 1–8, state the type of conic and give a focus and its corresponding directrix and the eccentricity.

1. $r = \dfrac{4}{1 + 2 \cos \theta}$.
2. $r = \dfrac{12}{1 - 3 \sin \theta}$.
3. $r = \dfrac{4}{3 + 2 \sin \theta}$.
4. $r = \dfrac{5}{4 - 4 \cos \theta}$.
5. $r = \dfrac{3}{1 + \sin \theta}$.
6. $r = \dfrac{10}{5 - 2 \cos \theta}$.
7. $r(3 + 2 \sin \theta) = 6$.
8. $r(2 - 4 \cos \theta) = 5$.

In Problems 9–14, sketch the given conic.

9. $r = \dfrac{2}{1 + \cos \theta}$.
10. $r = \dfrac{16}{5 - 3 \cos \theta}$.
11. $r = \dfrac{16}{4 - 5 \sin \theta}$.
12. $r(3 - 5 \cos \theta) = 9$.
13. $r(13 + 12 \sin \theta) = 25$.
14. $r(3 + 3 \sin \theta) = 4$.

In Problems 15–20, find a polar equation of the conic with focus at the origin and the given eccentricity and directrix.

15. Directrix: $x = 5$; $e = 2/3$.
16. Directrix: $y = -3$; $e = 2$.
17. Directrix: $y = 2$; $e = 1$.
18. Directrix: $x = -4$; $e = 1$.
19. Directrix: $x = 5$; $e = 5/4$.
20. Directrix: $y = 3$; $e = 3/4$.

21. Sketch $r = \dfrac{-2}{1 - \cos \theta}$. Compare it with the conic of Problem 9 (see the following problem).

22. Show that the conic section with focus at the origin, directrix $x = p$ (p positive), and eccentricity e has polar equation

$$r = \frac{-ep}{1 - e \cos \theta}.$$

23. Suppose, in the equation $r = \dfrac{ep}{1 + e \cos \theta}$, $e \to 0$ and $p \to +\infty$ in such a way that ep remains constant. What happens to the shape of the conic? What happens to the equation of the conic?

24. Find a polar equation of a circle with center (k, α) and radius a by using the law of

cosines (see Figure 9.14).

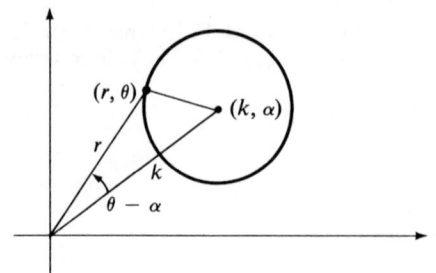

Figure 9.14   Figure 9.15

25. By using the trigonometry of right triangles, show that the line PQ (Figure 9.15) can be represented by the equation

$$x \cos \alpha + y \sin \alpha - p = 0.$$

This is called the *normal form* of the line, since it is expressed in terms of the polar coordinates of the point Q, which is the intersection of the original line and another perpendicular (or normal) to it and through the origin (see Problem 37, Section 3.3).

26. By using the identity

$$\sin^2 \alpha + \cos^2 \alpha = 1,$$

show that $Ax + By + C = 0$ can be put into the normal form by dividing through by $\pm\sqrt{A^2 + B^2}$ (see Problem 25 for the normal form).

27. Show that the distance from the point (x_1, y_1) to the line $Ax + By + C = 0$ is

$$d = \frac{|Ax_1 + By_1 + C|}{\sqrt{A^2 + B^2}}.$$

Hint: Put the original line and the one parallel to it and through (x_1, y_1) into the normal form (see Problems 25 and 26).

10

Parametric Equations

10.1

Parametric Equations

Up to now all of the equations we have dealt with have been in the form

$$y = f(x) \quad \text{or} \quad F(x, y) = 0.$$

In either case, a direct relationship between x and y is given. Another way of representing equations is to show x and y each as a function of a third variable or parameter; that is,

$$x = f(t), \quad y = g(t).$$

Each value of the parameter t gives a value of x and a value of y.

For instance, in the parametric equations

$$x = \sin t, \quad y = \cos t,$$

we see that, if $t = 0$, $x = 0$ and $y = 1$. Thus the point (0, 1) is a point of the graph. Note that we still have just the x and y axes; t does not appear on the graph. Let us continue with this process. The resulting graph is given in Figure 10.1. Of course, we could continue with values of t beyond 360°, but we would simply go over the same points again. Although the value of *t* need not appear anywhere on the graph, we have labeled several points with their corresponding values of *t*. Once the points are plotted, they are joined in the order of increasing (or decreasing) values of *t*.

The result seems to resemble a circle. How can we be *sure* it is a circle? If we had a single equation in *x* and *y*, we could easily see by the form of the equation whether

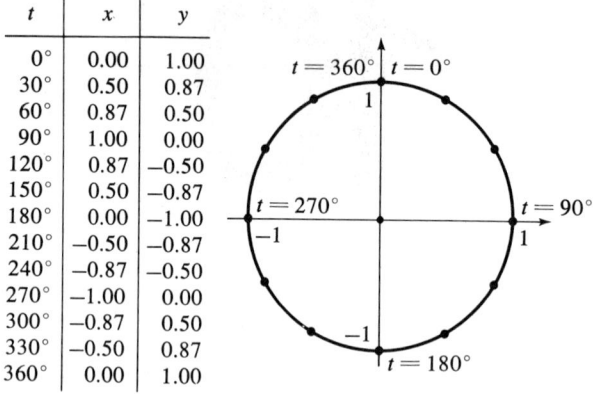

t	x	y
0°	0.00	1.00
30°	0.50	0.87
60°	0.87	0.50
90°	1.00	0.00
120°	0.87	−0.50
150°	0.50	−0.87
180°	0.00	−1.00
210°	−0.50	−0.87
240°	−0.87	−0.50
270°	−1.00	0.00
300°	−0.87	0.50
330°	−0.50	0.87
360°	0.00	1.00

Figure 10.1

or not we have a circle. Let us try to eliminate the parameter *t* between the equations $x = \sin t$ and $y = \cos t$:

$$\sin^2 t + \cos^2 t = 1,$$

$$x^2 + y^2 = 1.$$

We now see that we have a circle with center at the origin and radius 1.

Not only does elimination of the parameter assure us that this particular curve is a circle, it gives us a basis for sketching more rapidly that can be done by point-by-point plotting. However, we must be careful with the domain of the resulting equation. Let us illustrate this with some examples and see how the domain of $F(x, y) = 0$ plays an important role in sketching the graph.

Example 1

Graph the following two pairs of parameteric equations by eliminating the parameter.

$$x = t, \quad x = t^2,$$
$$y = t; \quad y = t^2.$$

Elimination of the parameter gives $y = x$ in both cases. But the graphs are not the same, as the domains in the two cases will show. The domain can be determined from the first of the two parametric equations in each case. In the first case, $x = t$ and, since there is no restriction on *t*, there is none on *x*; the

10.1 Parametric Equations

domain is the set of all real numbers. In the second case, $x = t^2$. The domain of $y = x$ is the range of $x = t^2$, which is $\{x \mid x \geq 0\}$. Thus we have a restricted domain here that we did not have in the first case. The graphs are given in Figure 10.2.

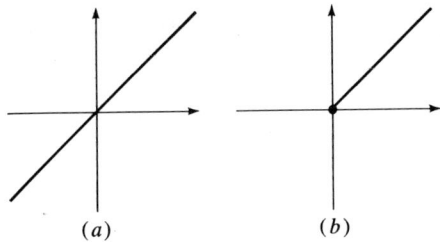

(a)   (b)

Figure 10.2

Example 2

Eliminate the parameter between $x = t + 1$ and $y = t^2 + 3t + 2$ and sketch.

Solving $x = t + 1$ for t, we have

$$t = x - 1.$$

If this is substituted into $y = t^2 + 3t + 2$, then

$$y = (x - 1)^2 + 3(x - 1) + 2$$
$$= x^2 + x.$$

Note that there is no restriction on x; the domain of $y = x^2 + x$ is the set of all real numbers. It is now a simple matter to sketch the curve; it is given in Figure 10.3.

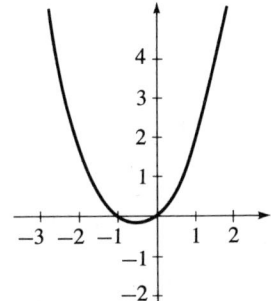

Figure 10.3

Occasionally it is difficult or impossible to eliminate the parameter. In such cases the curve must be plotted point by point. Closely related to parametric equations are vector valued functions. They are represented in the following way:

$$\mathbf{f}(t) = f_1(t)\mathbf{i} + f_2(t)\mathbf{j}.$$

Thus, when a value of t is substituted into the equation, the function takes on a vector value. For example, if

$$\mathbf{f}(t) = t\mathbf{i} + t^2\mathbf{j},$$

then

$$\mathbf{f}(1) = \mathbf{i} + \mathbf{j} \quad \text{and} \quad \mathbf{f}(2) = 2\mathbf{i} + 4\mathbf{j}.$$

Recall that, in graphing vectors, we graph only representatives. Thus, in graphing vector functions, let us graph representatives of the vectors, each having its tail at

the origin. Thus,

$$\mathbf{f}(t) = t\mathbf{i} + t^2\mathbf{j}$$

has the graphical representation shown in Figure 10.4(a). Normally we shall omit the directed line segments and show only their heads, as in Figure 10.4(b). Thus, the result is equivalent to graphing the curve represented parametrically by

$$x = t, \quad y = t^2.$$

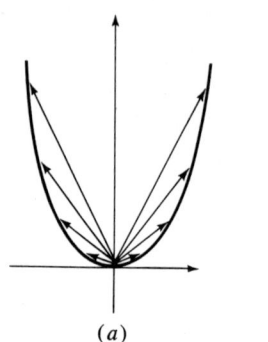
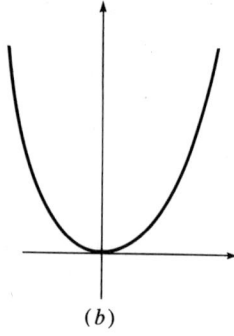

(a)   (b)

Figure 10.4

Example 3

Sketch the curve $\mathbf{f}(t) = (t+2)\mathbf{i} + (t^2 + 7t + 12)\mathbf{j}$.

This is equivalent to the parametric equations

$$x = t + 2, \quad y = t^2 + 7t + 12.$$

Eliminating the parameter, we have

$$y = x^2 + 3x + 2.$$

Its graph is given in Figure 10.5.

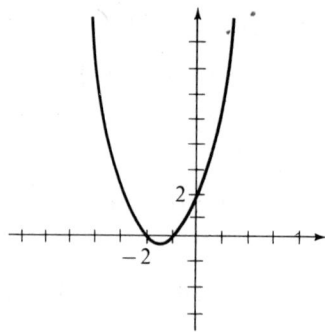

Figure 10.5

10.2 Parametric Equations of a Locus

Problems

In Problems 1–20, eliminate the parameter and sketch the curve.

1. $x = t^2 + 1$, $y = t + 1$.
2. $x = t^2 + t - 2$, $y = t + 2$.
3. $x = t - 1$, $y = t^2 - 2t$.
4. $x = 2t^2 + t - 3$, $y = t - 1$.
5. $x = t^2 + t$, $y = t^2 - t$.
6. $x = t^2 + 1$, $y = t^2 - 1$.
7. $x = t^3$, $y = t^2$.
8. $x = e^t$, $y = \sin t$.
9. $x = a \cos \theta$, $y = b \sin \theta$.
10. $x = \theta - \sin \theta$, $y = 1 - \cos \theta$.
11. $x = 2 + \cos \theta$, $y = -1 + \sin \theta$.
12. $x = 3 - \cos \theta$, $y = 2 + 4 \sin \theta$.
13. $x = 3 + \cosh \theta$, $y = 2 + \sinh \theta$. (*Hint*: Use the identity $\cosh^2 \theta - \sinh^2 \theta = 1$.)
14. $x = 4 + 2 \cosh \theta$, $y = 1 - 4 \sinh \theta$. (See Problem 13.)
15. $\mathbf{f}(t) = (t - 1)\mathbf{i} + t^2 \mathbf{j}$.
16. $\mathbf{f}(t) = (t^2 + 1)\mathbf{i} + t^2 \mathbf{j}$.
17. $\mathbf{f}(t) = t^2 \mathbf{i} + t^3 \mathbf{j}$.
18. $\mathbf{f}(t) = (3t + 1)\mathbf{i} + (t^3 - 1)\mathbf{j}$.
19. $\mathbf{f}(t) = \cos t \, \mathbf{i} + \sin t \, \mathbf{j}$.
20. $\mathbf{f}(t) = t^2 \mathbf{i} + e^t \mathbf{j}$.

In Problems 21–24, sketch the curve.

21. $x = \cos \theta + \theta \sin \theta$, $y = \sin \theta - \theta \cos \theta$.
22. $x = a \cos^3 \theta$, $y = a \sin^3 \theta$.
23. $x = 2a \cos \theta - a \cos 2\theta$, $y = 2a \sin \theta - a \sin 2\theta$.
24. $x = t - a \tanh \dfrac{t}{a}$, $y = a \operatorname{sech} \dfrac{t}{a}$.

25. Sketch each of the following parametric equations and note the similarities and differences.
   (a) $x = t$, $y = t$;
   (b) $x = t^2$, $y = t^2$;
   (c) $x = |t|$, $y = |t|$;
   (d) $x = \sin t$, $y = \sin t$;
   (e) $x = \ln(-t)$, $y = \ln(-t)$;
   (f) $x = \sqrt{(t-1)(t-2)}$, $y = \sqrt{(t-1)(t-2)}$.

10.2

Parametric Equations of a Locus

The principal advantage of parametric equations is in the determination of equations of a locus. It is frequently simpler to relate x and y to some third variable than to relate them to each other directly. Let us consider some examples.

Example 1

A ball is thrown horizontally with a velocity of 40 ft/sec and 10 ft above ground level. Neglecting air resistance, find parametric equations representing the path of the ball.

It is difficult to relate the x and y coordinates of the ball directly, but each can be related to the time rather easily. Suppose we take the starting position to be $(0, 10)$ and the ball to be thrown in the direction of the positive x axis. Since we are neglecting air resistance, there is nothing to change the horizontal velocity. Thus, using the formula $d = rt$, we have

$$x = 40t.$$

Now the vertical motion of the ball is governed by the law of falling bodies. The position y is given by

$$y = -16t^2 + v_0 t + x_0,$$

where v_0 is the initial, vertical velocity (or the vertical component of the velocity; see Problems 1–3) and x_0 is the initial position. In this case $v_0 = 0$, since the ball is given no upward or downward motion, and $x_0 = 10$. Thus,

$$y = -16t^2 + 10.$$

In the past we have found equations of curves from a geometric description. Again, this can often be accomplished by relating the x and y coordinates of a point on the curve to some third variable.

Example 2

Find parametric equations for the set of all points P which are determined as illustrated in Figure 10.6.

Since the ray OA is determined by the angle θ, θ is a convenient parameter. The x coordinate of P is the x coordinate of A; thus

$$x = a \cos \theta.$$

Similarly, the y coordinate of P is the y coordinate of B.

$$y = b \sin \theta.$$

Thus we have the curve in parametric form. It might be noted here that the parameter can easily be eliminated to give

$$\frac{x^2}{a^2} + \frac{y^2}{b^2} = 1.$$

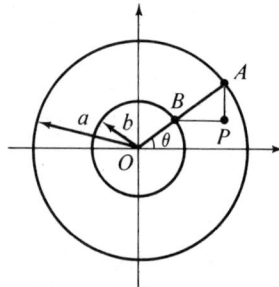

Figure 10.6

10.2 Parametric Equations of a Locus

Example 3

A wheel of radius a is rolling along a flat plane. Find the path traced by a point on the circumference if the plane is the x axis and the point starts at the origin.

Suppose we use the angle θ (see Figure 10.7) as the parameter. Since the wheel is rolling along the x axis,

$$\overline{OT} = \overline{PT} = a\theta.$$

Thus C is the point $(a\theta, a)$. Furthermore, from triangle CPQ,

$$\overline{PQ} = a \sin \theta \quad \text{and} \quad \overline{CQ} = a \cos \theta.$$

Thus,

$$\begin{aligned} x &= \overline{OT} - \overline{PQ} & y &= \overline{CT} - \overline{CQ} \\ &= a\theta - a \sin \theta & &= a - a \cos \theta \\ &= a(\theta - \sin \theta); & &= a(1 - \cos \theta). \end{aligned}$$

This curve is called a cycloid.

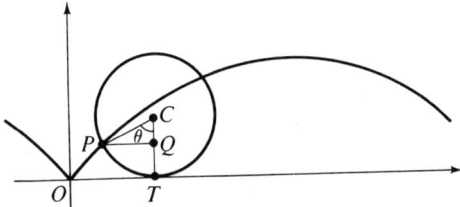

Figure 10.7

Problems

If, in Example 1, the ball is thrown at an angle θ with the horizontal, then the horizontal component of the velocity v_0 is $v_0 \cos \theta$, and the vertical component is $v_0 \sin \theta$. In Problems 1–3, find parametric equations representing the path of the ball when it is thrown from ground level at the indicated angle.

1. 30°.
2. 45°.
3. 60°.
4. Where does the ball hit the ground in Problems 1–3?
5. Find parametric equations for the set of all points P determined as shown in Figure 10.8. Eliminate the parameter.
6. Suppose, in Example 3, the point starts at $(0, 2a)$. Find parametric equations for the curve.
7. Find the path traced by a point P a distance b from the center of the circle of Example 3 if P starts at $(0, a - b)$.

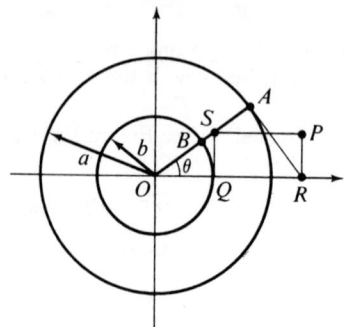

Figure 10.8

8. Sketch the curve of Problem 7 if
   (a) $a = 2$ and $b = 1$; (b) $a = 1$ and $b = 2$.
9. In Figure 10.9, $\overline{BA} = \overline{BP} = \overline{BP'}$, Find parametric equations for the set of all points P and P' determined as shown. Sketch. This curve is called a strophoid.

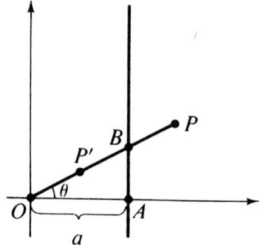

Figure 10.9

10. Find a polar representation for the curve of Problem 9.
11. Find parametric equations for the set of all points P and P' determined as shown in Figure 10.10. This curve is called a conchoid.

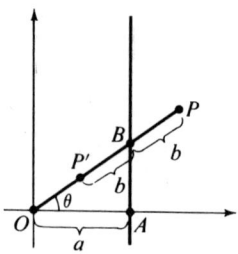

Figure 10.10

10.2 Parametric Equations of a Locus

12. Find a polar equation for the curve of Problem 11.
13. Sketch the conchoid (see Problems 11 and 12) if (a) $a=b=1$; (b) $a=1$ and $b=2$; (c) $a=2$ and $b=1$.
14. Find parametric equations for the set of all points P determined as shown in Figure 10.11. (*Hint*: Use a polar equation for the circle.) Sketch. This curve is called the witch of Agnesi.

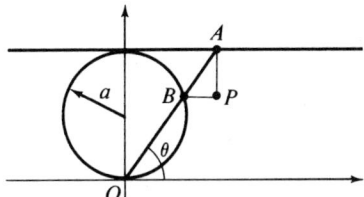

Figure 10.11   Figure 10.12

15. In Figure 10.12, $\overline{OP} = \overline{AB}$. Find parametric equations for the set of all points P determined as shown. (*Hint*: Use a polar equation for the circle.) Sketch. This curve is called the cissoid of Diocles.
16. If a string which is wound on a spool is unwound while the string is kept taut (see Figure 10.13), the curve traced by the end of the string is called the involute of the circle. Find parametric equations for the involute of a circle of radius a.

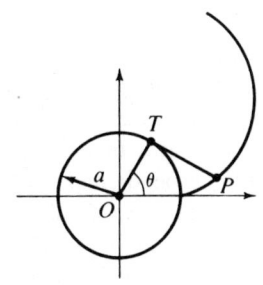
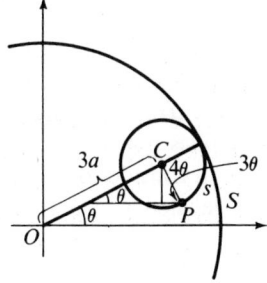

Figure 10.13   Figure 10.14

17. Find the path traced by a point P on the circumference of a circle of radius a which rolls inside a circle of radius $4a$ (see Figure 10.14). This curve is called a hypocycloid.
18. Find the path of the point P of Problem 17 if the smaller circle rolls outside the larger one. This curve is called an epicycloid.
19. What is the result if both of the circles of Problem 18 have radius a? This curve is called a cardioid.

11

Solid Analytic Geometry

11.1

Introduction: The Distance and Point-of-Division Formulas

Up to now we have been dealing almost exclusively with plane figures. Let us now consider solid figures. Forming the bridge between the algebra and geometry is a set of three *axes* concurrent at a point (the origin). The only requirement is that these three lines not be coplanar—that is, that they not all lie in the same plane. However, we shall consider only the case in which the axes are mutually perpendicular. The three axes, labeled x, y, and z, with a scale on each, determine a set of three numbers associated uniquely with any point in space. Since any pair of intersecting lines determines a plane, the three pairs of axes determine three *coordinate planes*, which we shall call the *xy plane*, the *xz plane*, and the *yz plane*. These planes separate space into eight *octants*. Although we shall not number all of them, the one in which all three coordinates are positive is called the *first octant*. Note that points of the xy plane have z coordinate 0, points of the xz plane have y coordinate 0, and points of the yz plane have x coordinate 0. Similarly, points of the x axis have y and z coordinates 0, and so on. Of course the origin has all of these coordinates 0.

The two basic geometric representatives of the axes are given in Figure 11.1; (a) shows a *right-hand system*, while (b) shows a *left-hand system*. Graphs of equations in the two systems are mirror images of each other. Since we shall normally represent

11.1 Introduction: The distance and Point-of-Division Formulas

space by a right-hand system, the axes will usually appear in the positions indicated in (a).

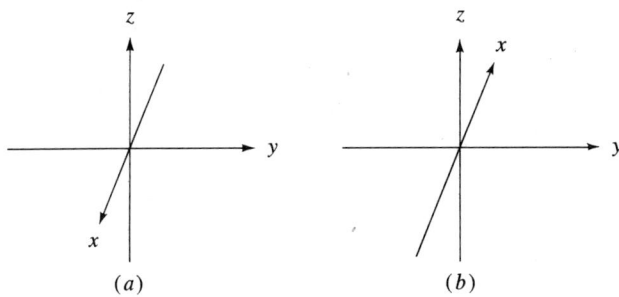

Figure 11.1

Many of the formulas of solid analytic geometry are simple extensions of plane analytic geometry. The one that follows is an example.

Theorem 11.1

The distance between two points (x_1, y_1, z_1) and (x_2, y_2, z_2) is

$$d = \sqrt{(x_1 - x_2)^2 + (y_1 - y_2)^2 + (z_1 - z_2)^2}.$$

The proof, which requires a double application of the Theorem of Pythagoras, is left to the student. Of course, if the line joining the two points is on or parallel to one of the coordinate planes, at least one of the three terms of this formula is zero, and it reduces to the plane case. Similarly, if the line joining the points is on or parallel to one of the axes, at least two terms are zero and the distance is the absolute value of the difference between the coordinates of the remaining pair.

Example 1

Find the distance between $(1, -2, 5)$ and $(-3, 6, 4)$.

$$\begin{aligned} d &= \sqrt{(x_1 - x_2)^2 + (y_1 - y_2)^2 + (z_1 - z_2)^2} \\ &= \sqrt{(1+3)^2 + (-2-6)^2 + (5-4)^2} \\ &= \sqrt{16 + 64 + 1} \\ &= \sqrt{81} \\ &= 9. \end{aligned}$$

Another easy extension from two dimensions is the point-of-division formula.

Theorem 11.2

If $P_1 = (x_1, y_1, z_1)$ and $P_2 = (x_2, y_2, z_2)$ and P is a point such that $r = P_1P/P_1P_2$, then the coordinates of P are

$$x = x_1 + r(x_2 - x_1),$$
$$y = y_1 + r(y_2 - y_1),$$
$$z = z_1 + r(z_2 - z_1).$$

Again the proof is similar to the one for the two-dimensional case, and it is left to the student.

Example 2

Find the point $1/3$ of the way from $(-2, 4, 1)$ to $(4, 1, 7)$.

$$x = x_1 + r(x_2 - x_1) = -2 + \frac{1}{3}(4 + 2) = 0,$$

$$y = y_1 + r(y_2 - y_1) = 4 + \frac{1}{3}(1 - 4) = 3,$$

$$z = z_1 + r(z_2 - z_1) = 1 + \frac{1}{3}(7 - 1) = 3.$$

The desired point is $(0, 3, 3)$.

The following theorem is a direct corollary of the point-of-division formulas.

Theorem 11.3

If $P_1 = (x_1, y_1, z_1)$ and $P_2 = (x_2, y_2, z_2)$, then the coordinates of the midpoint of the segment P_1P_2 are

$$x = \frac{x_1 + x_2}{2}, \quad y = \frac{y_1 + y_2}{2}, \quad z = \frac{z_1 + z_2}{2}.$$

Example 3

Find the midpoint of the segment with ends $(4, -3, 1)$ and $(-2, 5, 3)$.

$$x = \frac{x_1 + x_2}{2} = \frac{4 - 2}{2} = 1,$$

$$y = \frac{y_1 + y_2}{2} = \frac{-3 + 5}{2} = 1,$$

11.1 Introduction: The distance and Point-of-Division Formulas

$$z = \frac{z_1 + z_2}{2} = \frac{1+3}{2} = 2.$$

The point is $(1, 1, 2)$.

Problems

In Problems 1–10, find the distance between the pair of points given.

1. $(2, 5, 0), (-3, 1, 3)$.
2. $(4, -2, 1), (2, 2, -3)$.
3. $(5, 4, -1), (2, 0, -1)$.
4. $(2, 5, 3), (-2, 4, -1)$.
5. $(-5, 0, 2), (4, 1, -5)$.
6. $(-2, 5, 1), (-2, 8, 4)$.
7. $(3, -1, 4), (3, 4, 4)$.
8. $(2, 5, 0), (5, 5, 0)$.
9. $(4, 7, -1), (3, -1, 3)$.
10. $(5, 2, 3), (4, 5, -1)$.

In Problems 11–16, find the point P such that $AP/AB = r$.

11. $A = (4, 3, -2), B = (-5, 0, 4), r = 2/3$.
12. $A = (5, 2, 3), B = (-5, 7, -2), r = 2/5$.
13. $A = (-2, 0, 1), B = (10, 8, 5), r = 1/4$.
14. $A = (5, 5, 3), B = (2, -4, 0), r = 1/3$.
15. $A = (3, 1, 5), B = (-3, 4, 2), r = 2$.
16. $A = (-2, 5, 1), B = (4, -1, 2), r = 3/2$.

In Problems 17–20, find the midpoint of segment AB.

17. $A = (5, -2, 3), B = (-3, 4, 7)$.
18. $A = (4, 3, 5), B = (-2, -1, 2)$.
19. $A = (-3, 2, 0), B = (5, 4, 3)$.
20. $A = (4, 3, -1), B = (4, 8, -3)$.
21. Given $A = (5, -2, 3), P = (6, 0, 0)$, and $AP/AB = 1/3$, find B.
22. Given $B = (-4, 14, 4), P = (-1, 8, -4)$, and $AP/AB = 2/5$, find A.
23. Given $B = (6, 0, 9), P = (4, 1, 6)$, and $AP/AB = 3/4$, find A.
24. Given $A = (5, 3, -2), P = (1, 5, 2)$, and $AP/AB = 2/3$, find B.

In Problems 25–28, find the unknown quantity.

25. $A = (5, 1, 0), B = (1, y, 2), AB = 6$.
26. $A = (-2, 4, 3), B = (x, -4, 2), AB = 9$.
27. $A = (x, 4, -2), B = (-x, -6, 3), AB = 15$.
28. $A = (x, x, 5), B = (-1, -2, 0), AB = 5\sqrt{2}$.
29. The point $(-1, 5, 2)$ is a distance 6 from the midpoint of the segment joining $(1, 3, 2)$ and $(x, -1, 6)$. Find x.
30. The point $(1, -2, 9)$ is a distance $5\sqrt{5}$ from the midpoint of the segment joining $(1, y, 2)$ and $(5, -1, 6)$. Find y.
31. Prove Theorem 11.1   32. Prove Theorem 11.2.   33. Prove Theorem 11.3.

11.2

Vectors in Space

Vectors in three-dimensional space may be handled in much the same way as vectors in the plane. Vectors themselves, the sum and difference of two vectors, the absolute value of a vector, and scalar multiple of a vector are defined in the same way as they were in Chapter 2; and Theorem 2.2 (see page 31) holds for vectors in space as well as for vectors in the plane. The following theorems and definitions are the three-dimensional analogs of theorems and definitions of Chapter 2.

Definition

If $O = (0, 0, 0)$, $X = (1, 0, 0)$, $Y = (0, 1, 0)$ and $Z = (0, 0, 1)$, then the vectors represented by $\overrightarrow{OX}, \overrightarrow{OY},$ and $\overrightarrow{OZ}$ are **i, j,** and **k**, respectively, and are called **basis vectors**.

Theorem 11.4

Every vector in space can be written in the form
$$a\mathbf{i} + b\mathbf{j} + c\mathbf{k}.$$
The numbers a, b, and c are called the **components** of the vector.

Theorem 11.5

If $\overrightarrow{AB}$, where $A = (x_1, y_1, z_1)$ and $B = (x_2, y_2, z_2)$, represents a vector in space, then
$$\mathbf{v} = (x_2 - x_1)\mathbf{i} + (y_2 - y_1)\mathbf{j} + (z_2 - z_1)\mathbf{k}.$$

Theorem 11.6

$$(a_1\mathbf{i} + b_1\mathbf{j} + c_1\mathbf{k}) + (a_2\mathbf{i} + b_2\mathbf{j} + c_2\mathbf{k}) = (a_1 + a_2)\mathbf{i} + (b_1 + b_2)\mathbf{j} + (c_1 + c_2)\mathbf{k};$$
$$(a_1\mathbf{i} + b_1\mathbf{j} + c_1\mathbf{k}) - (a_2\mathbf{i} + b_2\mathbf{j} + c_2\mathbf{k}) = (a_1 - a_2)\mathbf{i} + (b_1 - b_2)\mathbf{j} + (c_1 - c_2)\mathbf{k};$$
$$d(a\mathbf{i} + b\mathbf{j} + c\mathbf{k}) = da\mathbf{i} + db\mathbf{j} + dc\mathbf{k};$$
$$|a\mathbf{i} + b\mathbf{j} + c\mathbf{k}| = \sqrt{a^2 + b^2 + c^2}.$$

Definition

The angle between two nonzero vectors **u** and **v** is the smallest angle between the representatives of **u** and **v** having their tails at the origin.

11.2 Vectors in Space

Theorem 11.7

If $\mathbf{u} = a_1\mathbf{i} + b_1\mathbf{j} + c_1\mathbf{k}$ and $\mathbf{v} = a_2\mathbf{i} + b_2\mathbf{j} + c_2\mathbf{k}$ ($\mathbf{u} \neq \mathbf{0}$ and $\mathbf{v} \neq \mathbf{0}$) and if θ is the angle between them, then

$$\cos\theta = \frac{a_1a_2 + b_1b_2 + c_1c_2}{|\mathbf{u}| \cdot |\mathbf{v}|}.$$

Definition

If $\mathbf{u} = a_1\mathbf{i} + b_1\mathbf{j} + c_1\mathbf{k}$ and $\mathbf{v} = a_2\mathbf{i} + b_2\mathbf{j} + c_2\mathbf{k}$, then the **dot product** (scalar product, inner product) of $\mathbf{u}$ and $\mathbf{v}$ is

$$\mathbf{u} \cdot \mathbf{v} = a_1a_2 + b_1b_2 + c_1c_2.$$

Theorems 2.9–2.12 (pages 44–45) still hold for three-dimensional vectors. They are restated here for convenience.

Theorem 11.8

The vectors $\mathbf{u}$ and $\mathbf{v}$ (not both $\mathbf{0}$) are orthogonal (perpendicular) if and only if $\mathbf{u} \cdot \mathbf{v} = 0$ (the zero vector is taken to be orthogonal to every vector).

Theorem 11.9

If $\mathbf{u}$ and $\mathbf{v}$ are vectors, then

$$\mathbf{u} \cdot \mathbf{v} = \mathbf{v} \cdot \mathbf{u},$$

and

$$(\mathbf{u} + \mathbf{v}) \cdot \mathbf{w} = \mathbf{u} \cdot \mathbf{w} + \mathbf{v} \cdot \mathbf{w}.$$

Theorem 11.10

If $\mathbf{u}$ and $\mathbf{v}$ are vectors and θ is the angle between them, then

$$\mathbf{u} \cdot \mathbf{v} = |\mathbf{u}| \cdot |\mathbf{v}| \cos\theta,$$

and

$$\mathbf{v} \cdot \mathbf{v} = |\mathbf{v}|^2.$$

Definition

The **projection** of $\mathbf{u}$ on $\mathbf{v}$ ($\mathbf{v} \neq \mathbf{0}$) is a vector $\mathbf{w}$ such that if $\overrightarrow{AB}$ is a representative of $\mathbf{u}$ and $\overrightarrow{CD}$ is a representative of $\mathbf{v}$, then a representative of $\mathbf{w}$ is a directed line segment $\overrightarrow{EF}$ such that $EF \perp AE$, $AE \perp CD$, and $BF \perp CD$ (see Figure 2.12, page 45).

Theorem 11.11

If **w** *is the projection of* **u** *on* **v**, *then*

$$|\mathbf{w}| = \frac{|\mathbf{u} \cdot \mathbf{v}|}{|\mathbf{v}|}.$$

Theorems 11.8–11.11 are not stated in terms of the dimensions of the vectors; their proofs are identical to those of Chapter 2. Proofs of the others are left to the student (see Problems 27–30).

Example 1

Given $\mathbf{u} = 2\mathbf{i} + \mathbf{j} - 3\mathbf{k}$ and $\mathbf{v} = \mathbf{i} - 2\mathbf{j} - \mathbf{k}$, find $\mathbf{u} + \mathbf{v}$, $\mathbf{u} - \mathbf{v}$, and $\mathbf{u} \cdot \mathbf{v}$.

$\mathbf{u} + \mathbf{v} = (2 + 1)\mathbf{i} + (1 - 2)\mathbf{j} + (-3 - 1)\mathbf{k} = 3\mathbf{i} - \mathbf{j} - 4\mathbf{k},$

$\mathbf{u} - \mathbf{v} = (2 - 1)\mathbf{i} + (1 + 2)\mathbf{j} + (-3 + 1)\mathbf{k} = \mathbf{i} + 3\mathbf{j} - 2\mathbf{k},$

$\mathbf{u} \cdot \mathbf{v} = 2 \cdot 1 + 1(-2) + (-3)(-1) = 3.$

Example 2

Give in component form the vector **v** that is represented by $\overrightarrow{AB}$ where $A = (4, 3, -1)$ and $B = (-1, 2, -3)$.

$\mathbf{v} = (-1 - 4)\mathbf{i} + (2 - 3)\mathbf{j} + (-3 + 1)\mathbf{k} = -5\mathbf{i} - \mathbf{j} - 2\mathbf{k}.$

Example 3

Find the end points of the representative $\overrightarrow{AB}$ of **v** if $\mathbf{v} = 2\mathbf{i} - 4\mathbf{j} + \mathbf{k}$ and $(2, -3, 5)$ is the midpoint of AB.

If $A = (x_1, y_1, z_1)$ and $B = (x_2, y_2, z_2)$, then, by Theorem 11.5,

$$x_2 - x_1 = 2, \qquad y_2 - y_1 = -4, \qquad z_2 - z_1 = 1.$$

By Theorem 11.3,

$$\frac{x_1 + x_2}{2} = 2, \qquad \frac{y_1 + y_2}{2} = -3, \qquad \frac{z_1 + z_2}{2} = 5.$$

Solving simultaneously, we have

$$A = \left(1, -1, \frac{9}{2}\right), \qquad B = \left(3, -5, \frac{11}{2}\right).$$

Example 4

Find the projection **w** of $\mathbf{u} = 2\mathbf{i} - \mathbf{j} + \mathbf{k}$ upon $\mathbf{v} = 3\mathbf{i} + \mathbf{j} - 4\mathbf{k}$.

11.2 Vectors in Space

By Theorem 11.11,

$$|\mathbf{w}| = \frac{|\mathbf{u} \cdot \mathbf{v}|}{|\mathbf{v}|} = \frac{|1|}{\sqrt{26}} = \frac{1}{\sqrt{26}}.$$

Since $\mathbf{w}$ and $\mathbf{v}$ have the same (or opposite) directions, there is a number c such that

$$\mathbf{w} = c\mathbf{v} = 3c\mathbf{i} + c\mathbf{j} - 4c\mathbf{k}.$$

Thus

$$|\mathbf{w}| = \frac{1}{\sqrt{26}} = \sqrt{26c^2},$$

$$c = \pm \frac{1}{26}.$$

Since $\mathbf{u} \cdot \mathbf{v}$ is positive, $\cos\theta$ is positive. Thus $\theta < 90°$ and $\mathbf{w}$ and $\mathbf{v}$ have the same direction. Therefore, $c = 1/26$ and

$$\mathbf{w} = \frac{3}{26}\mathbf{i} + \frac{1}{26}\mathbf{j} - \frac{2}{13}\mathbf{k}.$$

Problems

In Problems 1–4, give in component form the vector $\mathbf{v}$ that is represented by $\overrightarrow{AB}$.

1. $A = (2, 3, -5)$, $B = (-4, 1, 2)$.
2. $A = (3, -2, 4)$, $B = (5, 4, -1)$.
3. $A = (5, 0, -2)$, $B = (2, -4, 1)$.
4. $A = (2, -3, 8)$, $B = (2, 5, 2)$.

In Problems 5–8, give the unit vector in the direction of $\mathbf{v}$.

5. $\mathbf{v} = 4\mathbf{i} + \mathbf{j} - 2\mathbf{k}$.
6. $\mathbf{v} = \mathbf{i} - 2\mathbf{j} + 2\mathbf{k}$.
7. $\mathbf{v} = \mathbf{i} + 5\mathbf{j} - 3\mathbf{k}$.
8. $\mathbf{v} = 3\mathbf{i} - 4\mathbf{k}$.

In Problems 9–14, find the end points of the representative $\overrightarrow{AB}$ of $\mathbf{v}$ from the given information.

9. $\mathbf{v} = 2\mathbf{i} - \mathbf{j} + 3\mathbf{k}$, $A = (2, 1, 5)$.
10. $\mathbf{v} = 3\mathbf{i} + \mathbf{j} - 4\mathbf{k}$, $A = (1, 4, 3)$.
11. $\mathbf{v} = -\mathbf{i} + 2\mathbf{j} + 5\mathbf{k}$, $B = (2, 3, 8)$.
12. $\mathbf{v} = 2\mathbf{i} + 5\mathbf{j} - 3\mathbf{k}$, $B = (4, -2, 6)$.
13. $\mathbf{v} = 4\mathbf{i} - 2\mathbf{j} + \mathbf{k}$, $(2, 5, -1)$ is the midpoint of AB.
14. $\mathbf{v} = 6\mathbf{i} + \mathbf{j} - 4\mathbf{k}$, $(3, 2, -5)$ is the midpoint of AB.

In Problems 15–18, find the angle θ between the given vectors.

15. $\mathbf{u} = \mathbf{i} + \mathbf{j} + 2\mathbf{k}$, $\mathbf{v} = 2\mathbf{i} - \mathbf{j} + \mathbf{k}$.
16. $\mathbf{u} = 2\mathbf{i} - 2\mathbf{j} - \mathbf{k}$, $\mathbf{v} = -\mathbf{i} + 4\mathbf{j} + 2\mathbf{k}$.
17. $\mathbf{u} = 5\mathbf{i} - \mathbf{j} + 3\mathbf{k}$, $\mathbf{v} = 4\mathbf{i} + 5\mathbf{j} - 2\mathbf{k}$.
18. $\mathbf{u} = 2\mathbf{i} + 4\mathbf{j} + 4\mathbf{k}$, $\mathbf{v} = 4\mathbf{i} - 3\mathbf{k}$.

In Problems 19–22, find $\mathbf{u} + \mathbf{v}$, $\mathbf{u} - \mathbf{v}$, and $\mathbf{u} \cdot \mathbf{v}$. Indicate whether or not $\mathbf{u}$ and $\mathbf{v}$ are orthogonal.

19. $\mathbf{u} = \mathbf{i} - 2\mathbf{j} + 5\mathbf{k}$, $\mathbf{v} = 2\mathbf{i} + 4\mathbf{j} + \mathbf{k}$.
20. $\mathbf{u} = 3\mathbf{i} + \mathbf{j} - 4\mathbf{k}$, $\mathbf{v} = 2\mathbf{i} + 6\mathbf{j} + 3\mathbf{k}$.
21. $\mathbf{u} = 2\mathbf{i} - \mathbf{j} + 6\mathbf{k}$, $\mathbf{v} = \mathbf{i} - 4\mathbf{j} - \mathbf{k}$.
22. $\mathbf{u} = 4\mathbf{i} + 3\mathbf{j} - \mathbf{k}$, $\mathbf{v} = \mathbf{i} + 2\mathbf{j} + 3\mathbf{k}$.

In Problems 23–26, find the projection of **u** upon **v**.

23. $\mathbf{u} = 4\mathbf{i} + \mathbf{j} - \mathbf{k}$, $\mathbf{v} = \mathbf{i} + \mathbf{j} + \mathbf{k}$.
24. $\mathbf{u} = \mathbf{i} - 2\mathbf{j} + 4\mathbf{k}$, $\mathbf{v} = 2\mathbf{j} + 3\mathbf{k}$.
25. $\mathbf{u} = 4\mathbf{i} - 2\mathbf{j} - \mathbf{k}$, $\mathbf{v} = \mathbf{i} - 2\mathbf{j} + \mathbf{k}$.
26. $\mathbf{u} = 2\mathbf{i} + \mathbf{j}$, $\mathbf{v} = \mathbf{j} - 2\mathbf{k}$.
27. Prove Theorem 11.4.
28. Prove Theorem 11.5.
29. Prove Theorem 11.6.
30. Prove Theorem 11.7.

11.3

Direction Angles, Cosines, and Numbers

Direction angles, cosines, and numbers are defined for vectors in space in the same way as they were for vectors in the plane. However they are far more important in space, because they represent the most convenient way of giving the direction of a line. For this reason, the definitions and theorems of Section 2.2 are repeated here.

The angle between two vectors was given in the previous section. Note again that this is not a directed angle; in fact, we have given no convention for positive and negative angles in space. Thus the angle is never negative and never greater than 180°,

$$0° \leq \theta \leq 180°.$$

Definition

*If **v** is a vector, $\{\alpha, \beta, \gamma\}$ is the set of **direction angles** for **v**, where α is the angle between **v** and **i**, β is the angle between **v** and **j**, and γ is the angle between **v** and **k**.*

The direction angles for a vector **v** are illustrated in Figure 11.2. Note that the direction angles are not necessarily in the coordinate planes. The angle α is in the plane determined by **v** and **i**; β and γ are in planes determined by **v** and **j** and by **v** and **k**, respectively.

Definition

*If **v** is a vector, then $\{l, m, n\}$ is the set of **direction cosines** for **v**, where $l = \cos \alpha$, $m = \cos \beta$, and $n = \cos \gamma$, given that $\{\alpha, \beta, \gamma\}$ is the set of direction angles for **v**.*

Theorem 11.12

*If $\{l, m, n\}$ is the set of direction cosines for a vector **v**, then*

$$l^2 + m^2 + n^2 = 1.$$

This is easily proved by considering the projections of **v** on the coordinate axes.

11.3 Direction Angles, Cosines, and Numbers

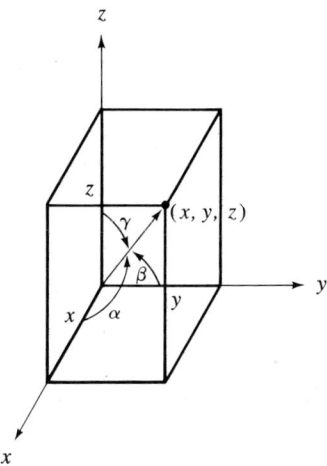

Figure 11.2

Definition

$\{a, b, c\}$ *is a set of* **direction numbers** *for the vector* **v** *if there is a nonzero constant* k *such that* $a = kl$, $b = km$, *and* $c = kn$, *where* $\{l, m, n\}$ *is the set of direction cosines for* **v**.

Theorem 11.12 may be used to find direction cosines of a vector if its direction numbers are known. Of course, additional information is needed to get the proper signs.

Example 1

Given that a vector **v** has direction numbers $\{4, 1, -2\}$ and is directed upward, find its direction cosines.

$$a^2 + b^2 + c^2 = k^2 l^2 + k^2 m^2 + k^2 n^2$$
$$= k^2(l^2 + m^2 + n^2) = k^2.$$

Thus,

$$k^2 = 16 + 1 + 4 = 21, \qquad k = \pm\sqrt{21}.$$

Since **v** is directed upward, $\gamma < 90°$ and $n > 0$. Thus $k = -\sqrt{21}$; and

$$l = -\frac{4}{\sqrt{21}}, \qquad m = -\frac{1}{\sqrt{21}}, \qquad n = \frac{2}{\sqrt{21}}.$$

Theorem 11.13

The vector $\mathbf{v} = a\mathbf{i} + b\mathbf{j} + c\mathbf{k}$ *has direction numbers* $\{a, b, c\}$.

Example 2

Give a set of direction numbers for the vector **v** represented by $\overrightarrow{AB}$, where $A = (3, -1, 2)$ and $B = (5, 2, -1)$.

By Theorem 11.5,
$$v = (5-3)i + (2+1)j + (-1-2)k = 2i + 3j - 3k.$$
By Theorem 11.13, one set of direction numbers for v is $\{2, 3, -3\}$.

Example 3

Are the vectors $u = 3i + j - 2k$ and $v = 2i - 4j + k$ orthogonal or not?
$$u \cdot v = 3 \cdot 2 + 1(-4) + (-2)1 = 0.$$
Thus u and v are orthogonal.

Direction angles, cosines, and numbers of vectors carry over directly to lines.

Definition

The statement, "A vector v is directed along a line l", means that a representative of v is on l.

Definition

A set of direction angles, cosines, or numbers for a line l is any set of direction angles, cosines, or numbers, respectively, for any vector v directed along l.

Note that every line has two sets of direction angles and two sets of direction cosines, corresponding to the two possible directions on the line. It is easily seen that if $\{l, m, n\}$ is one set of direction cosines for a line, then the other is $\{-l, -m, -n\}$.

Example 4

Find the two sets of direction cosines and direction angles for a line if $\{1, 2, 2\}$ is a set of direction numbers for it.

Since the three numbers given are a number k times the direction cosines, we have
$$kl = 1, \quad km = 2, \quad kn = 2;$$
$$k^2 l^2 + k^2 m^2 + k^2 n^2 = 1 + 4 + 4 = 9,$$
$$k^2(l^2 + m^2 + n^2) = 9,$$
$$k^2 = 9, \quad \text{(by Theorem 11.12)},$$
$$k = \pm 3.$$

Thus, the two possible sets of direction cosines are $\{1/3, 2/3, 2/3\}$ and $\{-1/3, -2/3, -2/3\}$, and they give approximate direction angles $\{71°, 48°, 48°\}$ and $\{109°, 132°, 132°\}$, respectively.

11.3 Direction Angles, Cosines, and Numbers

Example 5

Suppose a line has direction numbers $\{2, -4, 1\}$ and contains the point $(1, 3, 4)$. Find another point on the line.

By Theorems 11.5 and 11.13, we have

$$a = x_2 - x_1, \qquad b = y_2 - y_1, \qquad c = z_2 - z_1,$$

or

$$x_2 = x_1 + a, \qquad y_2 = y_1 + b, \qquad z_2 = z_1 + c.$$

Thus,

$$x_2 = 1 + 2 = 3, \qquad y_2 = 3 - 4 = -1, \qquad z_2 = 4 + 1 = 5,$$

which give the point $(3, -1, 5)$. Of course, any nonzero multiple of the given direction numbers gives another set of direction numbers; these may be used to find other points on the line.

Problems

In Problems 1–6, find the set of direction angles for the vector described.

1. Direction numbers $\{1, 4, 8\}$; directed to the right of the xz plane.
2. Direction numbers $\{4, -4, 2\}$; directed to the right of the xz plane.
3. Direction numbers $\{1, 2, -4\}$; directed behind the yz plane.
4. Direction numbers $\{2, -1, -3\}$; directed above the xy plane.
5. Direction numbers $\{1, 1, 1\}$; directed behind the yz plane.
6. Direction numbers $\{1, -1, 0\}$; directed to the right of the xz plane.

In Problems 7–12, find a set of direction numbers for the lines containing the two given points.

7. $(1, 4, 3)$ and $(5, 2, -1)$.
8. $(2, 0, -4)$ and $(-1, 2, 3)$.
9. $(2, 2, 1)$ and $(0, 0, 3)$.
10. $(3, 5, -2)$ and $(-1, 4, 4)$.
11. $(0, 0, 0)$ and $(5, 1, -2)$.
12. $(-1, 4, 5)$ and $(3, -4, 0)$.

In Problems 13–18, find two more points on the line.

13. Direction numbers $\{1, 5, 2\}$; containing $(2, 3, -1)$.
14. Direction numbers $\{1, 4, 0\}$; containing $(-2, 1, 1)$.
15. Direction numbers $\{2, 1, 2\}$; containing $(1, 3, 3)$.
16. Direction numbers $\{1, 1, 1\}$; containing $(2, 4, -1)$.
17. Direction numbers $\{4, 0, -1\}$; containing $(1, 3, -1)$.
18. Direction numbers $\{4, 4, 3\}$; containing $(-4, -4, -3)$.
19. Give the direction angles and direction cosines for the coordinate axes with their usual directions.
20. Give a set of formulas for finding all points on the line described in Example 5.

In Problems 21–30, two lines are described by a pair of points on each. Indicate whether the lines are parallel, perpendicular, coincident or none of these.

21. $(3, 4, 1), (4, 8, -1); (2, 3, -5), (0, -5, -1)$.
22. $(2, 1, 5), (3, 3, -1); (4, 2, 10), (1, -4, 5)$.
23. $(4, 1, -4), (3, 2, 1); (4, 1, -4), (11, 3, -3)$.
24. $(4, 2, -1), (7, 6, 2); (5, 10, 3), (-4, -2, -6)$.
25. $(2, 1, 4), (4, -3, 12); (1, 3, 0), (6, -7, 20)$.
26. $(4, 5, 1), (3, 2, -4); (4, 1, 2), (5, -1, 3)$.
27. $(2, 3, 1), (4, -2, 2); (1, 0, 3), (3, -3, 1)$.
28. $(3, 1, 4), (4, 3, 3); (5, 5, 2), (0, -5, 7)$.
29. $(2, 1, 3), (5, -1, 1); (3, 4, -1), (5, 3, 3)$.
30. $(4, 4, -3), (1, 3, -1); (2, 1, 5), (8, 3, 1)$.

11.4

The Line

We have seen that, given a set of direction numbers for a line and one point on that line, we can find another simply by adding. Furthermore, we can find other sets of direction numbers by taking a multiple of the original set. Thus if the point given is (x_0, y_0, z_0) and the set of direction numbers given is $\{a, b, c\}$, then any point (x, y, z) such that

$$x = x_0 + at,$$
$$y = y_0 + bt,$$
$$z = z_0 + ct,$$

where t is a real number, is on the given line. Furthermore, if (x, y, z) is a point on the line different from (x_0, y_0, z_0), then a set of direction numbers for the line is $\{x - x_0, y - y_0, z - z_0\}$. These must be a multiple of the given set of direction numbers $\{a, b, c\}$; that is, for some t,

$$x - x_0 = at,$$
$$y - y_0 = bt,$$
$$z - z_0 = ct.$$

These equations hold not only for every point on the line different from (x_0, y_0, z_0) but also for (x_0, y_0, z_0). Thus a point is on the given line if and only if it satisfies the set of equations given above.

11.4 The Line

Theorem 11.14

A parametric representation of the line containing (x_0, y_0, z_0) and having direction numbers $\{a, b, c\}$ is

$$x = x_0 + at, \qquad y = y_0 + bt, \qquad z = z_0 + ct.$$

Example 1

Find a parametric representation for the line containing $(1, 3, -2)$ and having direction numbers $\{3, 2, -1\}$.

$$x = 1 + 3t, \qquad y = 3 + 2t, \qquad z = -2 - t.$$

Example 2

Find a parametric representation of the line containing $(4, 2, -1)$ and $(0, 2, 3)$.

A set of direction numbers is $\{4 - 0, 2 - 2, -1 - 3\} = \{4, 0, -4\}$. Thus the line is

$$x = 4 + 4t, \qquad y = 2, \qquad z = -1 - 4t.$$

Once we have the direction numbers we may use them with either of the two given points. Thus, another representation is

$$x = 4s, \qquad y = 2, \qquad z = 3 - 4s.$$

Although this does not look much like the first representation, it is easily seen that they are the same. For instance, $t = 0$ gives the point $(4, 2, -1)$, as does $s = 1$; $t = -1$ gives the point $(0, 2, 3)$, as does $s = 0$, and so forth.

In fact,

$$\begin{aligned} x &= 4 + 4t & y &= 2, & z &= -1 - 4t \\ &= 4(t+1) & & & &= 3 - 4 - 4t \\ &= 4s, & & & &= 3 - 4(t+1) \\ & & & & &= 3 - 4s, \end{aligned}$$

where $s = t + 1$. Thus, whatever point we get using a value of t can be found by choosing $s = t + 1$.

A simpler set of direction numbers can also be found. Since the ones we have are all multiples of 4, we can multiply through by 1/4 to get another set of direction numbers, $\{1, 0, -1\}$. Using these with the first point gives

$$x = 4 + u, \qquad y = 2, \qquad z = -1 - u.$$

Again, we see that $4t = u$, so the two representations are equivalent.

Perhaps you wonder what is needed to be able to say that two parametric representations are equivalent. If a value of t and another of s both give the same point, then, for those values of t and s, the three coordinates must be equal. Eliminating x, y, and z between the two parametric representations gives three equations in t and s (in some of these, the parameters may both be absent, as they are in the representation of y here). If all give the same result when they are solved for one parameter in terms of the other, and if the domain and range are the same, then the representations are equivalent.

Suppose we eliminate the parameter in the representation given by Theorem 20.7. If none of the direction numbers is zero, we can solve each equation for t and set them equal to each other. This gives

$$\frac{x - x_0}{a} = \frac{y - y_0}{b} = \frac{z - z_0}{c}.$$

Actually, this is just a shorter way of writing the three equations

$$\frac{x - x_0}{a} = \frac{y - y_0}{b},$$

$$\frac{y - y_0}{b} = \frac{z - z_0}{c},$$

$$\frac{x - x_0}{a} = \frac{z - z_0}{c}.$$

But these three equations are not independent—the last can be found from the first two. Let us discard it and consider only the first two, which, as we shall see in the next section, represent planes. Any point that satisfies both equations is on both planes and therefore on the intersection of the two planes, which is a line. Thus this representation of a line gives it as the intersection of two planes. It might be noted that the equation we discarded is also a plane containing the same line.

What, now, if one of the direction numbers is zero. Let us suppose that $a = 0$. Then the line in parametric form is

$$x = x_0, \qquad y = y_0 + bt, \qquad z = z_0 + ct.$$

We do not have to eliminate the parameter from the first equation—it is already gone. By eliminating t between the last two equations as before, we have

$$\frac{y - y_0}{b} = \frac{z - z_0}{c}.$$

This, together with $x = x_0$ (or $x - x_0 = 0$) gives the line as the intersection of two planes.

If two of the direction numbers are zero, we have two equations in which the parameter is missing. The parameter in the third equation cannot be eliminated, because there is no second equation with which to combine it. But it is not necessary to eliminate it! The two equations without the parameter already give us the necessary two planes.

11.4 The Line

Theorem 11.15

If a line contains the point (x_0, y_0, z_0) and has direction numbers $\{a, b, c\}$, then it can be represented by

(a) $$\frac{x - x_0}{a} = \frac{y - y_0}{b} = \frac{z - z_0}{c}$$

if none of the direction numbers is zero;

(b) $$x - x_0 = 0 \quad \text{and} \quad \frac{y - y_0}{b} = \frac{z - z_0}{c}$$

if $a = 0$ and neither b nor c is zero (similar results follow if $b = 0$ or $c = 0$);

(c) $$x - x_0 = 0 \quad \text{and} \quad y - y_0 = 0$$

if $a = 0$ and $b = 0$ (again, similar results follow for some other pair of direction numbers equaling zero).

Example 3

Find a representation as the intersection of two planes of the line containing $(4, 1, -2)$ and having direction numbers $\{1, 3, -2\}$.

$$\frac{x - 4}{1} = \frac{y - 1}{3} = \frac{z + 2}{-2}.$$

Example 4

Find a representation as the intersection of two planes for the line containing $(4, 1, 3)$ and $(2, 1, -2)$.

A set of direction numbers is $\{4 - 2, 1 - 1, 3 + 2\} = \{2, 0, 5\}$. Since $b = 0$, we have (using the first point)

$$\frac{x - 4}{2} = \frac{z - 3}{5} \quad \text{and} \quad y - 1 = 0.$$

Example 5

Find the point of intersection (if any) of the lines

$$x = 3 + 2t, \quad y = 2 - t, \quad z = 5 + t$$

and

$$x = -3 - s, \quad y = 7 + s, \quad z = 16 + 3s.$$

Let us assume that there is a point of intersection. Then there is a value of t

and a value of s which yield the same values of x, y, and z. For these particular values of t and s, we have

$$x = 3 + 2t = -3 - s,$$
$$y = 2 - t = 7 + s,$$
$$z = 5 + t = 16 + 3s$$

or

$$2t + s = -6, \qquad t + s = -5, \qquad t - 3s = 11.$$

If we solve the first pair simultaneously, we get

$$t = -1 \quad \text{and} \quad s = -4.$$

We see that they also satisfy the third equation. Thus there is a point of intersection which corresponds to $t = -1$ (or $s = -4$). It is (1, 3, 4).

It might be noted that there are three possibilities. One is the situation in which there is a value of t and a value of s satisfying all three of the equations in t and s, as above. This results in a single point of intersection. In a second possibility, there is no value for t or s satisfying all three of the equations; that is, the values of t and s that satisfy the first two equations fail to satisfy the third. Thus, there is no point of intersection. The third possibility is that any two of the three equations in t and s are dependent; that is, any pair of values for t and s that satisfies one of them, satisfies all three. In this case we have two different representations for the same line (see the discussion following Example 2).

Problems

In Problems 1–16, represent the given line in parametric form and as the intersection of two planes.

1. Containing (5, 1, 3); direction numbers {3, −2, 4}.
2. Containing (2, −4, 2); direction numbers {2, 3, 1}.
3. Containing (5, −2, 1); direction numbers {4, 1, −2}.
4. Containing (2, 0, 3); direction numbers {4, −1, 3}.
5. Containing (1, 1, 1); direction numbers {2, 0, 1}.
6. Containing (1, 0, 5); direction numbers {3, 1, 0}.
7. Containing (4, 4, 1); direction numbers {0, 0, 1}.
8. Containing (3, 1, 2); direction numbers {1, 0, 0}.
9. Containing (4, 0, 5) and (2, 3, 1).
10. Containing (3, 3, 1) and (4, 0, 2).
11. Containing (8, 4, 1) and (−2, 0, 3).
12. Containing (−4, 2, 0) and (3, 1, 2).
13. Containing (5, 1, 3) and (5, 2, 4).
14. Containing (2, 2, 4) and (1, 2, 7).
15. Containing (1, −2, 3) and (1, 4, 3).

11.5 The Cross Product

16. Containing $(2, 4, -5)$ and $(5, 4, -5)$.

In Problems 17–24, find the point of intersection (if any) of the given lines.

17. $x = 4 + t, y = -8 - 2t, z = 12t;\quad x = 3 + 2s, y = -1 + s, z = -3 - 3s.$
18. $x = 2 - t, y = 3 + 2t, z = 4 + t;\quad x = 1 + t, y = -2 + t, z = 5 - 4t.$
19. $x = 3 + t, y = 4 - 2t, z = 1 + 5t;\quad x = 5 - t, y = 3 + 2t, z = 8 + 4t.$
20. $x = 3 - t, y = 5 + 3t, z = -1 - 4t;\quad x = 8 + 2s, y = -6 - 4s, z = 5 + s.$
21. $\dfrac{x-2}{1} = \dfrac{y-3}{-2} = \dfrac{z+1}{1};\quad \dfrac{x-3}{2} = \dfrac{y-1}{-4} = \dfrac{z}{2}.$
22. $\dfrac{x-5}{1} = \dfrac{y+2}{-2} = \dfrac{z-3}{5};\quad \dfrac{x-4}{-2} = \dfrac{y-2}{1} = \dfrac{z-4}{3}.$
23. $\dfrac{x-3}{1} = \dfrac{y+3}{-4}, z + 1 = 0;\quad \dfrac{x}{-2} = \dfrac{y-2}{1} = \dfrac{z-3}{4}.$
24. $\dfrac{x-2}{1} = \dfrac{y-3}{-2} = \dfrac{z}{4};\quad x - 4 = 0, \dfrac{y-2}{1} = \dfrac{z-3}{-1}.$

In Problems 25–30, indicate whether the two given lines are parallel, perpendicular, coincident, or none of these.

25. $x = 3 + 5t, y = -1 - 2t, z = 4 + t;\quad x = 3, y = 4 + 2s, z = -2 + 4s.$
26. $x = 4 - t, y = 3 + 2t, z = 1 + t;\quad x = 1 + 2t, y = 4 - 4t, z = 3 - 2t.$
27. $x = 2 + t, y = 5 - 3t, z = 1 + 4t;\quad x = 4 - t, y = 2 + 2t, z = 3t.$
28. $\dfrac{x-2}{1} = \dfrac{y-5}{-3} = \dfrac{z+1}{2};\quad \dfrac{x-4}{-3} = \dfrac{y+1}{9} = \dfrac{z-3}{-6}.$
29. $\dfrac{x+3}{1} = \dfrac{y-4}{3} = \dfrac{z+2}{-2};\quad \dfrac{x-5}{-3} = \dfrac{y+3}{-9} = \dfrac{z-1}{6}.$
30. $\dfrac{x-1}{2} = \dfrac{z+3}{4}, y - 5 = 0;\quad \dfrac{x+2}{6} = \dfrac{y-5}{3} = \dfrac{z}{2}.$
31. Give equations for each of the coordinates axes.

11.5

The Cross Product

Let us now look at the other product of two vectors—the cross product.

Definition

If $u = a_1 i + b_1 j + c_1 k$ *and* $v = a_2 i + b_2 j + c_2 k$, *then the* **cross product** *(vector product, outer product) of* **u** *and* **v** *is*

$$u \times v = (b_1 c_2 - c_1 b_2)i + (c_1 a_2 - a_1 c_2)j + (a_1 b_2 - b_1 a_2)k.$$

Some obvious questions arise. Why do we want to define a cross product this way? What is it good for? What are its properties? In some ways, all answers are the same. We define the cross product in this way to establish some interesting properties that are useful for certain applications. In a way, this is approaching the problem backward. It would be more logical to define the cross product of two vectors as that one having the desired properties and then show that such a vector must take the form given. The reason for our way of doing it is that it is by far the simpler approach. Before looking at some properties, let us consider a simpler form for the cross product.

Theorem 11.16

If $\mathbf{u} = a_1\mathbf{i} + b_1\mathbf{j} + c_1\mathbf{k}$ and $\mathbf{v} = a_2\mathbf{i} + b_2\mathbf{j} + c_2\mathbf{k}$, then

$$\mathbf{u} \times \mathbf{v} = \begin{vmatrix} \mathbf{i} & \mathbf{j} & \mathbf{k} \\ a_1 & b_1 & c_1 \\ a_2 & b_2 & c_2 \end{vmatrix}.$$

This theorem follows directly from the definition.

Example 1

If $\mathbf{u} = 3\mathbf{i} + \mathbf{j} - 2\mathbf{k}$ and $\mathbf{v} = \mathbf{i} + 2\mathbf{j} + \mathbf{k}$, find $\mathbf{u} \times \mathbf{v}$ and $\mathbf{v} \times \mathbf{u}$.

$$\mathbf{u} \times \mathbf{v} = \begin{vmatrix} \mathbf{i} & \mathbf{j} & \mathbf{k} \\ 3 & 1 & -2 \\ 1 & 2 & 1 \end{vmatrix} = 5\mathbf{i} - 5\mathbf{j} + 5\mathbf{k},$$

$$\mathbf{v} \times \mathbf{u} = \begin{vmatrix} \mathbf{i} & \mathbf{j} & \mathbf{k} \\ 1 & 2 & 1 \\ 3 & 1 & -2 \end{vmatrix} = -5\mathbf{i} + 5\mathbf{j} - 5\mathbf{k}.$$

Note that $\mathbf{u} \times \mathbf{v} \neq \mathbf{v} \times \mathbf{u}$!

Again we are not multiplying numbers; there is no reason to assume that the cross product of two vectors has the same properties as the product of two numbers. We have already seen one difference in Example 1. The cross product has the following properties.

Theorem 11.17

If $\mathbf{u}$, $\mathbf{v}$ and $\mathbf{w}$ are vectors and a is a scalar, then

(a) $\mathbf{u} \times \mathbf{v} = -\mathbf{v} \times \mathbf{u}$;
(b) $\mathbf{u} \times (\mathbf{v} + \mathbf{w}) = \mathbf{u} \times \mathbf{v} + \mathbf{u} \times \mathbf{w}$;
(c) $\mathbf{u} \times \mathbf{0} = \mathbf{0} \times \mathbf{u} = \mathbf{0}$;
(d) *if* $\mathbf{u} = a\mathbf{v}$, *then* $\mathbf{u} \times \mathbf{v} = \mathbf{0}$;
(e) $(\mathbf{u} \times \mathbf{v}) \cdot \mathbf{w} = \mathbf{u} \cdot (\mathbf{v} \times \mathbf{w})$.

11.5 The Cross Product

This follows from the definition of the cross product. The proof is left to the student. It might be noted that the definition was stated in terms of three-dimensional vectors. In fact, we must have a three-dimensional vector space, for **u** × **v** is not in the plane determined by **u** and **v**.

Theorem 11.18

*If **u** and **v** are nonzero vectors, then **u** × **v** is perpendicular to both **u** and **v**.*

Proof

$$\mathbf{u} \cdot (\mathbf{u} \times \mathbf{v}) = (\mathbf{u} \times \mathbf{u}) \cdot \mathbf{v} \quad \text{(why?)}$$
$$= \mathbf{0} \cdot \mathbf{v} \quad \text{(why?)}$$
$$= 0. \quad \text{(why?)}$$

Thus **u** and **u** × **v** are perpendicular. A similar argument shows that **u** × **v** and **v** are perpendicular.

This property of the cross product gives us its principal use. Certain problems in three-dimensional analytic geometry that were relatively difficult without the use of the cross product are easier now.

Example 2

Find a set of direction numbers for a line perpendicular to the plane containing

$$x = 1, \quad y = 3 + 2t, \quad z = 4 + t$$

and

$$x = 1 + 4s, \quad y = 3 + 2s, \quad z = 4 + 2s.$$

Any line perpendicular to a given plane is perpendicular to any line in that plane. This suggests the use of the cross product. Vectors directed along the given lines are

$$\mathbf{u} = 2\mathbf{j} + \mathbf{k} \quad \text{and} \quad \mathbf{v} = 4\mathbf{i} + 2\mathbf{j} + 2\mathbf{k}.$$

Since **u** × **v** = 2**i** + 4**j** − 8**k**, we have {2, 4, −8} as one set of direction numbers for the desired line; {1, 2, −4} is a simpler set.

Example 3

Find equations for the line containing (1, 4, 3) and perpendicular to

$$\frac{x-1}{2} = \frac{y+3}{1} = \frac{z-2}{4} \quad \text{and} \quad \frac{x+2}{3} = \frac{y-4}{2} = \frac{z+1}{-2}.$$

Again, vectors along the two given lines are

$$\mathbf{u} = 2\mathbf{i} + \mathbf{j} + 4\mathbf{k} \quad \text{and} \quad \mathbf{v} = 3\mathbf{i} + 2\mathbf{j} - 2\mathbf{k};$$

and **u** × **v** = −10**i** + 16**j** + **k** is perpendicular to both of them. The desired line is, therefore,

$$\frac{x-1}{-10} = \frac{y-4}{16} = \frac{z-3}{1}.$$

Example 4

Find the distance between the lines

$$x = 1 - 4t, \quad y = 2 + t, \quad z = 3 + 2t$$

and

$$x = 2 + s, \quad y = 2 - 2s, \quad z = 4 - s.$$

The desired distance is to be measured along a line perpendicular to both of the given lines. Again, vectors along the given lines are $\mathbf{u} = -4\mathbf{i} + \mathbf{j} + 2\mathbf{k}$ and $\mathbf{v} = \mathbf{i} - 2\mathbf{j} - \mathbf{k}$. Thus the distance is to be measured along $\mathbf{u} \times \mathbf{v} = 3\mathbf{i} - 2\mathbf{j} + 7\mathbf{k}$. The point $A: (1, 2, 3)$ is on the first line, and $B: (2, 2, 4)$ is on the second. The vector represented by $\overrightarrow{AB}$ is $\mathbf{w} = \mathbf{i} + \mathbf{k}$. We want a vector whose representatives are all perpendicular to both of the given lines and with one representative having its head on one line and its tail on the other. All of the representatives of $\mathbf{u} \times \mathbf{v}$ are perpendicular to both lines and one of the representatives of $\mathbf{w}$ has its end points on the given lines. Thus, the projection of $\mathbf{w}$ on $\mathbf{u} \times \mathbf{v}$ has the desired properties and its length is the distance between the given lines.

$$\frac{\mathbf{w} \cdot (\mathbf{u} \times \mathbf{v})}{|\mathbf{u} \times \mathbf{v}|} = \frac{1 \cdot 3 + 0(-2) + 1 \cdot 7}{\sqrt{9 + 4 + 49}} = \frac{10}{\sqrt{62}}.$$

Up to this point we have been dealing exclusively with the direction of $\mathbf{u} \times \mathbf{v}$. Its length also has some interesting properties.

Theorem 11.19

If $\mathbf{u}$ and $\mathbf{v}$ are vectors and θ is the angle between them, then

$$|\mathbf{u} \times \mathbf{v}| = |\mathbf{u}| |\mathbf{v}| \sin \theta.$$

Proof

Since $\cos \theta = \mathbf{u} \cdot \mathbf{v}/(|\mathbf{u}| |\mathbf{v}|)$ by Theorem 11.10,

$$|\mathbf{u}| |\mathbf{v}| \sin \theta = |\mathbf{u}| |\mathbf{v}| \sqrt{1 - \cos^2 \theta} \qquad \text{(See Note 1.)}$$

$$= |\mathbf{u}| |\mathbf{v}| \sqrt{1 - \frac{(\mathbf{u} \cdot \mathbf{v})^2}{|\mathbf{u}|^2 |\mathbf{v}|^2}}$$

$$= \sqrt{|\mathbf{u}|^2 |\mathbf{v}|^2 - (\mathbf{u} \cdot \mathbf{v})^2}.$$

If we let $\mathbf{u} = a_1 \mathbf{i} + b_1 \mathbf{j} + c_1 \mathbf{k}$ and $\mathbf{v} = a_2 \mathbf{i} + b_2 \mathbf{j} + c_2 \mathbf{k}$, then

$$|\mathbf{u}| |\mathbf{v}| \sin \theta = \sqrt{(a_1^2 + b_1^2 + c_1^2)(a_2^2 + b_2^2 + c_2^2) - (a_1 a_2 + b_1 b_2 + c_1 c_2)^2}$$

$$= \sqrt{(b_1 c_2 - c_1 b_2)^2 + (c_1 a_2 - a_1 c_2)^2 + (a_1 b_2 - b_1 a_2)^2} \qquad \text{(See Note 2.)}$$

$$= |\mathbf{u} \times \mathbf{v}|.$$

Note 1: By the definition of the angle between two vectors, $0° \leq \theta \leq 180°$; and $\sin \theta \geq 0$.

11.5 The Cross Product

Note 2: The algebra here is routine but tedious. It is left to the student.

Note the similarity between this theorem and the first part of Theorem 11.10. One consequence of this theorem is given in Problem 25.

Problems

In Problems 1–6, find $\mathbf{u} \times \mathbf{v}$.

1. $\mathbf{u} = 3\mathbf{i} - \mathbf{j} + 4\mathbf{k}$, $\mathbf{v} = 2\mathbf{i} + \mathbf{j} + \mathbf{k}$.
2. $\mathbf{u} = \mathbf{i} + \mathbf{j} + \mathbf{k}$, $\mathbf{v} = 2\mathbf{i} - \mathbf{j} - 4\mathbf{k}$.
3. $\mathbf{u} = 2\mathbf{i} + 3\mathbf{j} - \mathbf{k}$, $\mathbf{v} = -\mathbf{i} + 2\mathbf{j}$.
4. $\mathbf{u} = 4\mathbf{i} + 2\mathbf{j}$, $\mathbf{v} = 3\mathbf{i} - \mathbf{j}$.
5. $\mathbf{u} = 3\mathbf{i} + \mathbf{k}$, $\mathbf{v} = -\mathbf{i} + \mathbf{j}$.
6. $\mathbf{u} = 2\mathbf{i} + \mathbf{j} - \mathbf{k}$, $\mathbf{v} = -\mathbf{i} - \mathbf{j} + 3\mathbf{k}$.

In Problems 7–12, find direction numbers for the line described.

7. Perpendicular to the plane containing $(4, 1, 2)$, $(2, -1, 1)$, and $(3, 0, 4)$.
8. Perpendicular to the plane containing $(2, 2, 3)$, $(-1, 4, 1)$, and $(0, 1, 2)$.
9. Perpendicular to the plane containing $x = 2 + t$, $y = 3 - 2t$, $z = -t$ and $x = 2 - 2s$, $y = 3 + s$, $z = -s$.
10. Perpendicular to the plane containing $x = 3 + 4t$, $y = 1 - t$, $z = 3$ and $x = 3 - 2s$, $y = 1 + 2s$, $z = 3 - s$.
11. Perpendicular to the plane containing $x = 4 + t$, $y = -1 + 2t$, $z = 2t$ and $x = 2 + s$, $y = 4 + 2s$, $z = 1 + 2s$.
12. Perpendicular to the plane containing $x = 2 + 2t$, $y = 3 - t$, $z = -1 + t$ and $x = 4 + 2s$, $y = 2 - s$, $z = 4 + s$.

In Problems 13–18, find equations for the line described.

13. Containing $(3, 2, 1)$ and perpendicular to $x = 1 - 2t$, $y = 3 + t$, $z = 4 - t$ and $x = 2 + s$, $y = -1 + 2s$, $z = 3 - s$.
14. Containing $(4, -1, 0)$ and perpendicular to $x = 3 + t$, $y = 2 - t$, $z = 2t$ and $x = 4$, $y = 2 + s$, $z = -1 + s$.
15. Containing $(2, 3, 1)$ and perpendicular to the plane determined by $(2, 3, 1)$ and the line $x = 0$, $y = 2t$, $z = t$.
16. Containing $(0, 4, -2)$ and perpendicular to the plane determined by $(0, 4, -2)$ and the line $x = -2 + 2t$, $y = 8t$, $z = -1 + t$.
17. Containing $(2, 0, 5)$ and perpendicular to and containing a point of $x = 4 + t$, $y = 3 - 2t$, $z = 1 + t$.
18. Containing $(1, 1, 2)$ and perpendicular to and containing a point of $x = 1 - t$, $y = 2 + 2t$, $z = 4t$.

In Problems 19–24, find the distance between the given lines.

19. $x = 1 + t$, $y = -2 + 3t$, $z = 4 + t$ and $x = 2 - s$, $y = 3 + 2s$, $z = 1 + s$.
20. $x = 2 + t$, $y = 1 - t$, $z = 4t$ and $x = 2 + s$, $y = 4 - 2s$, $z = 1 + 3s$.
21. $x = 1 + t$, $y = 1 - 5t$, $z = 2 + t$ and $x = 4 + s$, $y = 5 + 2s$, $z = -3 + 4s$.
22. $x = 2 + t$, $y = -4 + t$, $z = 1 - 3t$ and $x = 3 - s$, $y = 4 + 2s$, $z = 2 + s$.
23. $x = 2 + 3t$, $y = 5 + t$, $z = -1 - 2t$ and $x = 2 + 3s$, $y = 3 + s$, $z = 5 - 2s$.
24. $x = 4t$, $y = 1 + t$, $z = -2 - t$ and $x = 9 + 4s$, $y = 1 + s$, $z = -2 - s$.
25. Suppose the vectors $\mathbf{u}$ and $\mathbf{v}$ are represented by $\overrightarrow{AB}$ and $\overrightarrow{AC}$, respectively. Show that the area of $\triangle ABC$ is $\frac{1}{2}|\mathbf{u} \times \mathbf{v}|$. (*Hint*: Use Theorem 11.19.)

In Problems 26–29, use the result of Problem 25 to find the area of the triangles with the given vertices.

26. (1, 0, 4), (2, −1, 2), (4, 4, 1).
27. (3, −2, 1), (−1, 2, 0), (4, 4, 2).
28. (2, 4, 3), (1, 0, 1), (−2, 2, 4).
29. (4, 2), (3, −1), (−1, 0).
30. Prove Theorem 11.17.
31. Complete the proof of Theorem 11.18.

11.6

The Plane

Suppose we are given a plane in space and a line l perpendicular to the plane and containing the origin O. Suppose this line intersects the plane at the point Q (see Figure 11.3). Then $\overrightarrow{OQ}$ represents a vector $\mathbf{v}$ normal (perpendicular) to the plane. If $\{\alpha, \beta, \gamma\}$ is the set of direction angles for $\mathbf{v}$ and $p = |\mathbf{v}|$, then

$$\mathbf{v} = p \cos \alpha \mathbf{i} + p \cos \beta \mathbf{j} + p \cos \gamma \mathbf{k}.$$

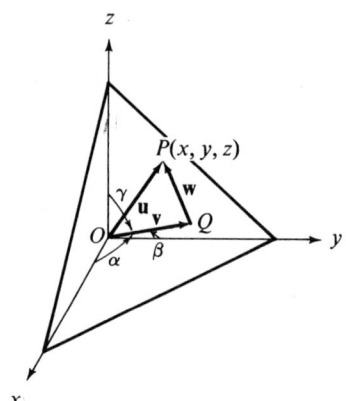

Figure 11.3

Let $P: (x, y, z)$ be any point of the plane; $\overrightarrow{OP}$ represents a vector $\mathbf{u} = x\mathbf{i} + y\mathbf{j} + z\mathbf{k}$. Finally $\overrightarrow{QP}$ represents a vector $\mathbf{w} = \mathbf{u} - \mathbf{v}$. Thus

$$\mathbf{w} = (x - p \cos \alpha)\mathbf{i} + (y - p \cos \beta)\mathbf{j} + (z - p \cos \gamma)\mathbf{k}.$$

Since $\overrightarrow{OQ}$ is perpendicular to the plane containing $\overrightarrow{QP}$, $\mathbf{v}$ and $\mathbf{w}$ are orthogonal.

11.6 The Plane

$$\mathbf{v} \cdot \mathbf{w} = p \cos \alpha (x - p \cos \alpha) + p \cos \beta (y - p \cos \beta) + p \cos \gamma (z - p \cos \gamma) = 0,$$
$$x \cos \alpha + y \cos \beta + z \cos \gamma - p(\cos^2 \alpha + \cos^2 \beta + \cos^2 \gamma) = 0,$$
$$x \cos \alpha + y \cos \beta + z \cos \gamma - p = 0.$$

If the above steps are reversed, we see that the last equation represents the plane with which we started.

Theorem 11.20

If a line perpendicular to a plane has direction angles $\{\alpha, \beta, \gamma\}$ and p is the distance from the origin to the plane, then a point is in the plane if and only if it satisfies the equation

$$x \cos \alpha + y \cos \beta + z \cos \gamma - p = 0.$$

This form for the equation of a plane is called the *normal form*. Compare it with the normal form for a line (see Problem 37, page 60; and Problem 25, page 182).

Theorem 11.21

Any plane can be represented by an equation of the form

$$Ax + By + Cz + D = 0,$$

where $\{A, B, C\}$ is a set of direction numbers for a line normal to (that is, perpendicular to) the plane. Conversely, an equation of the above form (where A, B, and C are not all zero) represents a plane with $\{A, B, C\}$ a set of direction numbers for a normal line.

The first part follows directly from Theorem 11.20, since any plane can be represented by an equation in normal form. If we are given an equation of the form

$$Ax + By + Cz + D = 0,$$

where A, B, and C are not all zero, let us multiply through by the constant

$$k = \frac{\pm 1}{\sqrt{A^2 + B^2 + C^2}},$$

where the sign is chosen opposite the sign of D. The coefficients of x, y, and z form a set of direction cosines,

$$\left\{ \frac{\pm A}{\sqrt{A^2 + B^2 + C^2}}, \frac{\pm B}{\sqrt{A^2 + B^2 + C^2}}, \frac{\pm C}{\sqrt{A^2 + B^2 + C^2}} \right\},$$

for some line, since the sum of their squares is 1 (see Problem 35). The resulting equation is, therefore, the normal form of a plane. Since the coefficients are a nonzero multiple of the set of direction cosines above, they are a set of direction numbers for the normal line.

Theorem 11.22

A point is on a plane containing (x_1, y_1, z_1) and perpendicular to a line with direction numbers $\{A, B, C\}$ if and only if it satisfies the equation

$$A(x - x_1) + B(y - y_1) + C(z - z_1) = 0.$$

The proof is left to the student.

Example 1

Sketch $x + 2y + 3z = 6$ and find its distance from the origin.

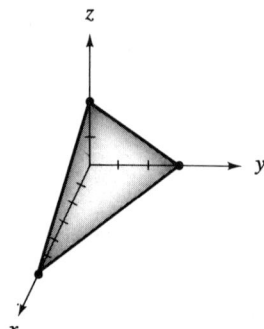

A simple way to sketch a plane is to find its three intercepts and identify its intersections with the coordinate planes. This has been done in Figure 11.4. To find its distance from the origin, we must put the equation into normal form by dividing through by $\sqrt{1^2 + 2^2 + 3^2} = \sqrt{14}$;

$$\frac{1}{\sqrt{14}}x + \frac{2}{\sqrt{14}}y + \frac{3}{\sqrt{14}}z - \frac{6}{\sqrt{14}} = 0.$$

The desired distance is $6/\sqrt{14}$.

Figure 11.4

Example 2

Find the distance between the parallel planes

$$x - 2y + 2z = 3 \quad \text{and} \quad x - 2y + 2z = 12.$$

Let us put them into normal form by dividing through by $\sqrt{1^2 + (-2)^2 + 2^2} = 3$:

$$\frac{1}{3}x - \frac{2}{3}y + \frac{2}{3}z - 1 = 0,$$

$$\frac{1}{3}x - \frac{2}{3}y + \frac{2}{3}z - 4 = 0.$$

They are at distances 1 and 4 from the origin, respectively, and both are on the same side, since their constant terms have the same sign. Thus, the distance between them is the difference, 3.

Example 3

Find the equation of the plane containing $(1, 3, -2)$ and perpendicular to the line through $(2, 5, 1)$ and $(0, 1, -3)$.

11.6 The Plane

A set of direction numbers for the given line is {2, 4, 4} or {1, 2, 2}. Thus the desired plane is
$$1(x-1)+2(y-3)+2(z+2)=0,$$
$$x+2y+2z-3=0.$$

Example 4

Find an equation of the plane containing the two lines
$$x=1, \quad y=3+2t, \quad z=4+t$$
and
$$x=1+4s, \quad y=3+2s, \quad z=4+2s.$$

These lines clearly intersect at (1, 3, 4). All we need, then, is a set of direction numbers for a line perpendicular to the desired plane. This was done in Example 2 of the previous section by using the cross product of two vectors. One such set is {1, 2, −4}. By Theorem 11.22, the corresponding plane is
$$1(x-1)+2(y-3)-4(z-4)=0,$$
$$x+2y-4z+9=0.$$

Example 5

Find the distance between the point (5, 1, 3) and the line $x=3$, $y=7+t$, $z=1+t$.

Of course the distance we seek is the minimum, or perpendicular, distance. To find it, we first find the plane containing the given point and perpendicular to the given line.
$$0(x-5)+1(y-1)+1(z-3)=0,$$
$$y+z-4=0.$$
Now the distance we want is the distance between the given point and the point of intersection of the given line with the plane we have just found. Substituting the x, y, and z of the line into the equation of the plane, we have
$$(7+t)+(1+t)-4=0,$$
$$t=-2.$$
Thus the point of intersection is (3, 5, −1) and the desired distance is
$$s=\sqrt{(5-3)^2+(1-5)^2+(3+1)^2}=6.$$

Problems

In Problems 1–6, sketch the plane and find its distance from the origin.

1. $2x+3y+z=6.$
2. $3x-y+z=9.$

3. $2x + y - 4z + 4 = 0$.
5. $x + 2y = 3$.

4. $x - y - 4z = 8$.
6. $y - 5 = 0$.

In Problems 7–12, find the distance between the parallel planes.

7. $2x - y + 2z = 9$, $2x - y + 2z = -12$.
8. $x - 4y - 2z = 5$, $x - 4y - 2z = 10$.
9. $3x + y - 4z = 3$, $6x + 2y - 8z = -5$.
10. $x + y + 4z = 6$, $2x + 2y + 8z = 9$.
11. $x + 2y = 6$, $x + 2y = 1$.
12. $x - y - z = 4$, $2x - 2y - 2z = -3$.

In Problems 13–30, find an equation(s) of the plane(s) satisfying the given conditions.

13. Containing $(4, 1, -3)$ and perpendicular to the line $x = 2 + 3t$, $y = 4 - t$, $z = 3 - 2t$.
14. Containing $(3, 2, 5)$ and perpendicular to the line $x = 1 + t$, $y = 3t$, $z = 4 + t$.
15. Containing $(3, 5, 1)$ and parallel to $3x - 4y + 2z = 3$.
16. Containing $(4, -1, 2)$ and parallel to $x + y - 2z = 4$.
17. Parallel to $3x + y - 4z + 2 = 0$ and twice as far from the origin.
18. Parallel to $5x - 2y + z - 2 = 0$ and equally distant from the origin but on the opposite side of the origin.
19. Parallel to $2x + y - z = 4$ and at a distance 2 from it.
20. Perpendicular to $x = 1 - t$, $y = 2 + 3t$, $z = 4 + t$ and at a distance 4 from the origin.
21. Containing $(1, 4, 2)$, $(2, 3, -1)$, and $(5, 0, 2)$.
22. Containing $(3, 1, -4)$, $(2, 3, 1)$, and $(7, 4, -2)$.
23. Containing $x = 4 + t$, $y = 2 - t$, $z = 1 + 2t$ and $x = 4 - 3s$, $y = 2 + 2s$, $z = 1 - s$.
24. Containing $x = 2 + 2t$, $y = -1 + t$, $z = 4 - t$ and $x = 2 - s$, $y = -1 - 2s$, $z = 4 + 3s$.
25. Containing $(4, 1, 2)$ and $x = 4 - t$, $y = 1 + 2t$, $z = 3 - t$.
26. Containing $(-2, 3, -4)$ and $x = 1 + t$, $y = 3 - 2t$, $z = -2 + t$.
27. Containing $x = 3 + 2t$, $y = 4 - t$, $z = 1 + t$ and $x = -1 + 2s$, $y = 3 - s$, $z = 4 + s$.
28. Containing $x = 4 + t$, $y = 2t$, $z = 5$ and $x = 1 + s$, $y = 3 + 2s$, $z = -2$.
29. Containing $(1, 5, -2)$ and perpendicular to $3x + 2y - z + 1 = 0$ and $x - y + 2z = 0$.
30. Containing $(3, 0, -4)$ and perpendicular to $2x - 5y + z = 1$ and $x - 2y - z = 3$.

In Problems 31–34, find the distance between the point and line given.

31. $(1, 3, -2)$; $x = 4$, $y = -3 + 4t$, $z = 11 + 5t$.
32. $(4, 3, 3)$; $x = 2 + 2t$, $y = 5 - 5t$, $z = -1 - t$.
33. $(2, 4, -1)$; $x = 5 + t$, $y = -2 + 3t$, $z = 3 + t$.
34. $(-1, 0, 5)$; $\dfrac{x-2}{2} = \dfrac{y-1}{1} = \dfrac{z+2}{3}$.

35. Show that if $A^2 + B^2 + C^2 = 1$, then $\{A, B, C\}$ is a set of direction cosines for some line. (*Hint:* Consider the point (A, B, C) together with the origin.)
36. Prove Theorem 11.22.

11.7

Distance between a Point and a Plane; Angles between Lines or Planes

The normal form for a plane is a very natural one for finding the distance between a point and a plane. Suppose we have a plane

$$Ax + By + Cz + D = 0$$

and a point (x_1, y_1, z_1) (See Figure 11.5). The plane parallel to the given plane and containing (x_1, y_1, z_1) is

$$A(x - x_1) + B(y - y_1) + C(z - z_1) = 0$$

or

$$Ax + By + Cz - (Ax_1 + By_1 + Cz_1) = 0.$$

Let us put both of these planes into normal form. The given plane is

$$\frac{Ax + By + Cz + D}{\sqrt{A^2 + B^2 + C^2}} = 0$$

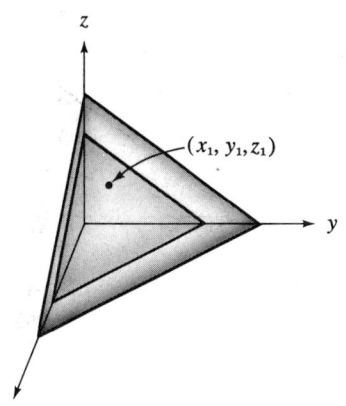

Figure 11.5

and the plane through the given point is

$$\frac{Ax + By + Cz - (Ax_1 + By_1 + Cz_1)}{\sqrt{A^2 + B^2 + C^2}} = 0.$$

Since the constant terms in these two equations give their signed distances from the origin, by the same line of reasoning that we used on page 216, the distance between these planes is the absolute value of the difference between their constant terms.

$$d = \frac{|Ax_1 + By_1 + Cz_1 + D|}{\sqrt{A^2 + B^2 + C^2}}.$$

Theorem 11.23

The distance between the point (x_1, y_1, z_1) and the plane $Ax + By + Cz + D = 0$ is

$$d = \frac{|Ax_1 + By_1 + Cz_1 + D|}{\sqrt{A^2 + B^2 + C^2}}.$$

Example 1

Find the distance between $(3, -4, 1)$ and $x - 2y + 2z + 4 = 0$.

$$d = \frac{|Ax_1 + By_1 + Cz_1 + D|}{\sqrt{A^2 + B^2 + C^2}}$$

$$= \frac{|1 \cdot 3 - 2(-4) + 2 \cdot 1 + 4|}{\sqrt{1^2 + (-2)^2 + 2^2}}$$

$$= \frac{17}{3}.$$

We have already considered the angle between two vectors in Section 11.2 (see page 197). The relationship between the two vectors and the angle between them was given in Theorem 11.7. This is restated here for convenience.

If $\mathbf{u} = a_1\mathbf{i} + b_1\mathbf{j} + c_1\mathbf{k}$ and $\mathbf{v} = a_2\mathbf{i} + b_2\mathbf{j} + c_2\mathbf{k}$ ($\mathbf{u} \neq 0$ and $\mathbf{v} \neq 0$) and if θ is the angle between them, then

$$\cos \theta = \frac{a_1 a_2 + b_1 b_2 + c_1 c_2}{|\mathbf{u}| \cdot |\mathbf{v}|}.$$

An angle between two lines can be found in much the same way, since the direction numbers of a line are the components of some vector directed along that line. However, an angle between two lines is the angle between two vectors directed along these lines. Since vectors directed along a line can be oriented in one of two directions, there are two angles θ_1 and θ_2 between any pair of lines. Since

$$\theta_1 + \theta_2 = 180°$$

and

$$\cos \theta_2 = -\cos \theta_1,$$

we can determine the absolute value of $\cos \theta$, where θ is an angle between two lines; but we need additional information to determine the sign of $\cos \theta$.

Theorem 11.24

If θ is an angle between two lines with direction numbers $\{a_1, b_1, c_1\}$ and $\{a_2, b_2, c_2\}$, then

$$|\cos \theta| = \frac{|a_1 a_2 + b_1 b_2 + c_1 c_2|}{\sqrt{a_1^2 + b_1^2 + c_1^2}\sqrt{a_2^2 + b_2^2 + c_2^2}}.$$

Example 2

Find the acute angle between the lines

$$x = 2 - 3t, \quad y = 4 + t, \quad z = t$$

11.7 Distance between a Point and a Plane; Angles between Lines or Planes

and
$$x = 3 + s, \quad y = 1 - s, \quad z = 2 + 2s.$$

Direction numbers for the lines are $\{-3, 1, 1\}$ and $\{1, -1, 2\}$. Thus

$$|\cos \theta| = \frac{|a_1 a_2 + b_1 b_2 + c_1 c_2|}{\sqrt{a_1^2 + b_1^2 + c_1^2} \sqrt{a_2^2 + b_2^2 + c_2^2}}$$

$$= \frac{|-3 \cdot 1 + 1(-1) + 1 \cdot 2|}{\sqrt{9 + 1 + 1}\sqrt{1 + 1 + 4}}$$

$$= \frac{|-2|}{\sqrt{11}\sqrt{6}} = \frac{2}{\sqrt{66}} \approx 0.2462.$$

Since the angle is acute, $\cos \theta$ is positive.

$$\cos \theta \approx 0.2462,$$
$$\theta \approx 76°.$$

If two lines are perpendicular, then $\cos \theta = 0$. This gives us the well-known test for perpendicularity

$$a_1 a_2 + b_1 b_2 + c_1 c_2 = 0,$$

where $\{a_1, b_1, c_1\}$ and $\{a_2, b_2, c_2\}$ are direction numbers for the lines.

Let us now consider an angle of intersection of two planes. This is defined to be an angle between their normal lines. With this definition, the problem is reduced to a familiar one. Note that we have again made no attempt to define *the* angle between two planes but only *an* angle. The particular one desired must be specified.

Theorem 11.25

If θ is an angle between the planes

$$A_1 x + B_1 y + C_1 z + D_1 = 0 \quad \text{and} \quad A_2 x + B_2 y + C_2 z + D_2 = 0,$$

then

$$|\cos \theta| = \frac{|A_1 A_2 + B_1 B_2 + C_1 C_2|}{\sqrt{A_1^2 + B_1^2 + C_1^2}\sqrt{A_2^2 + B_2^2 + C_2^2}}.$$

Example 3

Find the acute angle between the planes

$$2x + y - z + 3 = 0 \quad \text{and} \quad 4x - y + z + 1 = 0.$$

$$|\cos\theta| = \frac{|A_1A_2 + B_1B_2 + C_1C_2|}{\sqrt{A_1^2 + B_1^2 + C_1^2}\sqrt{A_2^2 + B_2^2 + C_2^2}}$$

$$= \frac{|2\cdot 4 + 1(-1) - 1\cdot 1|}{\sqrt{4+1+1}\sqrt{16+1+1}}$$

$$= \frac{|6|}{\sqrt{6}\sqrt{18}} = \frac{6}{6\sqrt{3}} = \frac{\sqrt{3}}{3} \approx 0.5774$$

Since we want the acute angle, $\cos\theta$ is positive.

$$\cos\theta \approx 0.5774,$$
$$\theta \approx 55°.$$

The conditions for parallel or perpendicular planes follow directly from the above discussion.

Theorem 11.26

The planes $A_1x + B_1y + C_1z + D_1 = 0$ and $A_2x + B_2y + C_2z + D_2 = 0$ are parallel if and only if $\{A_1, B_1, C_1\}$ and $\{A_2, B_2, C_2\}$ are proportional; they are perpendicular if and only if $A_1A_2 + B_1B_2 + C_1C_2 = 0$.

Problems

In Problems 1–8, find the distance between the plane and point given.

1. $2x - 4y + 4z + 3 = 0$; $(1, 3, -2)$.
2. $4x + y - 8z + 1 = 0$; $(2, 0, 3)$.
3. $x + y - 2z - 4 = 0$; $(3, 3, 1)$.
4. $2x - y + z + 5 = 0$; $(1, 0, 2)$.
5. $x + 3y + z - 2 = 0$; $(2, 1, -3)$.
6. $x - 2y + 4 = 0$; $(2, 2, 4)$.
7. $x + z - 5 = 0$; $(3, 3, 1)$.
8. $y + 7 = 0$; $(1, 3, 1)$.
9. If the distance between $(1, 4, z)$ and $8x - y + 4z - 3 = 0$ is 1, find z.
10. If the distance between $(2, y, 3)$ and $4x - 4y + 2z - 5 = 0$ is 3/2, find y.

In Problems 11–18, find the angle described.

11. The angle between $\mathbf{u} = \mathbf{i} + 3\mathbf{j} + 4\mathbf{k}$ and $\mathbf{v} = 3\mathbf{i} - \mathbf{j} + 4\mathbf{k}$.
12. The angle between $\mathbf{u} = \mathbf{i} - 2\mathbf{j} + 6\mathbf{k}$ and $\mathbf{v} = 4\mathbf{i} + 5\mathbf{j}$.
13. The acute angle between $x = 2t$, $y = 3t$, $z = -t$ and $x = 4t$, $y = -t$, $z = 2t$.
14. The obtuse angle between $x = 1 - 2t$, $y = 3 + t$, $z = 4t$, and $x = 2 + t$, $y = 3 - 2t$, $z = 1 + 3t$.
15. The angle between $x = 2 + t$, $y = 3 - 2t$, $z = 2 - t$ and $x = 4 - t$, $y = 2 + t$, $z = 2t$, both directed upward.
16. The angle between $x = 3 - t$, $y = 4 - 2t$, $z = -1 - t$ and $x = t$, $y = -4 + t$, $z = 2 + t$, both directed to the right.
17. The acute angle between $2x + y - z - 1 = 0$ and $x + y - 3z + 4 = 0$.
18. The obtuse angle between $x - y + z - 4 = 0$ and $2x + y + z = 0$.

In Problems 19–28, indicate whether the given lines (planes) are parallel, perpendicular, coincident, or none of these.

19. $x = 2 - t, y = 3 + t, z = -1 - 2t;\quad x = 4 + t, y = 6t, z = -1 + 2t$.
20. $x = 2 + t, y = -1 - 2t, z = 2 + 3t;\quad x = 1 + 2t, y = 4 - 4t, z = -2 + 6t$.
21. $\dfrac{x-2}{3} = \dfrac{y+1}{-2} = \dfrac{z-4}{2};\quad \dfrac{x}{2} = \dfrac{y-1}{4} = \dfrac{z+2}{1}$.
22. $\dfrac{x}{4} = \dfrac{y-1}{3} = \dfrac{z+2}{1};\quad \dfrac{x+1}{-2} = \dfrac{y+2}{1} = \dfrac{z-4}{5}$.
23. $2x + y - 4z + 2 = 0;\quad 6x + 3y + 4 = 0$.
24. $x - y - 5z + 1 = 0;\quad 2x + y - z = 0$.
25. $2x - y - 3z + 1 = 0;\quad 4x - 2y - 6z - 3 = 0$.
26. $3x + y - 4z + 1 = 0;\quad 5x - 3y + 3z - 1 = 0$.
27. $4x - 2y + z - 5 = 0;\quad 5x + 7y - 6z + 2 = 0$.
28. $2x - y + 3z - 1 = 0;\quad 4x - 2y + 6z - 2 = 0$.

11.8

Cylinders and Spheres

We now turn our attention to more complex surfaces, beginning with the cylinder. A cylinder is formed by a line (generatrix) moving along a curve (directrix) while remaining parallel to a fixed line. If the generatrix is parallel to one of the coordinate axes, the equation of the cylinder is quite simple.

Theorem 11.27

An equation of the form

$$f(x, y) = 0$$

is a cylinder with generatrix parallel to the z axis and directrix $f(x, y) = 0$ in the xy plane. Similar statements hold when one of the other variables is absent.

It is a simple matter to see why this is so. If $x = x_0$ and $y = y_0$ satisfies the equation $f(x, y) = 0$, then any point of the form (x_0, y_0, z), for *any* choice of z, is on the surface. But this is a line parallel to the z axis. Thus any point on the curve $f(x, y) = 0$ in the xy plane determines a line parallel to the z axis in space. The result is then a cylinder.

Example 1

Sketch $x^2 + y^2 = 4$.

The surface is a cylinder with generatrix parallel to the z axis and directrix a circle in the xy plane. A portion of the cylinder is given in Figure 11.6.

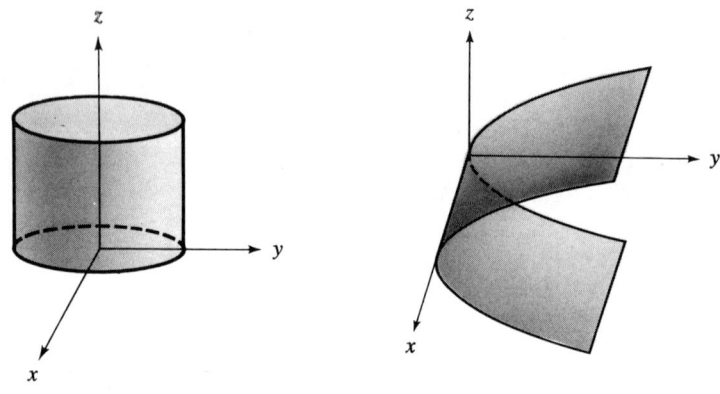

Figure 11.6   Figure 11.7

Example 2

Sketch $y = z^2$.

The surface is a cylinder with generatrix parallel to (or on) the x axis and directrix a parabola in the yz plane. A portion of the cylinder is given in Figure 11.7.

Another relatively simple surface is the sphere. The following theorems concerning the sphere are analogous to those for a circle and are proved in much the same way.

Theorem 11.28

A point (x, y, z) is on the sphere of radius r and center at (h, k, l) if and only if it satisfies the equation

$$(x - h)^2 + (y - k)^2 + (z - l)^2 = r^2.$$

Theorem 11.29

Any sphere can be represented by an equation of the form

$$Ax^2 + Ay^2 + Az^2 + Gx + Hy + Iz + J = 0,$$

where $A \neq 0$.

11.8 Cylinders and Spheres

Theorem 11.30

An equation of the form

$$Ax^2 + Ay^2 + Az^2 + Gx + Hy + Iz + J = 0,$$

where $A \neq 0$, represents a sphere, a point, or no locus.

The proofs are left to the student.

Example 3

Give an equation for the sphere with center $(1, 3, -2)$ and radius 3.

By Theorem 11.28, the equation is

$$(x-1)^2 + (y-3)^2 + (z+2)^2 = 9$$

or

$$x^2 + y^2 + z^2 - 2x - 6y + 4z + 5 = 0.$$

Example 4

Describe the locus of $x^2 + y^2 + z^2 + 2x - 4y - 8z + 5 = 0$.

Let us put the equation into the form of Theorem 11.28 by completing squares.

$$x^2 + 2x \quad + y^2 - 4y \quad + z^2 - 8z \quad = -5,$$
$$x^2 + 2x + 1 + y^2 - 4y + 4 + z^2 - 8z + 16 = -5 + 1 + 4 + 16,$$
$$(x+1)^2 + (y-2)^2 + (z-4)^2 = 16.$$

This represents a sphere with center $(-1, 2, 4)$ and radius 4.

Example 5

Describe the locus of

$$2x^2 + 2y^2 + 2z^2 - 2x + 6y - 4z + 7 = 0.$$

$$x^2 + y^2 + z^2 - x + 3y - 2z + \frac{7}{2} = 0,$$

$$x^2 - x \quad + y^2 + 3y \quad + z^2 - 2z \quad = -\frac{7}{2},$$

$$x^2 - x + \frac{1}{4} + y^2 + 3y + \frac{9}{4} + z^2 - 2z + 1 = -\frac{7}{2} + \frac{1}{4} + \frac{9}{4} + 1,$$

$$\left(x-\frac{1}{2}\right)^2 + \left(y+\frac{3}{2}\right)^2 + (z-1)^2 = 0.$$

The equation represents the point $\left(\frac{1}{2}, -\frac{3}{2}, 1\right)$.

Problems

In Problems 1–10, sketch the given surface.

1. $y^2 + z^2 = 1$.
2. $x^2 + z^2 = 4$.
3. $y = x^2$.
4. $x^2 - z^2 = 1$.
5. $xy = 4$.
6. $x^2 + z^2 + 2x = 0$.
7. $z = 4 - y^2$.
8. $x = \sin z$.
9. $y = \ln x$.
10. $z = e^y$.

In Problems 11–20, identify the equation as representing a sphere, a point, or no locus. If it is a sphere, give its center and radius. If it is a point give its coordinates.

11. $x^2 + y^2 + z^2 - 2x + 4z - 4 = 0$.
12. $x^2 + y^2 + z^2 + 6x - 10y + 2z + 19 = 0$.
13. $x^2 + y^2 + z^2 - 8x + 4y - 10z + 46 = 0$.
14. $x^2 + y^2 + z^2 + 6x - 8y - 2z + 22 = 0$.
15. $2x^2 + 2y^2 + 2z^2 + 2x - 6y + 4z - 1 = 0$.
16. $2x^2 + 2y^2 + 2z^2 - 2x + 2y - 10z + 13 = 0$.
17. $9x^2 + 9y^2 + 9z^2 - 6x + 6y + 12z - 2 = 0$.
18. $3x^2 + 3y^2 + 3z^2 + 4x - 2y - 8z + 7 = 0$.
19. $4x^2 + 4y^2 + 4z^2 - 8x - 4y + 16z + 21 = 0$.
20. $6x^2 + 6y^2 + 6z^2 - 6x - 4y - 3z = 0$.

In Problems 21–28, find an equation(s) in the general form of the sphere(s) described.

21. Center $(4, 1, -2)$ and radius 3.
22. Center $(3, 1, 1)$ and containing the origin.
23. Center $(2, 4, 7)$ and tangent to $4x - 8y + z = 1$.
24. Center $(4, 1, -3)$ and tangent to $2x - y - 2z = 4$.
25. Tangent to $x - 3y + 4z + 23 = 0$ with radius $\sqrt{26}$.
26. Tangent to $x + 2y + 2z - 17 = 0$ with radius 3.
27. Containing $(3, 1, -1)$, $(2, 5, 2)$, $(-3, 0, 1)$, and $(-1, 0, 0)$.
28. Containing $(4, 1, 0)$, $(-2, -1, 0)$, $(0, 2, 1)$, and $(1, 1, 1)$.
29. Prove Theorem 11.28.
30. Prove Theorem 11.29.
31. Prove Theorem 11.30.

11.9

Quadric Surfaces

In the plane, a second-degree equation represents a parabola, ellipse, hyperbola, or a degenerate case of one of them. There are far more variations in space, where we have seen that certain cylinders and the sphere are represented by second-degree equations.

The *ellipsoid* (see Figure 11.8) is represented by an equation of the form

$$\frac{x^2}{a^2} + \frac{y^2}{b^2} + \frac{z^2}{c^2} = 1.$$

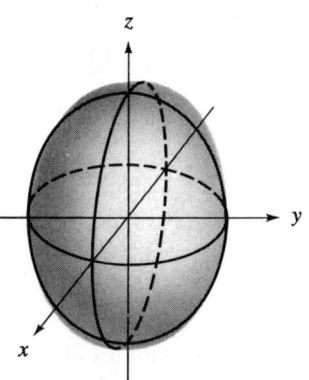

Figure 11.8

Its traces in the coordinate planes (that is, its intersections with these planes) are ellipses (or circles). We have already seen a special case of this, in which $a = b = c$. In that case, we have a sphere. There are two other special cases. One is the *prolate spheroid*. Here, two of the denominators are equal and both are less than the third. It has the shape of a football and may be generated by rotating an ellipse about its major axis. The other case is the *oblate spheroid*, in which two of the denominators are equal and both greater than the third. It has the shape of a doorknob and may be generated by rotating an ellipse about its minor axis.

The *hyperboloid of one sheet* (see Figure 11.9) is represented by an equation of the form

$$\frac{x^2}{a^2} + \frac{y^2}{b^2} - \frac{z^2}{c^2} = 1.$$

Its traces in the xz plane and yz plane are hyperbolas; in the xy plane, it is an ellipse. If $a = b$, it may be generated by rotating a hyperbola about its conjugate axis.

The *hyperboloid of two sheets* (see Figure 11.10) is represented by an equation of the form

$$\frac{x^2}{a^2} - \frac{y^2}{b^2} - \frac{z^2}{c^2} = 1.$$

Its traces in the xy plane and xz plane are hyperbolas. It has no trace in the yz plane; however, if $|x| > a$, its intersection with a plane parallel to the yz plane is an ellipse. If $b = c$, it may be generated by rotating a hyperbola about its transverse axis.

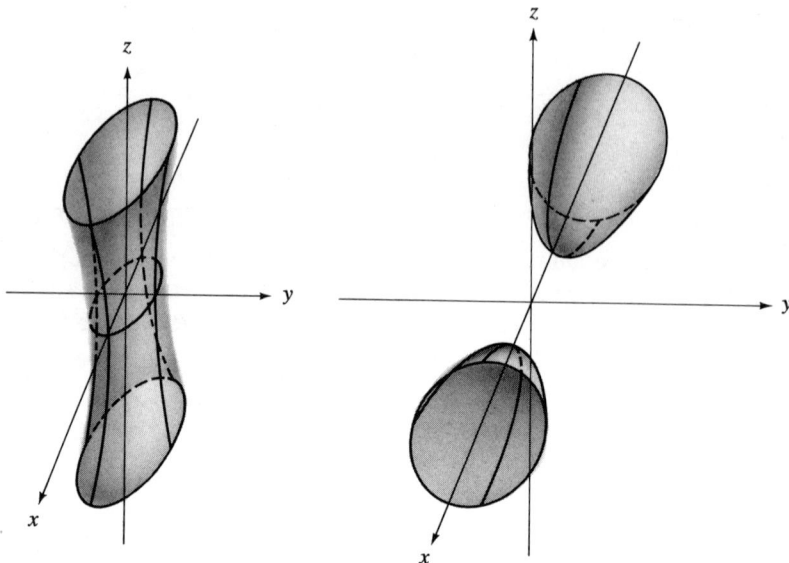

Figure 11.9

Figure 11.10

The *elliptic paraboloid* (see Figure 11.11) is represented by an equation of the form

$$\frac{x^2}{a^2} + \frac{y^2}{b^2} = \frac{z}{c}.$$

Its traces in the xz plane and yz plane are parabolas. Its trace in the xy plane is a single point. If $c > 0$, then its intersection with a plane parallel to and above the xy plane is an ellipse; below the xy plane there is no intersection. This situation is reversed if $c < 0$. In Figure 11 11 $c > 0$. If $a = b$, the elliptic paraboloid is generated by rotating a parabola about its axis.

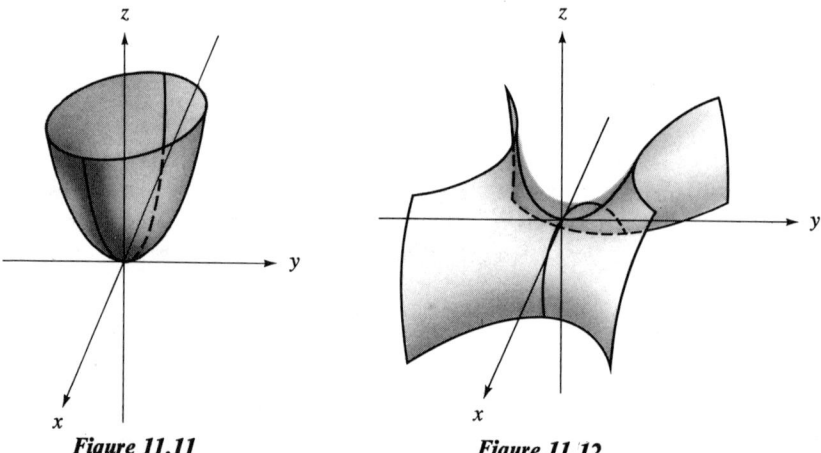

Figure 11.11

Figure 11.12

11.9 Quadric Surfaces

The *hyperbolic paraboloid* (see Figure 11.12) is represented by an equation of the form

$$\frac{x^2}{a^2} - \frac{y^2}{b^2} = \frac{z}{c}.$$

Its traces in the xz plane and yz plane are parabolas, one opening upward and the other down. Its trace in the xy plane is a pair of lines intersecting at the origin (a degenerate hyperbola). Its intersection with a plane parallel to the xy plane is a hyperbola. If $c > 0$, those hyperbolas above the xy plane have the transverse axis parallel to the x axis, while those below have it parallel to the y axis. If $c < 0$, this situation is reversed. In Figure 11.12, $c < 0$.

The *elliptic cone* (see Figure 11.13) is represented by an equation of the form

$$\frac{x^2}{a^2} + \frac{y^2}{b^2} - \frac{z^2}{c^2} = 0.$$

Its trace in the xz plane is a pair of lines intersecting at the origin. Its trace in the yz plane is also a pair of lines intersecting at the origin. Its trace in the xy plane is a single point at the origin. Its intersection with a plane parallel to the xy plane is an ellipse. If $a = b$, it is a circular cone.

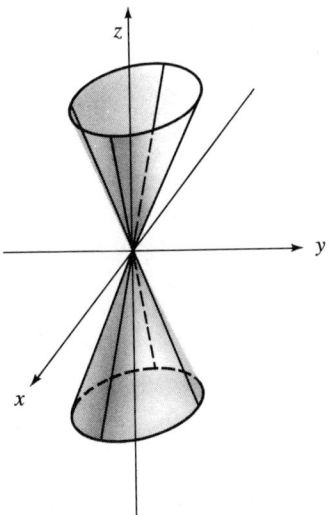

Figure 11.13

Example 1

Describe and sketch

$$9x^2 + 9y^2 - 4z^2 = 36.$$

Dividing by 36, we have a hyperboloid of one sheet in the standard form:

$$\frac{x^2}{4} + \frac{y^2}{4} - \frac{z^2}{9} = 1.$$

Since the denominators of the x^2 and y^2 terms are equal, the trace in the xy plane, as well as in any plane parallel to it, is a circle. The surface is given in Figure 11.14.

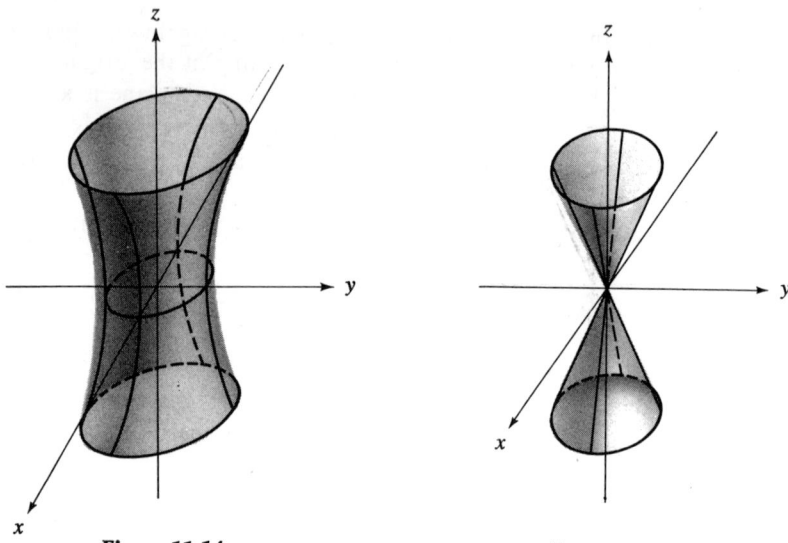

Figure 11.14   Figure 11.15

Example 2

Describe and sketch

$$9x^2 + 9y^2 - 4z^2 = 0.$$

Again dividing by 36, we have

$$\frac{x^2}{4} + \frac{y^2}{4} - \frac{z^2}{9} = 0,$$

which is a circular cone with its axis the z axis. The cone is given in Figure 11.15.

If the given equation does not fit any of these forms, determine its traces in the coordinate planes and planes parallel to them in order to have some idea of the shape.

Problems

Describe and sketch the following quadric surfaces.

1. $36x^2 + 9y^2 + 4z^2 = 36.$
2. $4x^2 + 9y^2 + 9z^2 = 36.$
3. $x - y^2 - z^2 = 0.$
4. $4x^2 + 4y^2 - z^2 + 16 = 0.$

11.10 Cylindrical and Spherical Coordinates

5. $x^2 - y^2 + z^2 = 0$.
6. $x^2 - y - z^2 = 0$.
7. $x^2 + 2y + z^2 = 0$.
8. $x^2 - 4y^2 - 4z^2 = 0$.
9. $25x^2 - 4y^2 + 25z^2 + 100 = 0$.
10. $16x^2 - 9y^2 - 9z^2 + 144 = 0$.
11. $x^2 + 4y - z^2 = 0$.
12. $25x^2 + 16y^2 + 25z^2 = 400$.
13. $36x^2 - 9y^2 + 16z^2 + 144 = 0$.
14. $16x^2 - 9y + 16z^2 = 0$.
15. $16x^2 - 9y^2 - 9z^2 = 0$.
16. $36x^2 - 4y - 9z^2 = 0$.
17. $9x^2 - 36y^2 + 16z^2 + 144 = 0$.
18. $x^2 + y^2 - 4z = 0$.
19. $x^2 + y^2 - 4z^2 = 4$.
20. $x^2 + y^2 + 4z^2 = 4$.
21. $x^2 - y^2 - 9z = 0$.
22. $9x^2 - y^2 - z^2 = 9$.
23. $9x^2 - y^2 - z^2 = 0$.
24. $x^2 + y^2 + 2z = 0$.
25. $25x^2 - 4y^2 + 25z^2 = 100$.

11.10

Cylindrical and Spherical Coordinates

Here we look at two other coordinate systems in space that are useful. The first of these is called a *cylindrical coordinate system*. In this system, a point P with projection Q on the xy plane (see Figure 11.16) is represented by (r, θ, z), where (r, θ) is a polar representation of Q and z is the (directed) distance of P from the xy plane. The relations between rectangular and cylindrical coordinates are the same as the relations between rectangular and polar coordinates; that is,

$$x = r \cos \theta, \quad y = r \sin \theta, \quad z = z$$

or

$$r = \pm\sqrt{x^2 + y^2}, \quad \theta = \arctan \frac{y}{x}, \quad z = z.$$

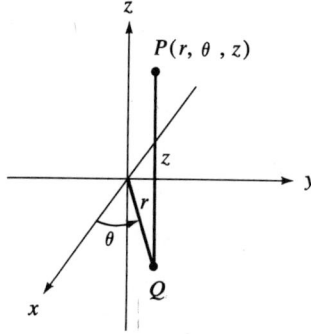

Figure 11.16

Example 1

Express $(4, 30°, -2)$ in rectangular coordinates.

$$x = r\cos\theta = 4\cos 30° = 4 \cdot \frac{\sqrt{3}}{2} = 2\sqrt{3},$$

$$y = r\sin\theta = 4\sin 30° = 4 \cdot \frac{1}{2} = 2,$$

$$z = -2.$$

Thus, the point is $(2\sqrt{3}, 2, -2)$.

Example 2

Express $(2, 2, 4)$ in cylindrical coordinates.

$$r = \pm\sqrt{x^2 + y^2} = \pm\sqrt{8} = \pm 2\sqrt{2},$$

$$\theta = \arctan\frac{y}{x} = \arctan 1 = 45° + 180° \cdot n,$$

$$z = 4.$$

There are two choices for r and infinitely many for θ. The choices we make are not independent of each other—the choice of one of them puts restrictions on the other. If we choose θ to be $45°$ (or any first-quadrant angle), we must choose $r = 2\sqrt{2}$; if we choose θ to be $225°$ (or any third-quadrant angle), we must choose $r = -2\sqrt{2}$. Thus, two possible representations are

$$(2\sqrt{2}, 45°, 4) = (-2\sqrt{2}, 225°, 4).$$

Example 3

Express $x + y + z = 1$ in cylindrical coordinates.

Substituting $x = r\cos\theta$ and $y = r\sin\theta$, we have

$$r\cos\theta + r\sin\theta + z = 1.$$

Example 4

Express $r = z\sin\theta$ in rectangular coordinates.

Multiplying both sides by r, we have

$$r^2 = z \cdot r\sin\theta, \qquad x^2 + y^2 = yz.$$

Another useful system for representing points in space uses *spherical coordinates*. In this system, a point is represented by (ρ, θ, φ), where ρ is the distance of the point

11.10 Cylindrical and Spherical Coordinates

from the origin, θ has the same meaning as in cylindrical coordinates, and φ is the angle between the positive end of the z axis and the segment joining the origin to the given point (see Figure 11.17). Since φ is undirected, it is never negative. Furthermore

$$0° \le \varphi \le 180°.$$

Likewise, we restrict ρ: $\rho \ge 0$.

From Figure 11.17, we see that $OQ = \rho \sin \varphi$. Thus

$$x = OQ \cos \theta = \rho \sin \varphi \cos \theta,$$
$$y = OQ \sin \theta = \rho \sin \varphi \sin \theta,$$
$$z = \rho \cos \varphi.$$

We can easily see from these that

$$\rho^2 = x^2 + y^2 + z^2.$$

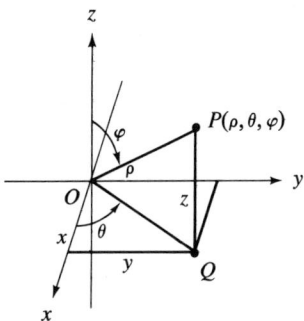

Figure 11.17

Example 5

Express $(2, 30°, 120°)$ in rectangular coordinates.

$$x = \rho \sin \varphi \cos \theta$$
$$= 2 \sin 120° \cdot \cos 30°$$
$$= 2 \cdot \frac{\sqrt{3}}{2} \cdot \frac{\sqrt{3}}{2} = \frac{3}{2},$$

$$y = \rho \sin \varphi \sin \theta$$
$$= 2 \sin 120° \sin 30°$$
$$= 2 \cdot \frac{\sqrt{3}}{2} \cdot \frac{1}{2} = \frac{\sqrt{3}}{2},$$

$$z = \rho \cos \varphi$$
$$= 2 \cos 120°$$
$$= 2\left(-\frac{1}{2}\right) = -1.$$

Thus the point is $(3/2, \sqrt{3}/2, -1)$.

Example 6

Express $(\sqrt{3}, \sqrt{3}, -\sqrt{2})$ in spherical coordinates.

$$\rho^2 = x^2 + y^2 + z^2$$
$$= 3 + 3 + 2 = 8,$$
$$\rho = 2\sqrt{2};$$

$$z = \rho \cos \varphi,$$
$$-\sqrt{2} = 2\sqrt{2} \cos \varphi,$$
$$\cos \varphi = -\frac{1}{2},$$
$$\varphi = 120°;$$
$$x = \rho \sin \varphi \cos \theta,$$
$$\sqrt{3} = 2\sqrt{2} \cdot \frac{\sqrt{3}}{2} \cos \theta,$$
$$\cos \theta = \frac{1}{\sqrt{2}},$$
$$\theta = 45°.$$

The point is $(2\sqrt{2}, 45°, 120°)$.

Example 7

Express $x^2 + y^2 - z^2 = 0$ in spherical coordinates.

$$\rho^2 \sin^2 \varphi \cos^2 \theta + \rho^2 \sin^2 \varphi \sin^2 \theta - \rho^2 \cos^2 \varphi = 0,$$
$$\rho^2 [\sin^2 \varphi (\cos^2 \theta + \sin^2 \theta) - \cos^2 \varphi] = 0,$$
$$\rho^2 (\sin^2 \varphi - \cos^2 \varphi) = 0,$$
$$\sin^2 \varphi - \cos^2 \varphi = 0 \quad \text{or} \quad \rho = 0;$$
$$\cos 2\varphi = 0,$$
$$2\varphi = 90°, 270°$$
$$\varphi = 45°, 135°.$$

Thus, we have $\varphi = 45°$, $\varphi = 135°$, and $\rho = 0$. $\varphi = 45°$ gives the top half of the cone, $\varphi = 135°$ gives the bottom half, and $\rho = 0$ gives a single point—the origin. Since $\rho = 0$ is included in both of the others, we may drop it. The final result is

$$\varphi = 45° \quad \text{and} \quad \varphi = 135°.$$

Example 8

Express $\rho^2 \sin \varphi \cos \varphi \cos \theta = 1$ in rectangular coordinates.

$$\rho^2 \sin \varphi \cos \varphi \cos \theta = 1,$$
$$(\rho \sin \varphi \cos \theta)(\rho \cos \varphi) = 1,$$
$$xz = 1.$$

Problems

1. The following points are given in cylindrical coordinates. Express them in rectangular coordinates.

11.10 Cylindrical and Spherical Coordinates

   a. $(2, 45°, 1)$,   b. $(3, 2\pi/3, -2)$,
   c. $(1, 0°, 2)$,   d. $(0, \pi/4, -3)$.

2. The following points are given in rectangular coordinates. Express them in cylindrical coordinates.

   a. $(1, 1, 3)$,   b. $(0, 2, -2)$,
   c. $(-1, \sqrt{3}, 4)$, d. $(-2\sqrt{3}, -2, 3)$.

3. The following points are given in spherical coordinates. Express them in rectangular coordinates.

   a. $(3, 45°, 30°)$,   b. $(1, \pi/6, 0)$,
   c. $(1, 90°, 45°)$,   d. $(2, 5\pi/6, 3\pi/4)$.

4. The following points are given in rectangular coordinates. Express them in spherical coordinates.

   a. $(2, 2, 0)$,   b. $(2, 1, -2)$,
   c. $(2, -\sqrt{3}, 4)$, d. $(1, 1, \sqrt{2})$.

5. The following points are given in cylindrical coordinates. Express them in spherical coordinates.

   a. $(3, 30°, 4)$,   b. $(2, \pi/4, -2)$,
   c. $(0, 45°, 3)$,   d. $(2, \pi/2, -4)$.

6. The following points are given in spherical coordinates. Express them in cylindrical coordinates.

   a. $(4, 45°, 30°)$,   b. $(2, 2\pi/3, \pi/2)$,
   c. $(2, 210°, 135°)$, d. $(3, \pi/6, 2\pi/3)$.

In Problems 7–14, express the given equations in cylindrical and spherical coordinates.

7. $x^2 + y^2 = 4$.
8. $x^2 + y^2 + z^2 = 4$.
9. $x^2 + y^2 = z$.
10. $x^2 - y^2 = z$.
11. $x^2 - y^2 - z^2 = 1$.
12. $x^2 - y^2 + z^2 = 1$.
13. $x^2 + y^2 - z^2 = 1$.
14. $4x^2 + 9y^2 + 9z^2 = 1$.

In Problems 15–22, express the given equations in rectangular coordinates.

15. $z = r^2 \sin 2\theta$.
16. $z = r^2 \cos 2\theta$.
17. $z = r^2$.
18. $z = 1 + \sin \theta$.
19. $\rho \sin \varphi \tan \varphi \sin 2\theta = 2$.
20. $\rho \sin \varphi = 1$.
21. $\rho = \sin \varphi \cos \theta$.
22. $\rho^2 \sin \varphi \cos \varphi = 1$.

Table 1 Trigonometric Functions

Degrees	Radians	Sin	Cos	Tan	Cot		
0	0.0000	0.0000	1.0000	0.0000		1.5708	90
1	0.0175	0.0175	0.9998	0.0175	57.290	1.5533	89
2	0.0349	0.0349	0.9994	0.0349	28.636	1.5359	88
3	0.0524	0.0523	0.9986	0.0524	19.081	1.5184	87
4	0.0698	0.0698	0.9976	0.0699	14.301	1.5010	86
5	0.0873	0.0872	0.9962	0.0875	11.430	1.4835	85
6	0.1047	0.1045	0.9945	0.1051	9.5144	1.4661	84
7	0.1222	0.1219	0.9925	0.1228	8.1443	1.4486	83
8	0.1396	0.1392	0.9903	0.1405	7.1154	1.4312	82
9	0.1571	0.1564	0.9877	0.1584	6.3138	1.4137	81
10	0.1745	0.1736	0.9848	0.1763	5.6713	1.3963	80
11	0.1920	0.1908	0.9816	0.1944	5.1446	1.3788	79
12	0.2094	0.2079	0.9781	0.2126	4.7046	1.3614	78
13	0.2269	0.2250	0.9744	0.2309	4.3315	1.3439	77
14	0.2443	0.2419	0.9703	0.2493	4.0108	1.3265	76
15	0.2618	0.2588	0.9659	0.2679	3.7321	1.3090	75
16	0.2793	0.2756	0.9613	0.2867	3.4874	1.2915	74
17	0.2967	0.2924	0.9563	0.3057	3.2709	1.2741	73
18	0.3142	0.3090	0.9511	0.3249	3.0777	1.2566	72
19	0.3316	0.3256	0.9455	0.3443	2.9042	1.2392	71
20	0.3491	0.3420	0.9397	0.3640	2.7475	1.2217	70
21	0.3665	0.3584	0.9336	0.3839	2.6051	1.2043	69
22	0.3840	0.3746	0.9272	0.4040	2.4751	1.1868	68
23	0.4014	0.3907	0.9205	0.4245	2.3559	1.1694	67
24	0.4189	0.4067	0.9135	0.4452	2.2460	1.1519	66
25	0.4363	0.4226	0.9063	0.4663	2.1445	1.1345	65
26	0.4538	0.4384	0.8988	0.4877	2.0503	1.1170	64
27	0.4712	0.4540	0.8910	0.5095	1.9626	1.0996	63
28	0.4887	0.4695	0.8829	0.5317	1.8807	1.0821	62
29	0.5061	0.4848	0.8746	0.5543	1.8040	1.0647	61
30	0.5236	0.5000	0.8660	0.5774	1.7321	1.0472	60
31	0.5411	0.5150	0.8572	0.6009	1.6643	1.0297	59
32	0.5585	0.5299	0.8480	0.6249	1.6003	1.0123	58
33	0.5760	0.5446	0.8387	0.6494	1.5399	0.9948	57
34	0.5934	0.5592	0.8290	0.6745	1.4826	0.9774	56
35	0.6109	0.5736	0.8192	0.7002	1.4281	0.9599	55
36	0.6283	0.5878	0.8090	0.7265	1.3764	0.9425	54
37	0.6458	0.6018	0.7986	0.7536	1.3270	0.9250	53
38	0.6632	0.6157	0.7880	0.7813	1.2799	0.9076	52
39	0.6807	0.6293	0.7771	0.8098	1.2349	0.8901	51
40	0.6981	0.6428	0.7660	0.8391	1.1918	0.8727	50
41	0.7156	0.6561	0.7547	0.8693	1.1504	0.8552	49
42	0.7330	0.6691	0.7431	0.9004	1.1106	0.8378	48
43	0.7505	0.6820	0.7314	0.9325	1.0724	0.8203	47
44	0.7679	0.6947	0.7193	0.9657	1.0355	0.8029	46
45	0.7854	0.7071	0.7071	1.0000	1.0000	0.7854	45
		Cos	Sin	Cot	Tan	Radians	Degrees

Table 2   Common Logarithms

x	0	1	2	3	4	5	6	7	8	9
1.0	.0000	.0043	.0086	.0128	.0170	.0212	.0253	.0294	.0334	.0374
1.1	.0414	.0453	.0492	.0531	.0569	.0607	.0645	.0682	.0719	.0755
1.2	.0792	.0828	.0864	.0899	.0934	.0969	.1004	.1038	.1072	.1106
1.3	.1139	.1173	.1206	.1239	.1271	.1303	.1335	.1367	.1399	.1430
1.4	.1461	.1492	.1523	.1553	.1584	.1614	.1644	.1673	.1703	.1732
1.5	.1761	.1790	.1818	.1847	.1875	.1903	.1931	.1959	.1987	.2014
1.6	.2041	.2068	.2095	.2122	.2148	.2175	.2201	.2227	.2253	.2279
1.7	.2304	.2330	.2355	.2380	.2405	.2430	.2455	.2480	.2504	.2529
1.8	.2553	.2577	.2601	.2625	.2648	.2672	.2695	.2718	.2742	.2765
1.9	.2788	.2810	.2833	.2856	.2878	.2900	.2923	.2945	.2967	.2989
2.0	.3010	.3032	.3054	.3075	.3096	.3118	.3139	.3160	.3181	.3201
2.1	.3222	.3243	.3263	.3284	.3304	.3324	.3345	.3365	.3385	.3404
2.2	.3424	.3444	.3464	.3483	.3502	.3522	.3541	.3560	.3579	.3598
2.3	.3617	.3636	.3655	.3674	.3692	.3711	.3729	.3747	.3766	.3784
2.4	.3802	.3820	.3838	.3856	.3874	.3892	.3909	.3927	.3945	.3962
2.5	.3979	.3997	.4014	.4031	.4048	.4065	.4082	.4099	.4116	.4133
2.6	.4150	.4166	.4183	.4200	.4216	.4232	.4249	.4265	.4281	.4298
2.7	.4314	.4330	.4346	.4362	.4378	.4393	.4409	.4425	.4440	.4456
2.8	.4472	.4487	.4502	.4518	.4533	.4548	.4564	.4579	.4594	.4609
2.9	.4624	.4639	.4654	.4669	.4683	.4698	.4713	.4728	.4742	.4757
3.0	.4771	.4786	.4800	.4814	.4829	.4843	.4857	.4871	.4886	.4900
3.1	.4914	.4928	.4942	.4955	.4969	.4983	.4997	.5011	.5024	.5038
3.2	.5051	.5065	.5079	.5092	.5105	.5119	.5132	.5145	.5159	.5172
3.3	.5185	.5198	.5211	.5224	.5237	.5250	.5263	.5276	.5289	.5302
3.4	.5315	.5328	.5340	.5353	.5366	.5378	.5391	.5403	.5416	.5428
3.5	.5441	.5453	.5465	.5478	.5490	.5502	.5514	.5527	.5539	.5551
3.6	.5563	.5575	.5587	.5599	.5611	.5623	.5635	.5647	.5658	.5670
3.7	.5682	.5694	.5705	.5717	.5729	.5740	.5752	.5763	.5775	.5786
3.8	.5798	.5809	.5821	.5832	.5843	.5855	.5866	.5877	.5888	.5899
3.9	.5911	.5922	.5933	.5944	.5955	.5966	.5977	.5988	.5999	.6010
4.0	.6021	.6031	.6042	.6053	.6064	.6075	.6085	.6096	.6107	.6117
4.1	.6128	.6138	.6149	.6160	.6170	.6180	.6191	.6201	.6212	.6222
4.2	.6232	.6243	.6253	.6263	.6274	.6284	.6294	.6304	.6314	.6325
4.3	.6335	.6345	.6355	.6365	.6375	.6385	.6395	.6405	.6415	.6425
4.4	.6435	.6444	.6454	.6464	.6474	.6484	.6493	.6503	.6513	.6522
4.5	.6532	.6542	.6551	.6561	.6571	.6580	.6590	.6599	.6609	.6618
4.6	.6628	.6637	.6646	.6656	.6665	.6675	.6684	.6693	.6702	.6712
4.7	.6721	.6730	.6739	.6749	.6758	.6767	.6776	.6785	.6794	.6803
4.8	.6812	.6821	.6830	.6839	.6848	.6857	.6866	.6875	.6884	.6893
4.9	.6902	.6911	.6920	.6928	.6937	.6946	.6955	.6964	.6972	.6981
5.0	.6990	.6998	.7007	.7016	.7024	.7033	.7042	.7050	.7059	.7067
5.1	.7076	.7084	.7093	.7101	.7110	.7118	.7126	.7135	.7143	.7152
5.2	.7160	.7168	.7177	.7185	.7193	.7202	.7210	.7218	.7226	.7235
5.3	.7243	.7251	.7259	.7267	.7275	.7284	.7292	.7300	.7308	.7316
5.4	.7324	.7332	.7340	.7348	.7356	.7364	.7372	.7380	.7388	.7396
x	0	1	2	3	4	5	6	7	8	9

Table 2 (*Continued*)

x	0	1	2	3	4	5	6	7	8	9
5.5	.7404	.7412	.7419	.7427	.7435	.7443	.7451	.7459	.7466	.7474
5.6	.7482	.7490	.7497	.7505	.7513	.7520	.7528	.7536	.7543	.7551
5.7	.7559	.7566	.7574	.7582	.7589	.7597	.7604	.7612	.7619	.7627
5.8	.7634	.7642	.7649	.7657	.7664	.7672	.7679	.7686	.7694	.7701
5.9	.7709	.7716	.7723	.7731	.7738	.7745	.7752	.7760	.7767	.7774
6.0	.7782	.7789	.7796	.7803	.7810	.7818	.7825	.7832	.7839	.7846
6.1	.7853	.7860	.7868	.7875	.7882	.7889	.7896	.7903	.7910	.7917
6.2	.7924	.7931	.7938	.7945	.7952	.7959	.7966	.7973	.7980	.7987
6.3	.7993	.8000	.8007	.8014	.8021	.8028	.8035	.8041	.8048	.8055
6.4	.8062	.8069	.8075	.8082	.8089	.8096	.8102	.8109	.8116	.8122
6.5	.8129	.8136	.8142	.8149	.8156	.8162	.8169	.8176	.8182	.8189
6.6	.8195	.8202	.8209	.8215	.8222	.8228	.8235	.8241	.8248	.8254
6.7	.8261	.8267	.8274	.8280	.8287	.8293	.8299	.8306	.8312	.8319
6.8	.8325	.8331	.8338	.8344	.8351	.8357	.8363	.8370	.8376	.8382
6.9	.8388	.8395	.8401	.8407	.8414	.8420	.8426	.8432	.8439	.8445
7.0	.8451	.8457	.8463	.8470	.8476	.8482	.8488	.8494	.8500	.8506
7.1	.8513	.8519	.8525	.8531	.8537	.8543	.8549	.8555	.8561	.8567
7.2	.8573	.8579	.8585	.8591	.8597	.8603	.8609	.8615	.8621	.8627
7.3	.8633	.8639	.8645	.8651	.8657	.8663	.8669	.8675	.8681	.8686
7.4	.8692	.8698	.8704	.8710	.8716	.8722	.8727	.8733	.8739	.8745
7.5	.8751	.8756	.8762	.8768	.8774	.8779	.8785	.8791	.8797	.8802
7.6	.8808	.8814	.8820	.8825	.8831	.8837	.8842	.8848	.8854	.8859
7.7	.8865	.8871	.8876	.8882	.8887	.8893	.8899	.8904	.8910	.8915
7.8	.8921	.8927	.8932	.8938	.8943	.8949	.8954	.8960	.8965	.8971
7.9	.8976	.8982	.8987	.8993	.8998	.9004	.9009	.9015	.9020	.9025
8.0	.9031	.9036	.9042	.9047	.9053	.9058	.9063	.9069	.9074	.9079
8.1	.9085	.9090	.9096	.9101	.9106	.9112	.9117	.9122	.9128	.9133
8.2	.9138	.9143	.9149	.9154	.9159	.9165	.9170	.9175	.9180	.9186
8.3	.9191	.9196	.9201	.9206	.9212	.9217	.9222	.9227	.9232	.9238
8.4	.9243	.9248	.9253	.9258	.9263	.9269	.9274	.9279	.9284	.9289
8.5	.9294	.9299	.9304	.9309	.9315	.9320	.9325	.9330	.9335	.9340
8.6	.9345	.9350	.9355	.9360	.9365	.9370	.9375	.9380	.9385	.9390
8.7	.9395	.9400	.9405	.9410	.9415	.9420	.9425	.9430	.9435	.9440
8.8	.9445	.9450	.9455	.9460	.9465	.9469	.9474	.9479	.9484	.9489
8.9	.9494	.9499	.9504	.9509	.9513	.9518	.9523	.9528	.9533	.9538
9.0	.9542	.9547	.9552	.9557	.9562	.9566	.9571	.9576	.9581	.9586
9.1	.9590	.9595	.9600	.9605	.9609	.9614	.9619	.9624	.9628	.9633
9.2	.9638	.9643	.9647	.9652	.9657	.9661	.9666	.9671	.9675	.9680
9.3	.9685	.9689	.9694	.9699	.9703	.9708	.9713	.9717	.9722	.9727
9.4	.9731	.9736	.9741	.9745	.9750	.9754	.9759	.9763	.9768	.9773
9.5	.9777	.9782	.9786	.9791	.9795	.9800	.9805	.9809	.9814	.9818
9.6	.9823	.9827	.9832	.9836	.9841	.9845	.9850	.9854	.9859	.9863
9.7	.9868	.9872	.9877	.9881	.9886	.9890	.9894	.9899	.9903	.9908
9.8	.9912	.9917	.9921	.9926	.9930	.9934	.9939	.9943	.9948	.9952
9.9	.9956	.9961	.9965	.9969	.9974	.9978	.9983	.9987	.9991	.9996
x	0	1	2	3	4	5	6	7	8	9

Table 3 Squares, Square Roots, and Prime Factors

No.	Sq.	Sq. Rt.	Factors	No.	Sq.	Sq. Rt.	Factors
1	1	1.000		51	2,601	7.141	$3 \cdot 17$
2	4	1.414	2	52	2,704	7.211	$2^2 \cdot 13$
3	9	1.732	3	53	2,809	7.280	53
4	16	2.000	2^2	54	2,916	7.348	$2 \cdot 3^3$
5	25	2.236	5	55	3,025	7.416	$5 \cdot 11$
6	36	2.449	$2 \cdot 3$	56	3,136	7.483	$2^3 \cdot 7$
7	49	2.646	7	57	3,249	7.550	$3 \cdot 19$
8	64	2.828	2^3	58	3,364	7.616	$2 \cdot 29$
9	81	3.000	3^2	59	3,481	7.681	59
10	100	3.162	$2 \cdot 5$	60	3,600	7.746	$2^2 \cdot 3 \cdot 5$
11	121	3.317	11	61	3,721	7.810	61
12	144	3.464	$2^2 \cdot 3$	62	3,844	7.874	$2 \cdot 31$
13	169	3.606	13	63	3,969	7.937	$3^2 \cdot 7$
14	196	3.742	$2 \cdot 7$	64	4,096	8.000	2^6
15	225	3.873	$3 \cdot 5$	65	4,225	8.062	$5 \cdot 13$
16	256	4.000	2^4	66	4,356	8.124	$2 \cdot 3 \cdot 11$
17	289	4.123	17	67	4,489	8.185	67
18	324	4.243	$2 \cdot 3^2$	68	4,624	8.246	$2^2 \cdot 17$
19	361	4.359	19	69	4,761	8.307	$3 \cdot 23$
20	400	4.472	$2^2 \cdot 5$	70	4,900	8.367	$2 \cdot 5 \cdot 7$
21	441	4.583	$3 \cdot 7$	71	5,041	8.426	71
22	484	4.690	$2 \cdot 11$	72	5,184	8.485	$2^3 \cdot 3^2$
23	529	4.796	23	73	5,329	8.544	73
24	576	4.899	$2^3 \cdot 3$	74	5,476	8.602	$2 \cdot 37$
25	625	5.000	5^2	75	5,625	8.660	$3 \cdot 5^2$
26	676	5.099	$2 \cdot 13$	76	5,776	8.718	$2^2 \cdot 19$
27	729	5.196	3^3	77	5,929	8.775	$7 \cdot 11$
28	784	5.292	$2^2 \cdot 7$	78	6,084	8.832	$2 \cdot 3 \cdot 13$
29	841	5.385	29	79	6,241	8.888	79
30	900	5.477	$2 \cdot 3 \cdot 5$	80	6,400	8.944	$2^4 \cdot 5$
31	961	5.568	31	81	6,561	9.000	3^4
32	1,024	5.657	2^5	82	6,724	9.055	$2 \cdot 41$
33	1,089	5.745	$3 \cdot 11$	83	6,889	9.110	83
34	1,156	5.831	$2 \cdot 17$	84	7,056	9.165	$2^2 \cdot 3 \cdot 7$
35	1,225	5.916	$5 \cdot 7$	85	7,225	9.220	$5 \cdot 17$
36	1,296	6.000	$2^2 \cdot 3^2$	86	7,396	9.274	$2 \cdot 43$
37	1,369	6.083	37	87	7,569	9.327	$3 \cdot 29$
38	1,444	6.164	$2 \cdot 19$	88	7,744	9.381	$2^3 \cdot 11$
39	1,521	6.245	$3 \cdot 13$	89	7,921	9.434	89
40	1,600	6.325	$2^3 \cdot 5$	90	8,100	9.487	$2 \cdot 3^2 \cdot 5$
41	1,681	6.403	41	91	8,281	9.539	$7 \cdot 13$
42	1,764	6.481	$2 \cdot 3 \cdot 7$	92	8,464	9.592	$2^2 \cdot 23$
43	1,849	6.557	43	93	8,649	9.644	$3 \cdot 31$
44	1,936	6.633	$2^2 \cdot 11$	94	8,836	9.695	$2 \cdot 47$
45	2,025	6.708	$3^2 \cdot 5$	95	9,025	9.747	$5 \cdot 19$
46	2,116	6.782	$2 \cdot 23$	96	9,216	9.798	$2^5 \cdot 3$
47	2,209	6.856	47	97	9,409	9.849	97
48	2,304	6.928	$2^4 \cdot 3$	98	9,604	9.899	$2 \cdot 7^2$
49	2,401	7.000	7^2	99	9,801	9.950	$3^2 \cdot 11$
50	2,500	7.071	$2 \cdot 5^2$	100	10,000	10.000	$2^2 \cdot 5^2$

Answers to Selected Problems

Chapter 1

Section 1.2, page 6

1. $\sqrt{65}$. 3. 2. 5. $\sqrt{37}/2$. 7. $\sqrt{6}$. 9. $-3, 5$. 11. 2, 3.
13. Collinear. 15. Not collinear. 17. Collinear. 19. Right triangle.
21. Not a right triangle.
27. $(0, -1)$ inside; $(1, 7), (2, 0), (-5, 7), (-5, -1), (-6, 6)$ on; $(-3, 8), (4, 2)$ outside.

Section 1.3, page 11

1. $(4, 3)$. 3. $(16/5, -9/5)$. 5. $(17, 22)$. 7. $(2, 1)$. 9. $(7/2, 1)$.
11. $(12, -4)$. 13. $(0, -7)$. 15. $(3, 0)$. 17. $(14/5, 7/5)$. 19. $(6, -7)$.
21. $(2, -1/2)$. 23. $7, -9$.

Section 1.5, page 15

1. $5/3, 59°$. 3. $2/3, 34°$. 5. No slope, $90°$. 7. $1, 45°$. 9. Parallel.
11. None. 13. Coincident. 15. Perpendicular. 17. $14/3$. 19. $10/7, 11$.
21. 9.

Section 1.6, page 18

1. $135°$. 3. $135°$. 5. $6°$. 7. $12°$. 9. $60°$. 11. $112°$.
13. $-7 - 5\sqrt{2}$. 15. $1 + \sqrt{2}$. 17. $(128 + \sqrt{17{,}170})/23$. 19. $37°, 72°, 72°$.
21. $1/5$.

Section 1.8, page 24

1. 3. 5.

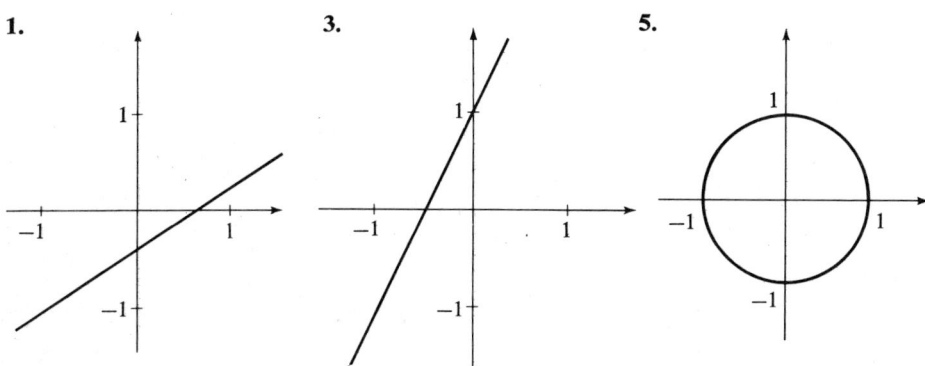

242 Answers

7.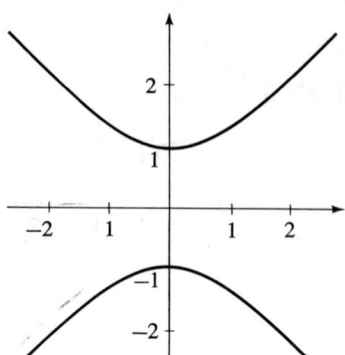

9.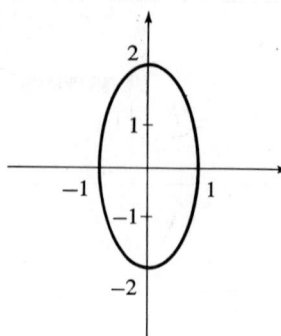

11.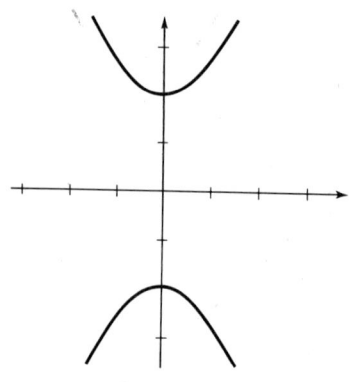

13.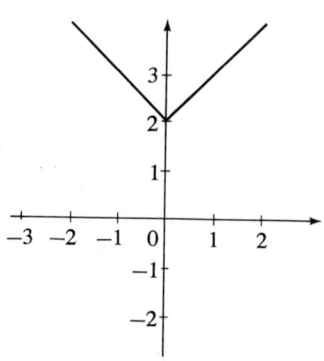

15.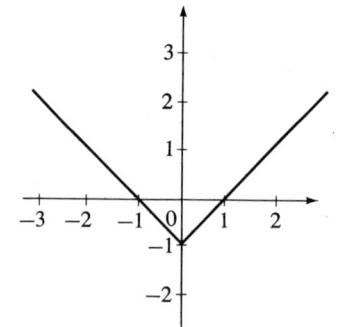

17.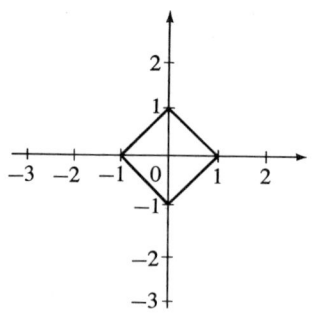

Answers 243

19.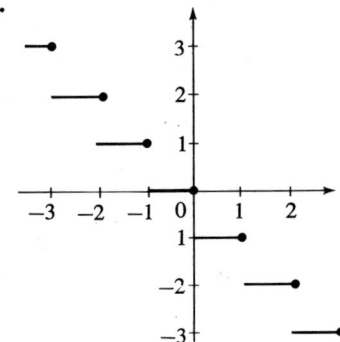

21.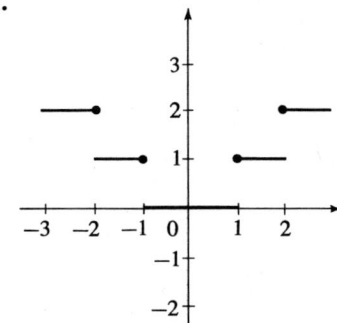

23.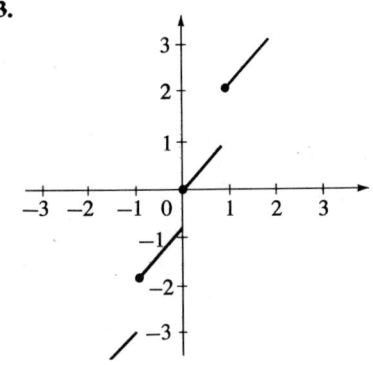

25.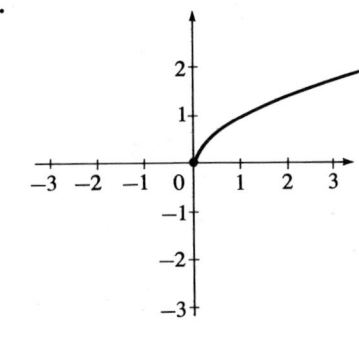

27.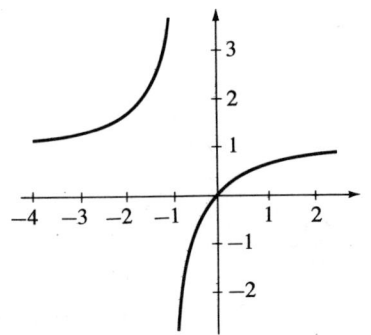

29.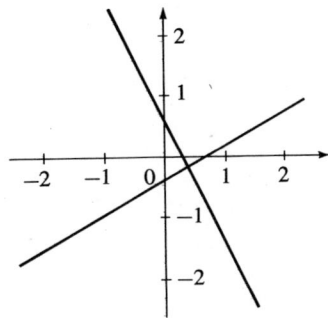

31.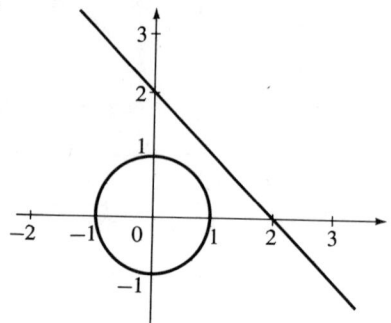

33.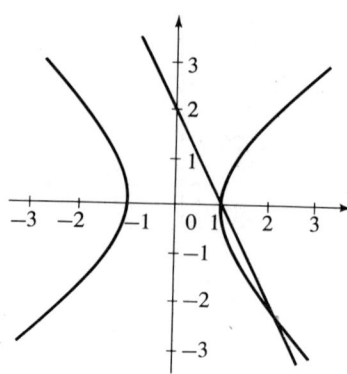

35.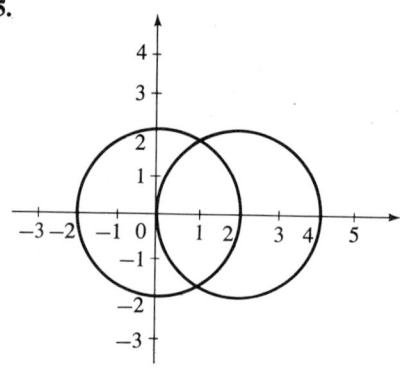

37.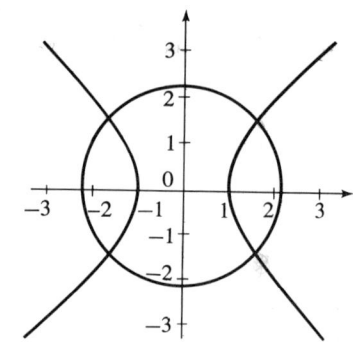

39.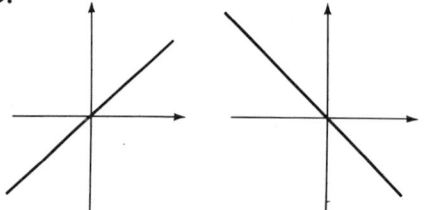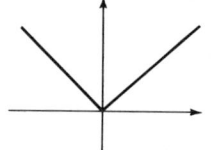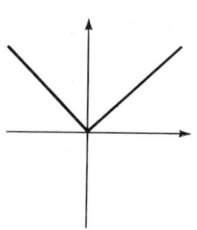

41. 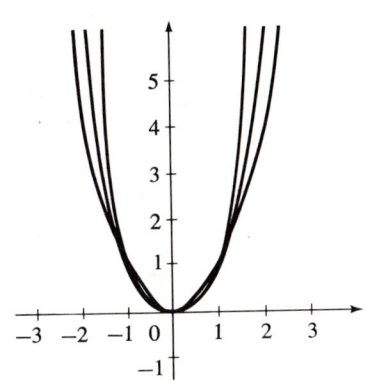   **43.**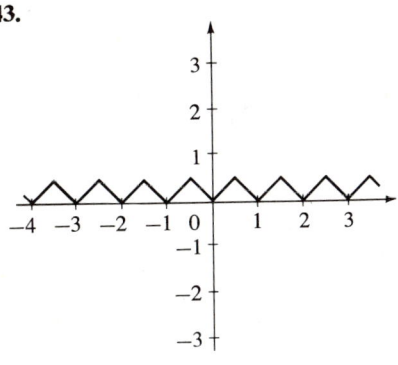

Section 1.9, page 27

1. $14x + 8y - 69 = 0$.   **3.** $2x - y - 8 = 0$.   **5.** $5x - 2y - 19 = 0$.
7. $x^2 + y^2 - 10x - 16y + 80 = 0$.   **9.** $x^2 + y^2 - 9x + 18 = 0$.
11. $x^2 + y^2 - x - 9y + 18 = 0$.   **13.** $y^2 - 8x + 16 = 0$.   **15.** $3x^2 - y^2 + 6x - 9 = 0$.
17. $25x^2 + 9y^2 = 225$.   **19.** $x^2 + y^2 = 16$.   **21.** $8x^2 - y^2 = 8$.   **23.** $xy = \pm 4$.

Chapter 2

Section 2.1, page 34

1. $-6\mathbf{i} - 2\mathbf{j}$.   **3.** $7\mathbf{i} + \mathbf{j}$.   **5.** $-\mathbf{i} + 5\mathbf{j}$.   **7.** $4\mathbf{i} - 3\mathbf{j}$.   **9.** $(1/\sqrt{10})\mathbf{i} - (1/\sqrt{10})\mathbf{j}$.
11. $-(1/\sqrt{5})\mathbf{i} + (2/\sqrt{5})\mathbf{j}$.   **13.** $(1/\sqrt{5})\mathbf{i} + (2/\sqrt{5})\mathbf{j}$.   **15.** $-(1/\sqrt{5})\mathbf{i} + (2/\sqrt{5})\mathbf{j}$.
17. $B = (4, 3)$.   **19.** $A = (5, 0)$.   **21.** $A = (5/2, -3/2)$, $B = (11/2, 7/2)$.
23. $B = (6, 0)$.   **25.** $A = (1, 3)$.   **27.** $A = (3/2, -1/2)$, $B = (5/2, 1/2)$.
29. $4\mathbf{i} + \mathbf{j}$.   **31.** $-\mathbf{i} - 3\mathbf{j}$.

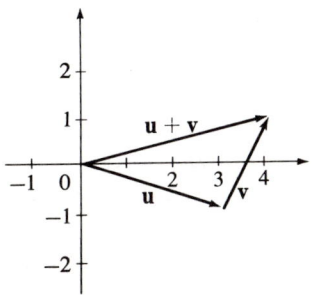   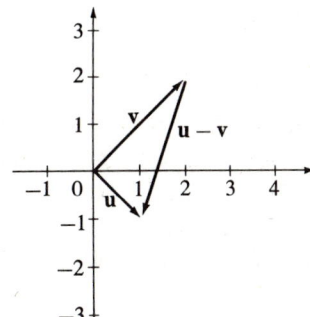

246 Answers

33. $4i - 2j$. **35.** $5i - 3j$, $i + j$. **37.** $14i + 11j$.

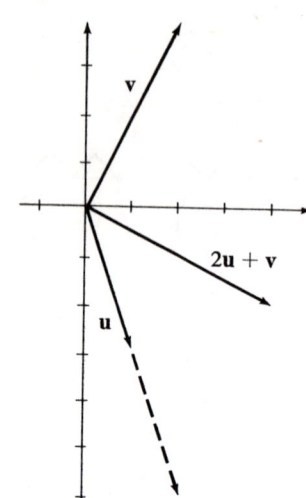

Section 2.2, page 40

1. $150°$, $-1/\sqrt{3}$. **3.** $120°$, $-\sqrt{3}$. **5.** $45°$, 1. **7.** $3/4$, $37°$. **9.** $5/2$, $68°$.
11. -3, $108°$. **13.** $\dfrac{2\sqrt{2}}{\sqrt{5}}i + \dfrac{6\sqrt{2}}{\sqrt{5}}j$. **15.** $-3i$. **17.** $-\dfrac{5}{\sqrt{37}}i - \dfrac{30}{\sqrt{37}}j$.
19. $5\sqrt{5}$ lb to the right and inclined upward at an angle of $63°$ with the horizontal.
21. $\sqrt{25 - 12\sqrt{2}}$ lb to the right and inclined downward at an angle of $3°$ with the horizontal.
23. $6i + 2j$.
25. $\sqrt{29}$ lb to the right and inclined downward at an angle of $68°$ with the horizontal.
27. $\sqrt{34 + 15\sqrt{2}}$ lb to the left and inclined downward at an angle of $28°$ with the horizontal.
29. j.

Section 2.3, page 46

1. Arccos $1/5\sqrt{2} = 82°$. **3.** $90°$. **5.** $90°$.
7. Arccos $3/\sqrt{34} = 59°$. **9.** 1, not orthogonal. **11.** 4, not orthogonal.
13. -1, not orthogonal. **15.** 3, not orthogonal. **17.** $(1/2)i + (1/2)j$.
19. $-(6/5)i + (12/5)j$. **21.** $(2/5)i + (1/5)j$. **23.** $(6/5)i - (3/5)j$. **25** 3.
27. $-1/2$. **29.** $4 \pm 2\sqrt{3}$. **31.** -8. **33.** -1. **35.** $(8 + 5\sqrt{3})/11$.
37. $-(2/17)i - (8/17)j$, $-(19/17)i - (76/17)j$. **39.** $-(2/5)i - (4/5)j$, $(3/5)i + (6/5)j$.

Answers 247

Chapter 3

Section 3.1, page 50

1.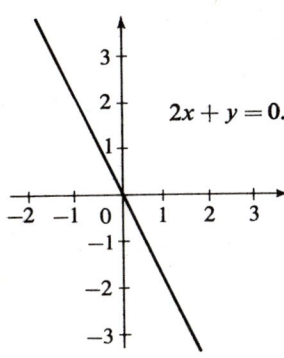

3.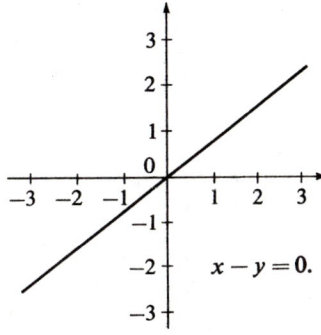

5.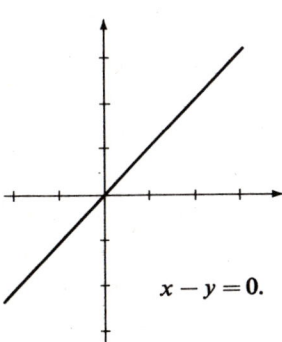

7.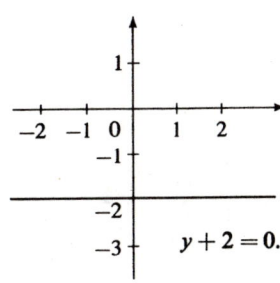

9.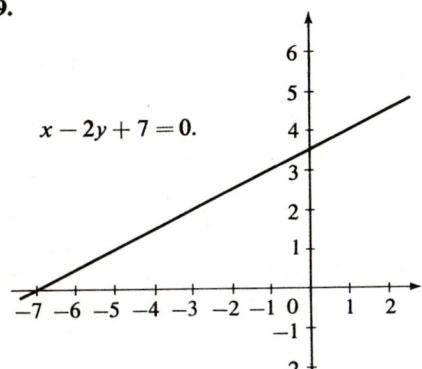

11.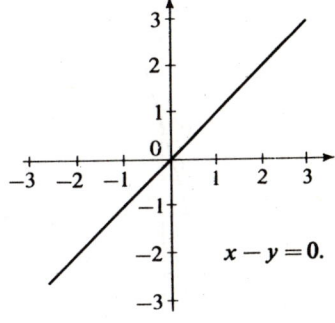

13.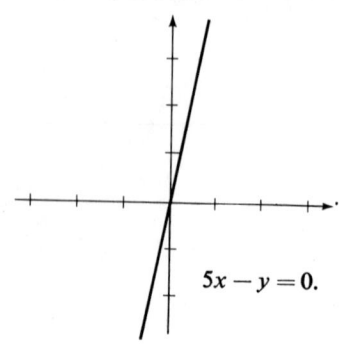

15.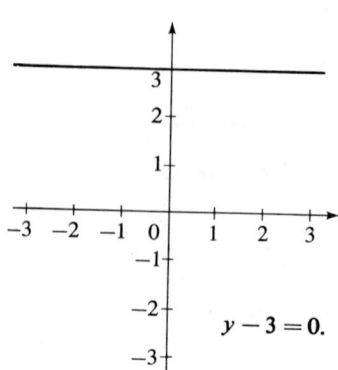

17. $2x+y-6=0$, $x-2y-3=0$, $3x-y+1=0$.

19. $x-2y-3=0$, $2x+y-6=0$, $x+3y-3=0$.

21. $5x-6y+2=0$, $2x+9y-3=0$, $7x+3y-1=0$.

23.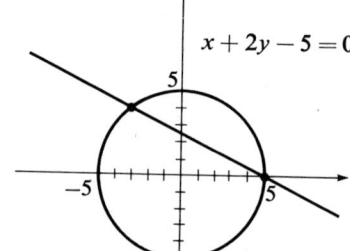

25. $3x-2y+5=0$. **29.** $x-3y+3=0$.
31. $7x-y-20=0$, $7x-y-20=0$. **33.** $9C-5F+160=0$.

Section 3.2, page 55

1.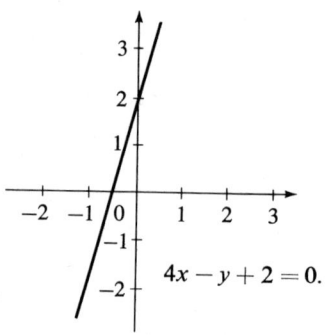

3.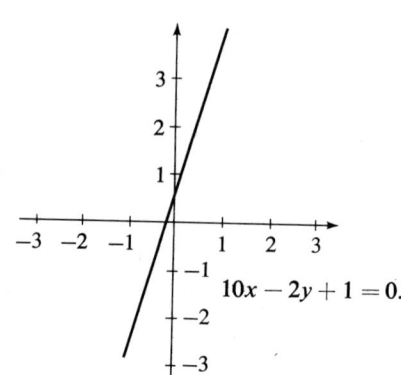

Answers 249

5.
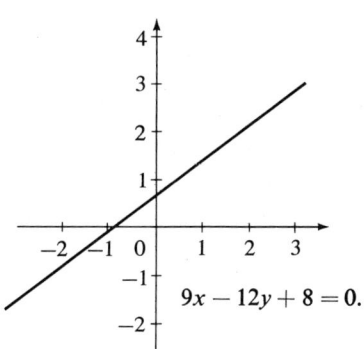

7.
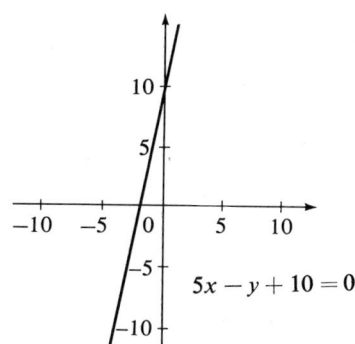

9.
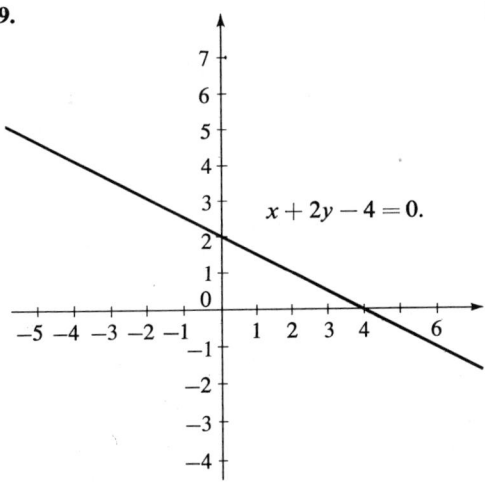

11.
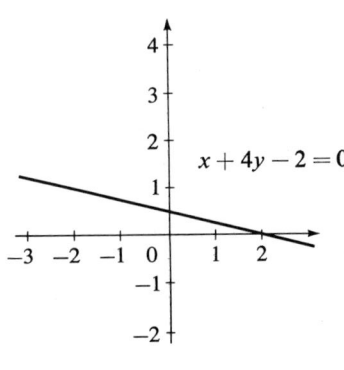

13.
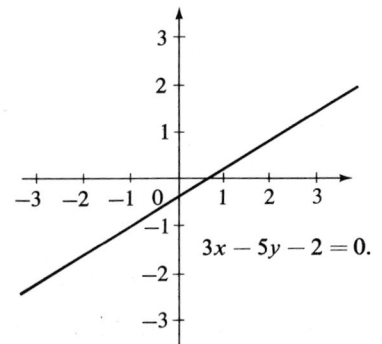

15.
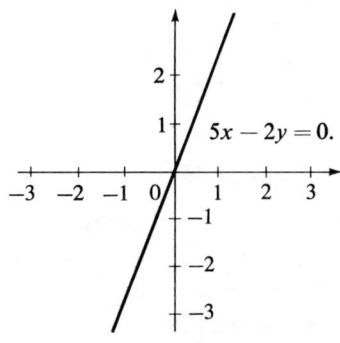

250 Answers

17.

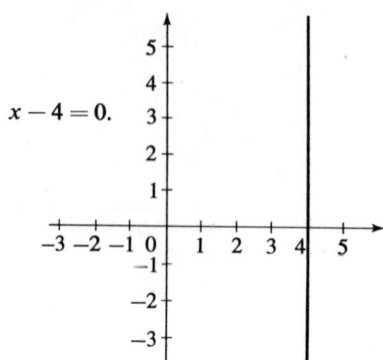

19. $2x - 5y + 11 = 0$.
21. $2x - y - 7 = 0$.
23. $(15/14, -5/14)$.
25. $(119/41, 99/41)$.
27. $5/2$.
29. 1.

31.

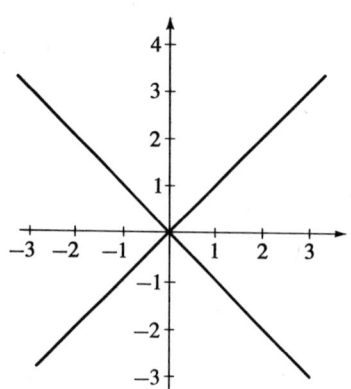

33.

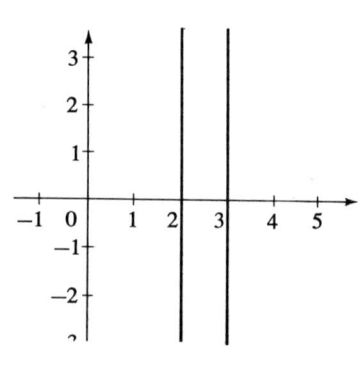

Section 3.3, page 59

1. $\sqrt{2}$.   **3.** $9/\sqrt{41}$.   **5.** $2/5$.   **7.** $18/\sqrt{29}$.   **9.** $10/3$.
11. $13/\sqrt{10}$, $26/\sqrt{29}$, $26/5$.   **13.** $7x - 7y + 2 = 0$.   **15.** $39x + 13y - 24 = 0$.
17. $2(\sqrt{2} - 1)x - 2y + (4 - 3\sqrt{2}) = 0$.   **19.** $4/\sqrt{29}$.   **21.** $7/2\sqrt{5}$.   **23.** $8/\sqrt{5}$.
25. 13.
27. $(21/10, 31/10)$, $(-9/4, -5/4)$, $(113/30, 31/10)$, $(-9/4, 159/28)$.
29. $\pm 3/\sqrt{7}$.   **31.** $\pm 3/2\sqrt{2}$.

Section 3.4, page 64

1. Lines through $(-1, 4)$. Does not include $x = -1$.
3. Lines with x intercept 2. Does not include $x = 2$.
5. All lines through the origin.
7. Lines having both an x intercept and a y intercept. Does not include any line through the origin.

9. Lines containing the point of intersection of $2x+3y+1=0$ and $4x+2y-5=0$. Does not include $4x+2y-5=0$.
11. Lines with y intercept twice the x intercept. Does not include any line through the origin.
13. All lines with y intercept equal to their slope.
15. $\{3x-5y=k \mid k \text{ real}\}$. 17. $\{y-5=m(x-2) \mid m \text{ real}\} \cup \{x=2\}$.
19. $\{3x-5y+1+k(2x+3y-7)=0 \mid k \text{ real}\} \cup \{2x+3y-7=0\}$.
21. $\{Ax+By=0 \mid A, B \text{ real}\}$. 23. $\{Ax+By+C=0 \mid |6A+C|/\sqrt{A^2+B^2}=5\}$.
25. (a) $3x-5y+25=0$, (b) $5x+3y-49=0$.
27. (a) $2x+y-5=0$, (b) $x-2y+10=0$. 29. $5x-y\pm 3\sqrt{26}=0$.
31. $15x-8y-43=0$, $3x-4y+1=0$. 33. $11x-60y-17=0$, $x-7=0$.
35. $16x-8y-27=0$. 37. $x-2y-10=0$, $3x+2y-6=0$.
39. $x+2y-8=0$, $9x+2y-24=0$, $(11\pm 4\sqrt{7})x-2y-(16\pm 8\sqrt{7})=0$.
41. $x+y-(4+2\sqrt{2})=0$.
43. $\sqrt{3}x-y+(6-4\sqrt{3})=0$, $\sqrt{3}x+y-(6+4\sqrt{3})=0$.

Section 3.5, page 69

1. $m=4.17$. 3. $k=0.412$. 5. $m=2.14$, $b=4.18$. 7. $p=2.40$, $q=4.13$.
9. $k=11.2$. 11. $\Delta H = 0.0703$ cal/mole.

Chapter 4

Section 4.1, page 75

1. $(x-1)^2+(y-3)^2=25$,
   $x^2+y^2-2x-6y-15=0$.

3. $(x-5)^2+(y+2)^2=4$,
   $x^2+y^2-10x+4y+25=0$.

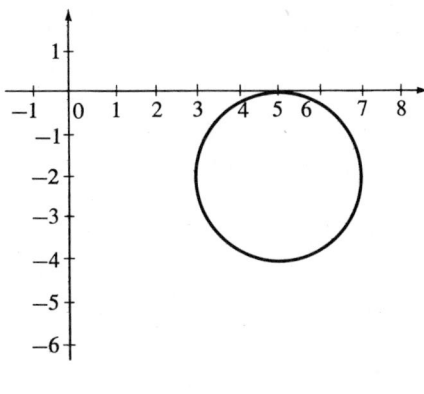

252 Answers

5. $(x - 1/2)^2 + (y + 3/2)^2 = 4$,
$2x^2 + 2y^2 - 2x + 6y - 3 = 0$.

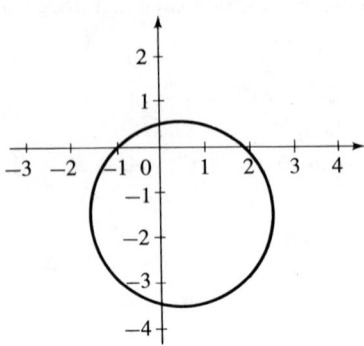

7. $(x - 4)^2 + (y + 2)^2 = 26$,
$x^2 + y^2 - 8x + 4y - 6 = 0$.

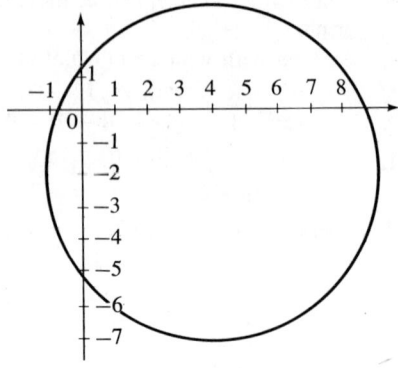

9. $x^2 + (y + 3/2)^2 = 25/4$,
$x^2 + y^2 + 3y - 4 = 0$.

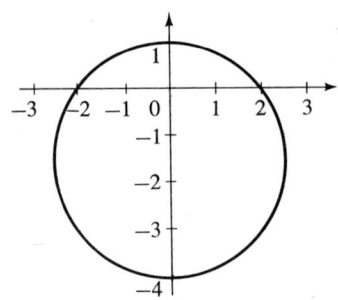

11. $(x - 3)^2 + (y - 3)^2 = 9$,
$x^2 + y^2 - 6x - 6y + 9 = 0$.

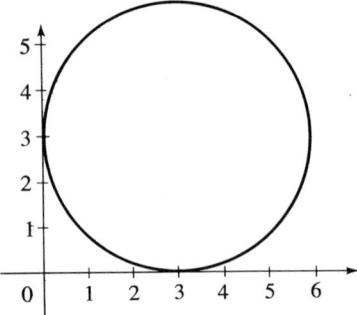

13. $(x - 4)^2 + (y - 1)^2 = 4$,
$x^2 + y^2 - 8x - 2y + 13 = 0$.

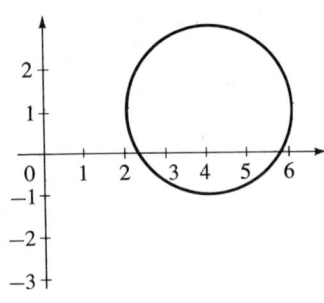

15. $(x - 4)^2 + (y + 4)^2 = 16$,
$x^2 + y^2 - 8x + 8y + 16 = 0$.

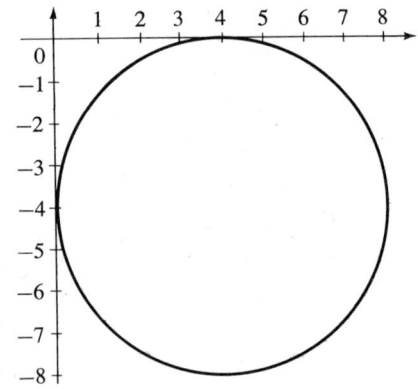

17. $(x-1)^2 + (y-2)^2 = 4$;   19. $(x+3)^2 + y^2 = 25$.

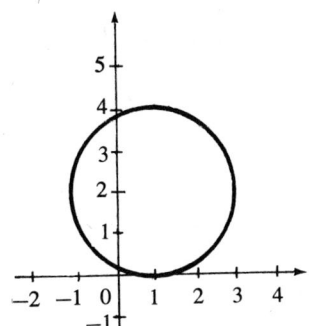   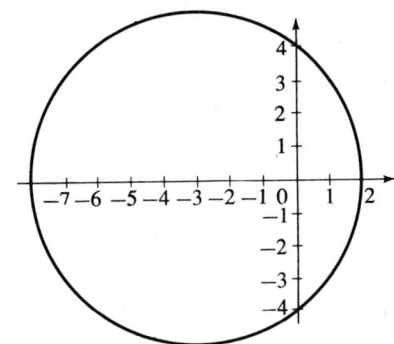

21. $(x-1/2)^2 + (y-3/2)^2 = 9/4$.   23. $(x-4/5)^2 + (y-2/5)^2 = 25$.

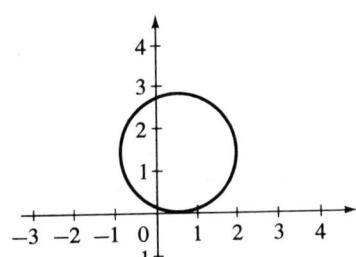   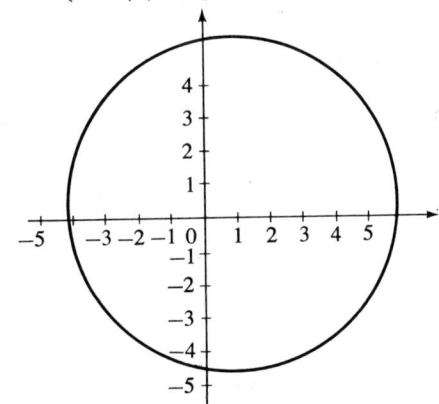

25. $(x-1/3)^2 + (y+1)^2 = -1/9$.   27. $(x-2/3)^2 + (y-1/2)^2 = 0$.

29. $(2, -1), (3, 3)$.
31. $(1, 4), (3, 1)$.
35. $x + y + 1 = 0$.
39. (a) $D^2 + E^2 - 4AF > 0$,
    (b) $= 0$,
    (c) < 0.

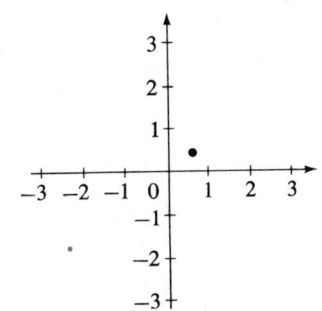

Section 4.2, page 82

1. $3x^2 + 3y^2 - 10x - 10y - 5 = 0$.   3. $5x^2 + 5y^2 + 13x - 17y - 40 = 0$.
5. $x^2 + y^2 - 13x - y = 0$.
7. $x^2 + y^2 + 4x - 10y + 4 = 0$, $x^2 + y^2 - 6x + 10y + 9 = 0$.

254 Answers

9. $x^2 + y^2 - 4x + 8y - 5 = 0$.
11. $2x^2 + 2y^2 + 8x + 12y + 1 = 0$, $2x^2 + 2y^2 - 12x - 48y + 81 = 0$.
13. $x^2 + y^2 - 2x + 6y = 0$.   15. $x^2 + y^2 + 4x + 6y - 37 = 0$.
17. $13x^2 + 13y^2 + 4x + 19y - 49 = 0$.
19. $x^2 + y^2 + 6x + 6y + 9 = 0$, $x^2 + y^2 + 6x - 6y + 9 = 0$, $x^2 + y^2 - 6x + 6y + 9 = 0$, $x^2 + y^2 - 6x - 6y + 9 = 0$.
21. $5x^2 + 5y^2 + 52x - 56y + 47 = 0$, $5x^2 + 5y^2 - 32x + 56y - 37 = 0$.
23. $x^2 + y^2 + 4x - 21 = 0$.   27. $x - 3y - 7 \pm 13\sqrt{10} = 0$.

Section 4.3, page 85

1. $7x^2 + 7y^2 + 26x + 32y - 17 = 0$.   3. $3x^2 + 3y^2 + 14x - 11y - 23 = 0$.
5. $3x^2 + 3y^2 - 2x - 10y - 2 = 0$.   7. $x^2 + y^2 - 6x + 2y + 3 = 0$.
9. $13x^2 + 13y^2 + 20x - 22y - 35 = 0$, $x^2 + y^2 - 4x + 2y + 1 = 0$.
11. $4x + 8y - 11 = 0$.
15. The family of all circles of radius 1, with center on the line $y = 1$.
17. The family of all circles in the first and third quadrants, which are tangent to both axes, together with the origin.
19. The family of all circles of radius 1 and containing the origin.
25. $2\sqrt{11}$.   27. $\sqrt{34}$.

Chapter 5

Section 5.2, page 91

1. Axis: x axis, $V(0, 0)$, $F(4, 0)$,
   D: $x = -4$, lr $= 16$.

3. Axis: y axis, $V(0, 0)$,
   $F(0, 1)$, D: $y = -1$, lr $= 4$.

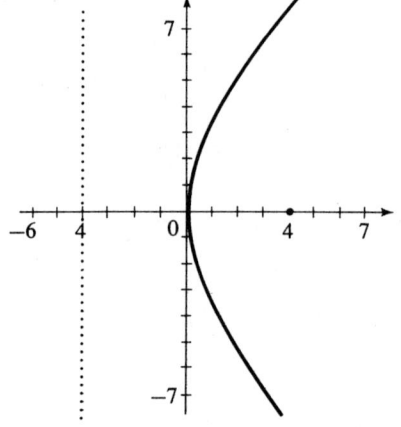
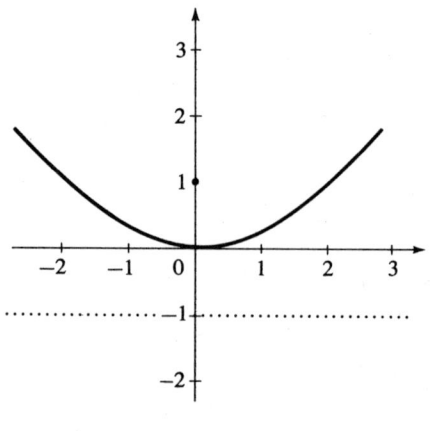

Answers 255

5. Axis: x axis, $V(0, 0)$, $F(5/2, 0)$,
D: $x = -5/2$, lr $= 10$.

7. Axis: y axis, $V(0, 0)$, $F(0, 5/4)$,
D: $y = -5/4$, lr $= 5$.

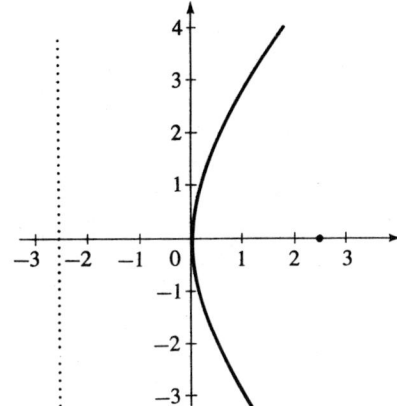

9. Axis: y axis, $V(0, 0)$,
$F(0, -1/2)$, D: $y = 1/2$, lr $= 2$.

11. Axis: y axis,
$V(0, 0)$, $F(0, 3/2)$,
D: $y = -3/2$, lr $= 6$.

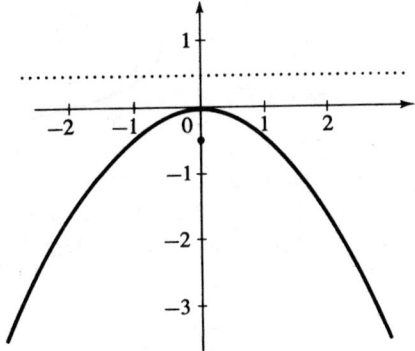 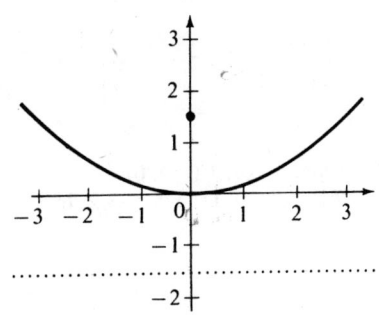

13. $y^2 = 25x$. **15.** $y^2 = 5x$, $y^2 = -5x$. **17.** $y^2 = -12x$. **19.** $x^2 = 4y/3$.
21. $y^2 = 4x - 4$. **23.** $x^2 - 2xy + y^2 + 8x + 8y - 16 = 0$. **25.** $2x - y - 1 = 0$.
27. $x + y - 4$.

Section 5.3, page 95

1. Axis: $x = 3$, $V(3, 2)$,
   $F(3, 4)$, D: $y = 0$, lr $= 8$.

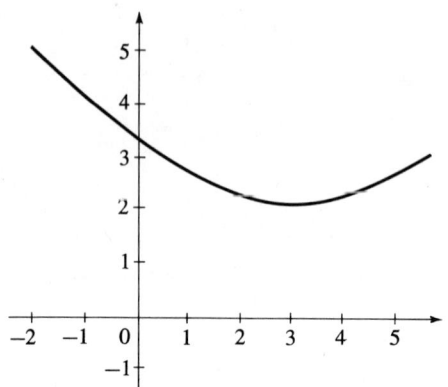

3. Axis: $x = -5$, $V(-5, -2)$,
   $F(-5, -1/2)$, D: $y = -7/2$, lr $= 6$.

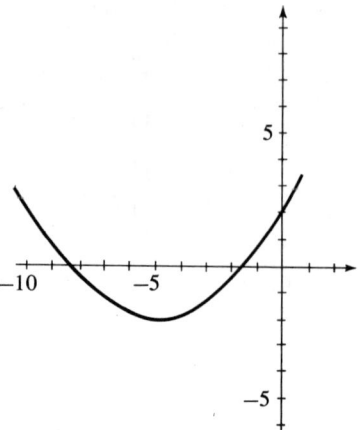

5. Axis: $x = -2$, $V(-2, 1)$,
   $F(-2, 2)$, D: $y = 0$, lr $= 4$.

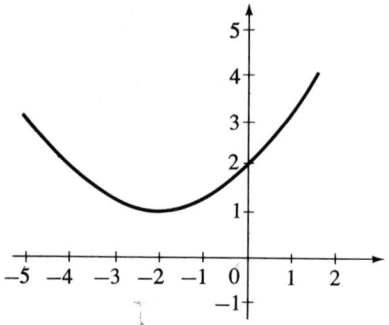

7. Axis: $y = -1$, $V(3, -1)$,
   $F(13/4, -1)$, D: $x = 11/4$, lr $= 1$.

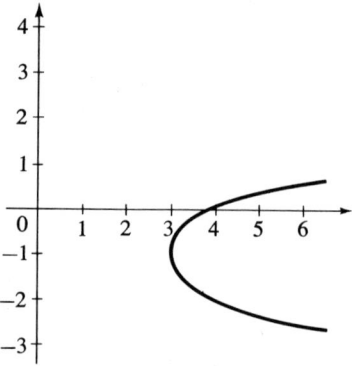

9. Axis: $y = 0$, $V(-2, 0)$,
   $F(-5/2, 0)$, D: $x = -3/2$, lr $= 2$.

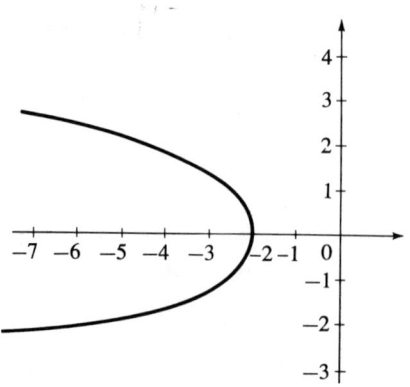

11. No locus.
13. Axis: $y = 2/5$, $V(-3/5, 2/5)$,
    $F(7/5, 2/5)$, D: $x = -13/5$, lr $= 8$.

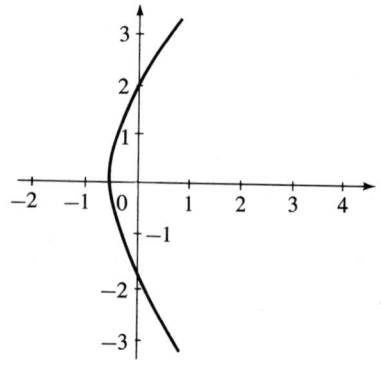

15. Axis: $x = 1/2$, $V(1/2, 1/3)$,
    $F(1/2, 5/6)$, $D: y = -1/6$, $lr = 2$.

17. $y^2 - 8x - 10y + 33 = 0$.
19. $x^2 - 8x - 12y + 28 = 0$.
21. $y^2 - 8x - 8y = 0$,
    $y^2 + 8x - 8y + 32 = 0$.
23. $x^2 - 7x - y + 6 = 0$.
25. $x + 3y - 2 = 0$.
27. $10x + y - 12 = 0$,
    $2x + y - 4 = 0$.

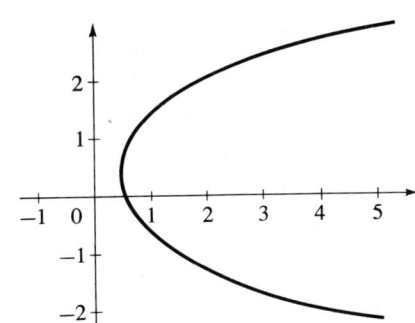

Section 5.4, page 101

1. $C(0, 0)$, $V(\pm 13, 0)$, $CV(0, \pm 5)$,
   $F(\pm 12, 0)$, $lr = 50/13$.

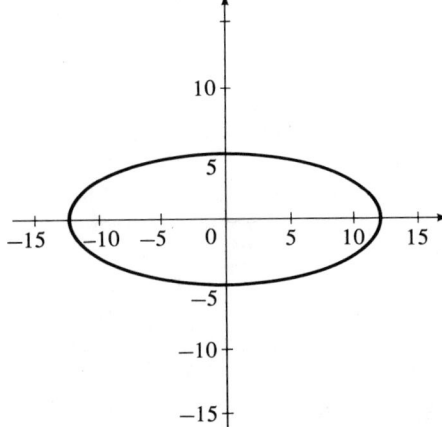

3. $C(0, 0)$, $V(\pm 5, 0)$, $CV(0, \pm 2)$,
   $F(\pm \sqrt{21}, 0)$, $lr = 8/5$.

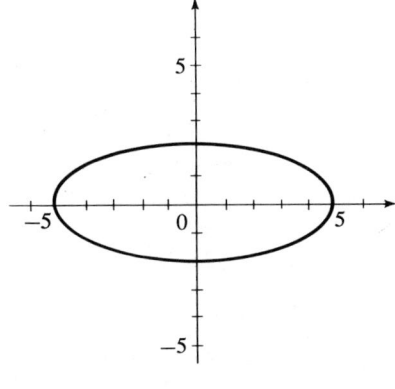

5. $C(0, 0)$, $V(0, \pm 7)$, $CV(\pm 5, 0)$,
   $F(0, \pm 2\sqrt{6})$, $lr = 50/7$.

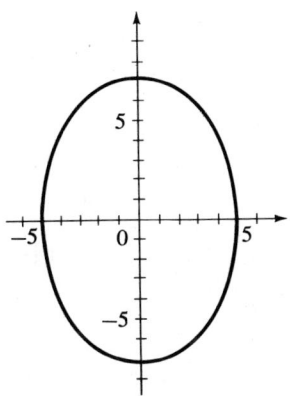

7. $C(0, 0)$, $V(0, \pm 3)$, $CV(\pm 2, 0)$,
   $F(0, \pm \sqrt{5})$, $lr = 8/3$.

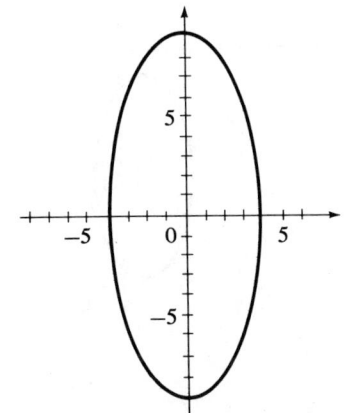

258 Answers

9. $C(0, 0)$, $V(0, \pm 4)$, $CV(\pm 3, 0)$, $F(0, \pm \sqrt{7})$, lr $= 9/2$.

11. $x^2/144 + y^2/169 = 1$.
13. $x^2/25 + y^2/10 = 1$.
15. $x^2/36 + y^2/9 = 1$.
17. $x^2/100 + y^2/64 = 1$.
21. $4x + 3y - 11 = 0$.
23. $x - 4y + 14 = 0$,
    $11x + 12y - 70 = 0$.
29. 0.0167.

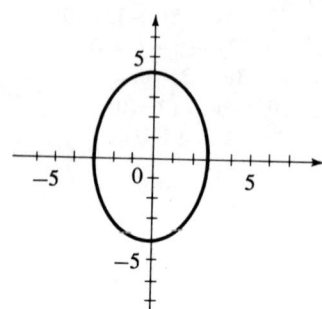

Section 5.5, page 105

1. $C(0, 3)$, $V(\pm 1, 3)$, $CV(0, 3 \pm 1/2)$, $F(\pm \sqrt{3}/2, 3)$, lr $= 1/2$.

3. $C(-4, 1)$, $V(-4 \pm 5, 1)$, $CV(-4, 1 \pm 3)$, $F(-4 \pm 4, 1)$, lr $= 18/5$.

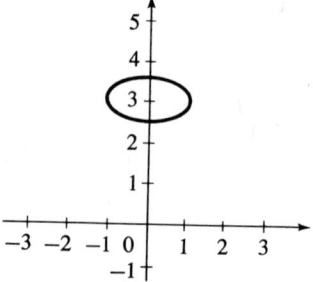
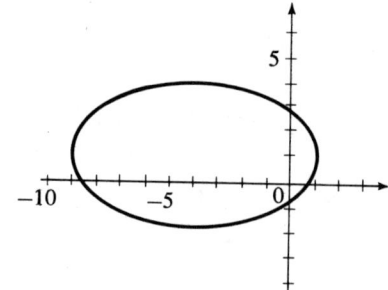

5. $C(-1, -5)$, $V(-1, -5 \pm 4)$, $CV(-1 \pm 2, -5)$, $F(-1, -5 \pm 2\sqrt{3})$, lr $= 2$.

7. $(-4, 3)$.

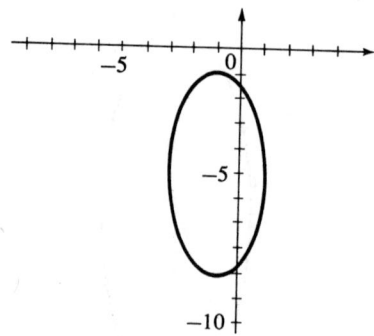
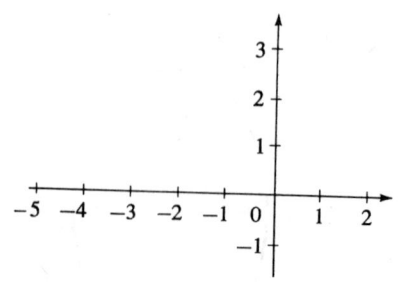

Answers 259

9. $C(-1, 2)$, $V(-1 \pm 3, 2)$, $CV(-1, 2 \pm 2)$,
   $F(-1 \pm \sqrt{5}, 2)$, $\text{lr} = 8/3$.

11. No locus.

13. $C(-6, 8)$, $V(-6 \pm 3, 8)$, $CV(-6, 8 \pm 2)$,
    $F(-6 \pm \sqrt{5}, 8)$, $\text{lr} = 8/3$.

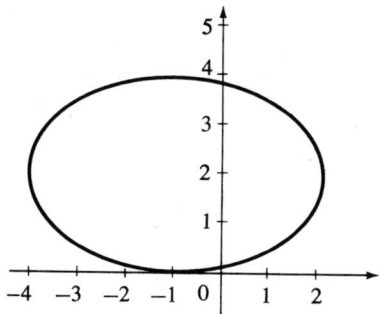
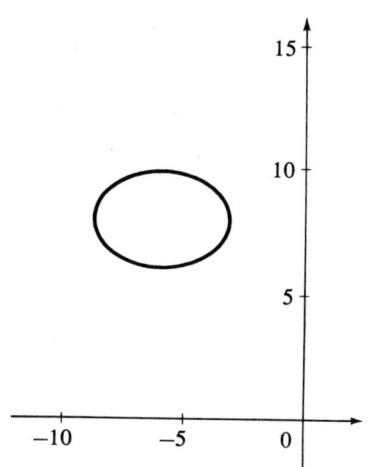

15. $16x^2 + y^2 - 32x + 8y + 16 = 0$.   17. $4x^2 + y^2 + 8x - 6y - 12 = 0$.
19. $4x^2 + 9y^2 + 16x - 20 = 0$.   21. $x^2 + 36y^2 - 2x + 216y + 156 = 0$.
23. $16x^2 + 25y^2 - 96x + 50y - 231 = 0$.   25. $9x^2 + 8y^2 + 54x - 16y - 559 = 0$.
27. $5x + 4y - 9 = 0$.   29. $3x + 8y + 40 = 0$, $3x - 8y + 24 = 0$.

Section 5.6, page 111

1. $C(0, 0)$, $V(\pm 4, 0)$, $F(\pm 5, 0)$,
   A: $y = \pm 3x/4$, $\text{lr} = 9/2$.

3. $C(0, 0)$, $V(0, \pm 3)$, $F(0, \pm \sqrt{13})$,
   A: $y = \pm 3x/2$, $\text{lr} = 8/3$.

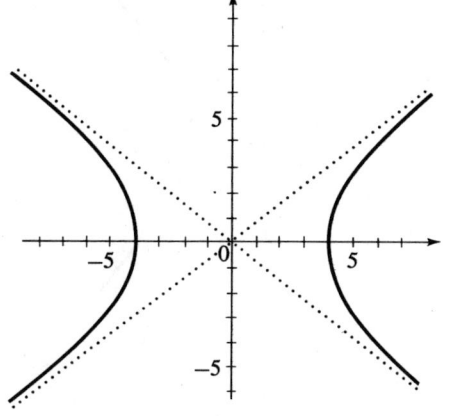
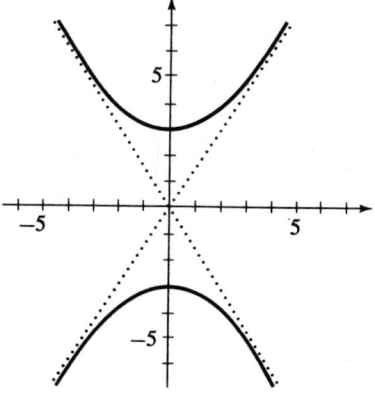

5. $C(0, 0)$, $V(\pm 12, 0)$, $F(\pm 13, 0)$,
   A: $y = \pm 5x/12$, lr $= 25/6$.

7. $C(0, 0)$, $V(0, \pm 5)$, $F(0, \pm \sqrt{34})$,
   A: $y = \pm 5x/3$, lr $= 18/5$.

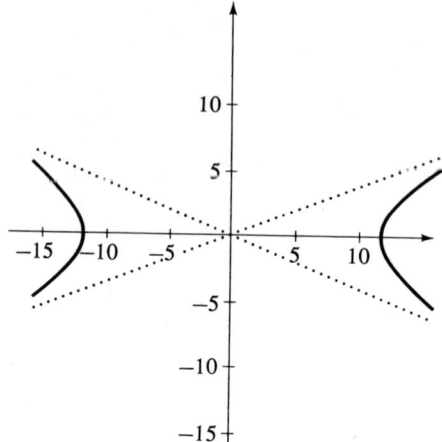
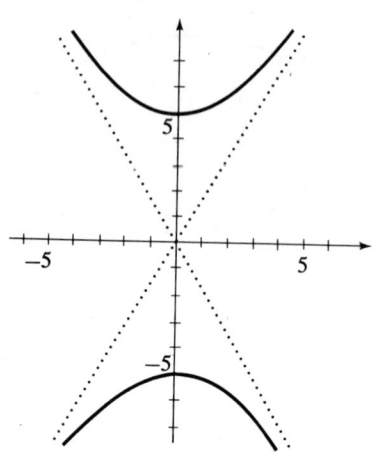

9. $C(0, 0)$, $V(\pm 1, 0)$, $F(\pm \sqrt{5}, 0)$,
   A: $y = \pm 2x$, lr $= 8$.

11. $C(0, 0)$, $V(\pm 3, 0)$, $F(\pm 3\sqrt{2}, 0)$,
    A: $y = \pm x$, lr $= 6$.

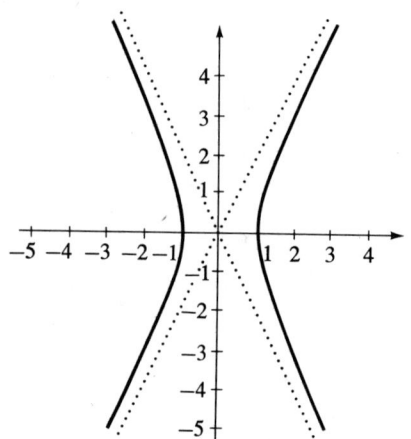
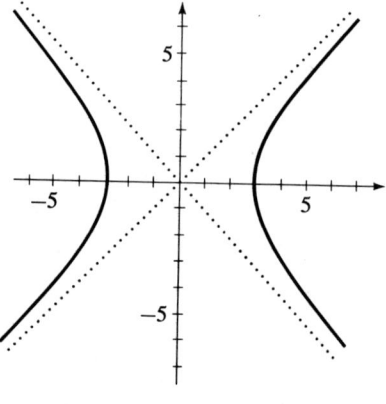

13. $C(0, 0)$, $V(0, \pm 5/2)$, $F(0, \pm \sqrt{34}/2)$,
    A: $y = \pm 5x/3$, lr $= 9/5$.

15. $x^2/4 - y^2/12 = 1$.
17. $x^2/36 - y^2/16 = 1$.
19. $16y^2 - 9x^2 = 144$.
21. None.
23. $y^2/9 - x^2/16 = 1$.
25. $x^2/9 - y^2/16 = 1$.
27. $52x - 15y - 144 = 0$.
29. $5x - 4y - 9 = 0$.

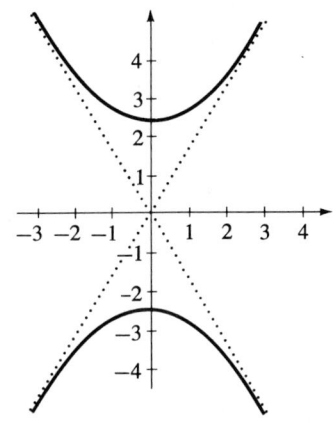

Section 5.7, page 114

1. $C(0, 3)$, $V(\pm 3, 3)$, $F(\pm 5, 3)$,
   A: $y = 3 \pm 4x/3$, lr $= 32/3$.

3. $C(5, -1)$, $V(5 \pm 8, -1)$, $F(5 \pm 8\sqrt{2}, -1)$,
   A: $y + 1 = \pm(x - 5)$, lr $= 16$.

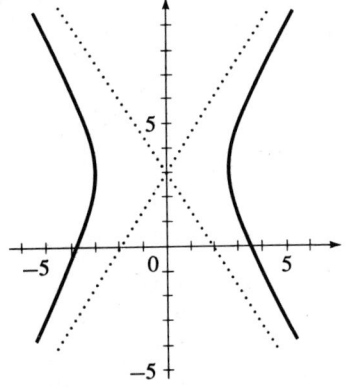
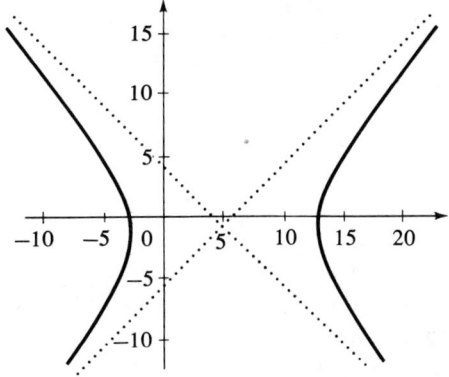

5. $C(-3/2, 1/2)$, $V(-3/2, 1/2 \pm 2)$,
$F(-3/2, 1/2 \pm 2\sqrt{5})$,
$A: y = 1/2 \pm (x + 3/2)/2$, $\text{lr} = 16$.

7. $y = -1 \pm 3(x-2)/2$.

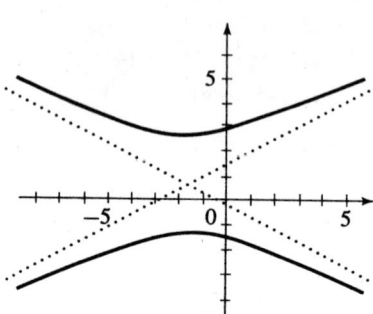
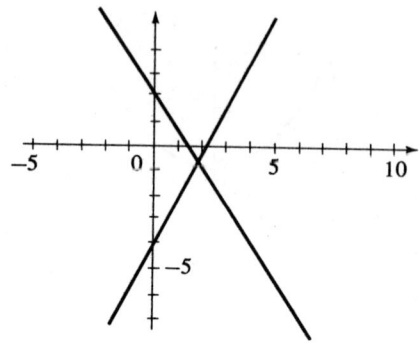

9. $C(1/3, -2/3)$, $V(1/3 \pm 2, -2/3)$,
$F(1/3 \pm 2\sqrt{2}, -2/3)$,
$A: y = -2/3 \pm (x - 1/3)$, $\text{lr} = 4$.

11. $C(-1/3, 1/2)$, $V(-1/3, 1/2 \pm 1)$,
$F(-1/3, 1/2 \pm \sqrt{2})$,
$A: y - 1/2 = \pm(x + 1/3)$, $\text{lr} = 2$.

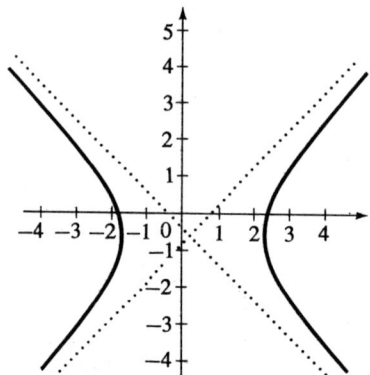
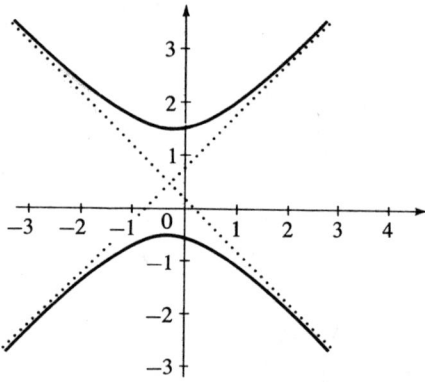

13. $3x^2 - y^2 - 12x + 2y - 1 = 0$. **15.** $9x^2 - 4y^2 - 24x - 8y - 184 = 0$.
17. $4x^2 - 21y^2 - 16x + 210y - 425 = 0$. **19.** $3x^2 - 2y^2 + 3x + 12y + 14 = 0$.
21. $21x^2 - 4y^2 + 42x - 63 = 0$. **23.** $4x^2 - y^2 - 40x + 2y + 63 = 0$.
25. $4x - 7y + 3 = 0$. **27.** $x + y + 4 = 0$, $x - y - 2 = 0$.

Answers 263

Chapter 6

Section 6.1, page 123

1. $y'^2 = 4x'$;

3. $4x'^2 + y'^2 = 16$.

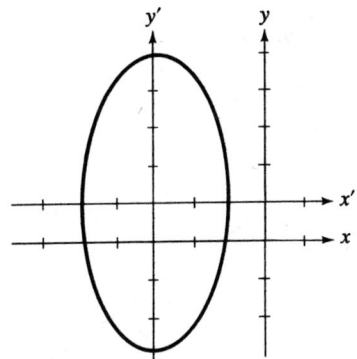

5. $9x'^2 - 4y'^2 = 36$.

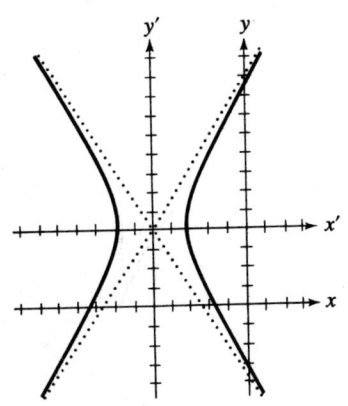

7. $9x'^2 + 4y'^2 = 0$.

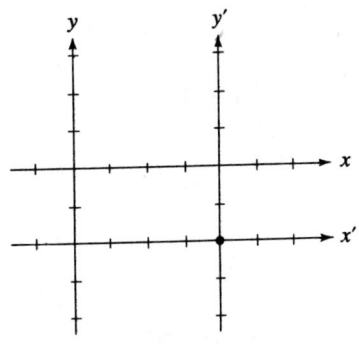

264 Answers

9. $x'^2 = y'$.

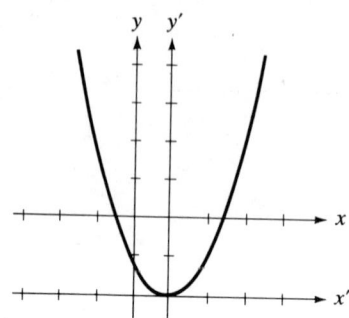

11. $x'^2 - 2x'y' + 4y'^2 - 5 = 0$.
13. $x'^2 + 4x'y' - y'^2 + 1 = 0$.
15. $x'y' + 16 = 0$.
17. $y' = x'^3 - x'$.
19. $y' = x'^4 - 4x'^3 + 6x'^2$.
21. $y' = x'^5 - 2x'^3 + 2x'$.
23. $x'^2 y' - 2x'^2 - 4 = 0$.

Section 6.2, page 126

1. $\sqrt{13}\,x' = 6$.

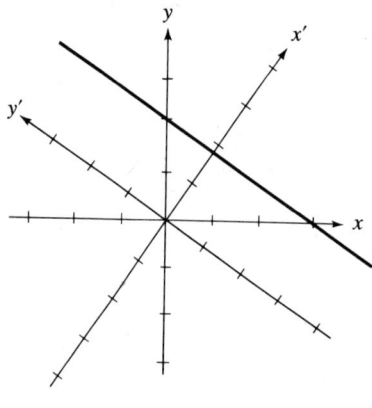

3. $x'^2 - y'^2 = 8$.

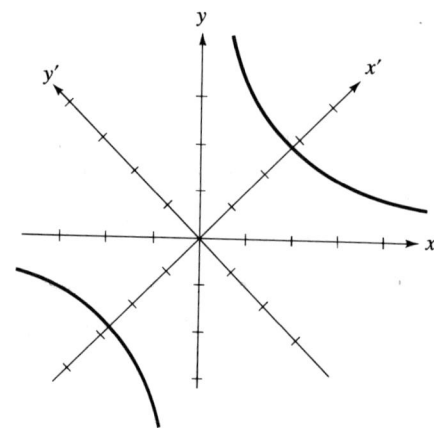

5. $\sqrt{2}\,y'^2 + x' = 0$.

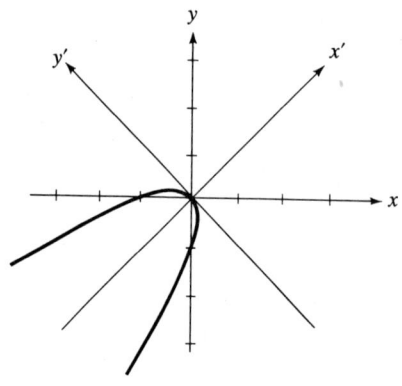

7. $x'^2 - 4y' = 0$.

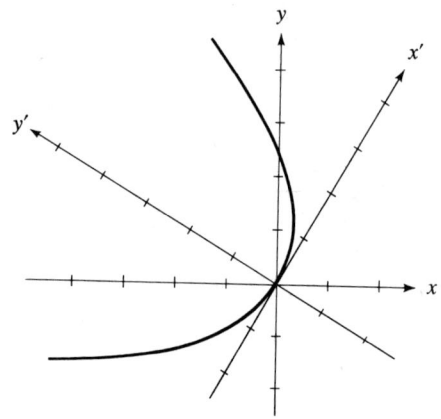

9. $17x'^2 - 9y'^2 = 8$.

11. $17x'^2 + 7x'y' - 7y'^2 + 20 = 0$.
13. $7x'^2 - 8x'y' + y'^2 - 10 = 0$.
15. $3\sqrt{3}x'^2 - (6 - 8\sqrt{3})x'y'$
$\qquad - (8 + 3\sqrt{3})y'^2 - 16 = 0$.

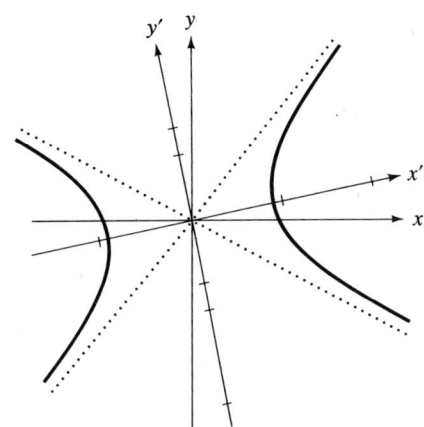

Section 6.3, page 131

1. $3x'^2 + y'^2 - 16y' = 0$.

3. $4x'^2 - y'^2 - 16 = 0$.

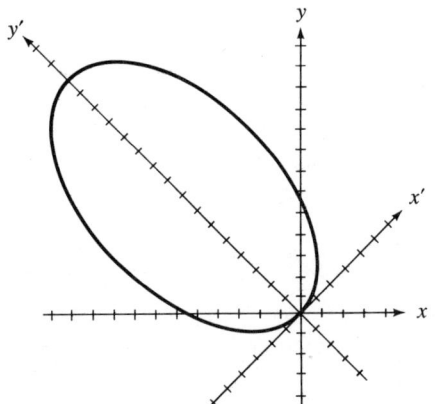

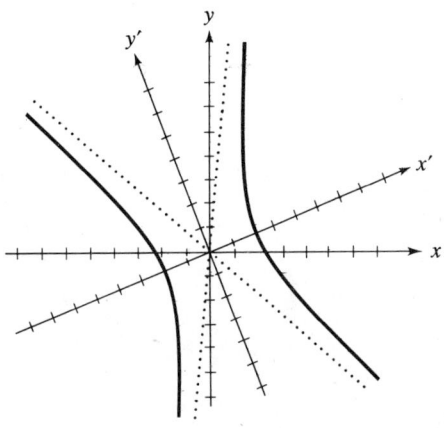

5. $25x'^2 + 4y'^2 = 100$.

7. $4x'^2 + 9y'^2 - 36 = 0$.

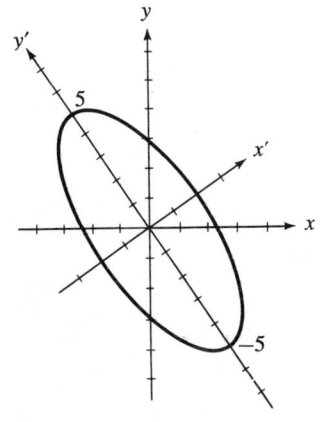

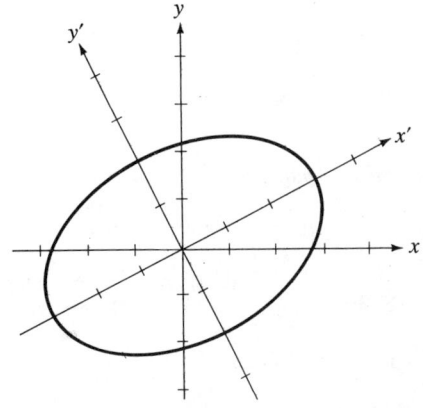

9. $x'^2 + 4x' - 5 = 0$.

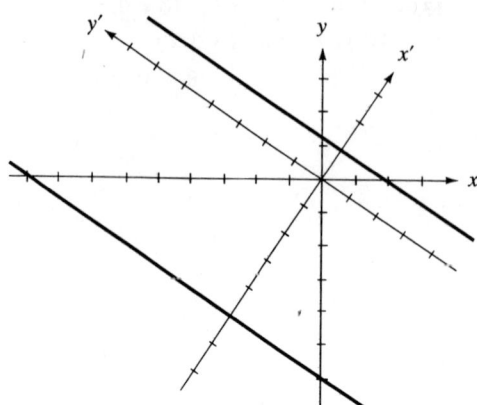

11. $y'^2 - 12x' = 0$.

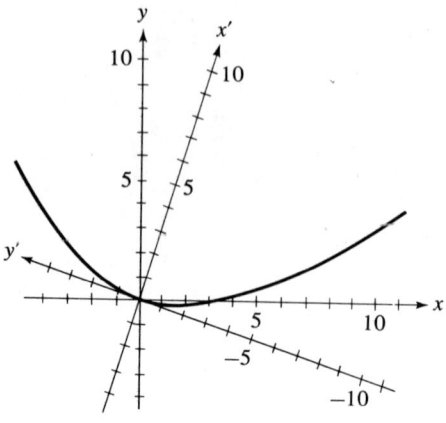

13. $x'^2 - y'^2 + 2x' + 4y' - 4 = 0$.

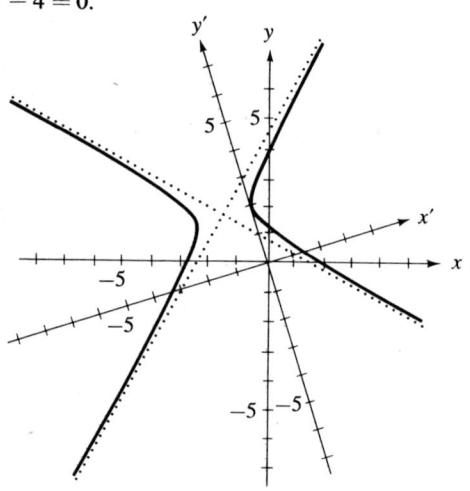

Chapter 7

Section 7.1, page 137

1. $(0, 0)$, $(-3, 0)$. **3.** $(1, 0)$, $(-1, 0)$, $(0, -1)$.
5. $(-1/4, 0)$, $(2, 0)$, $(-3/2, 0)$, $(0, -18)$. **7.** $(1, 0)$, $(0, -1)$.
9. $(1/3, 0)$, $(-1/2, 0)$, $(0, 8)$. **11.** $(0, 0)$. **13.** $(3, 0)$, $(0, 9)$. **15.** None.
17. None. **19.** $x = 1$, $y = 0$. **21.** $x = -3$, $y = 1$. **23.** $x = -1$, $y = 2$.
25. $x = -1$, $x = 3$, $y = 0$. **27.** $y = 1$. **29.** $x = -3/2$, $x = -1$, $y = 0$.
31. $y = 0$. **33.** $y = 0$. **35.** (a) $y = 0$, (b) $y = a_n/b_m$, (c) none.

Section 7.2, page 143

1. y axis. **3.** None. **5.** x axis. **7.** Origin. **9.** Origin.

11.
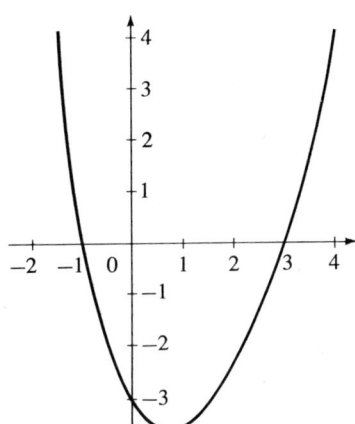

13.
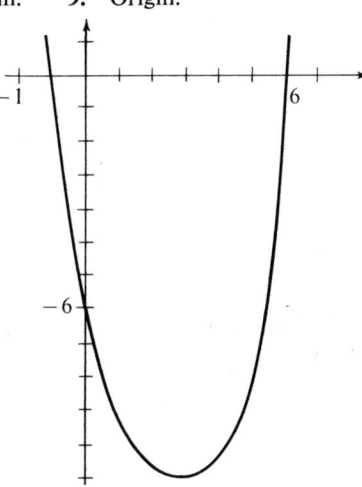

15.
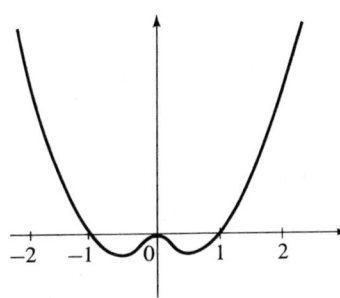

17.
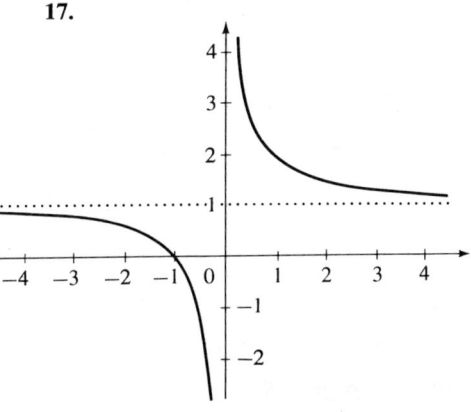

19.
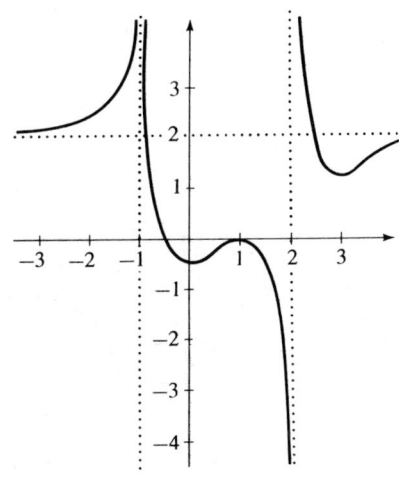

21.
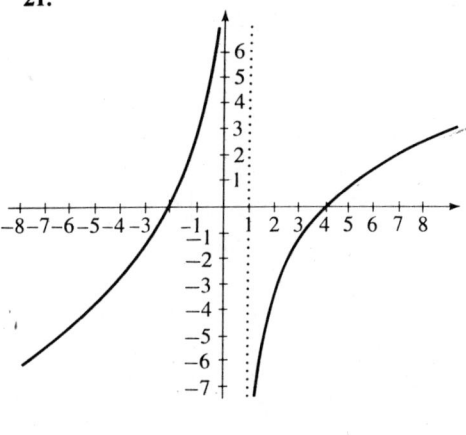

268 Answers

23.

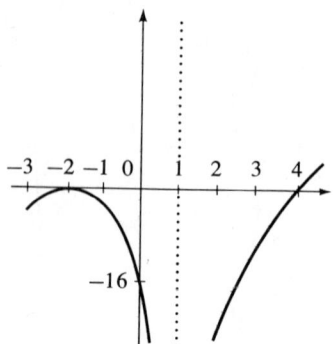

25.

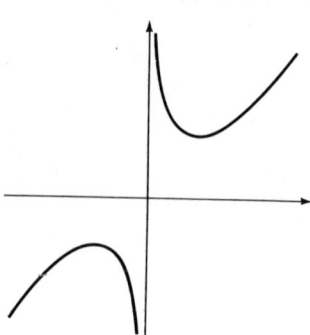

27.

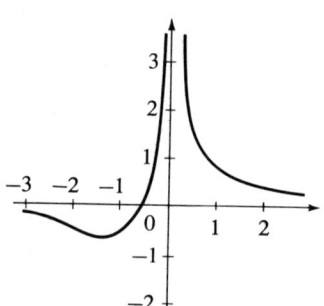

29.

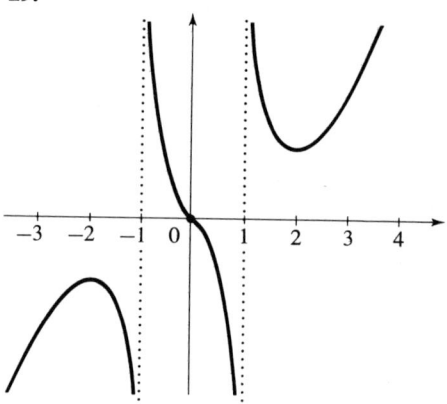

33. Yes.

Section 7.3, page 148

1.

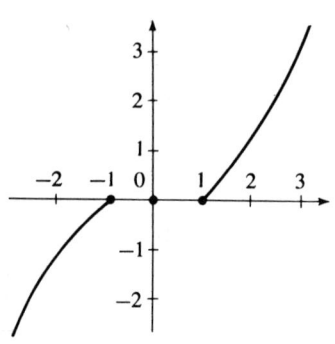

3.

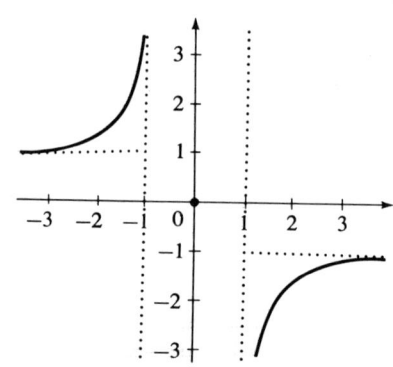

Answers

5.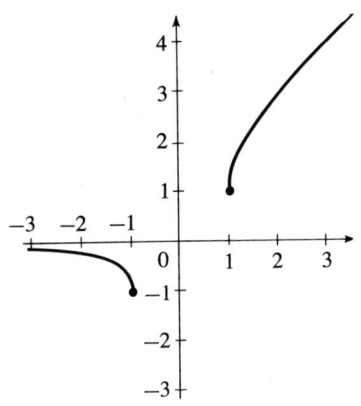

7.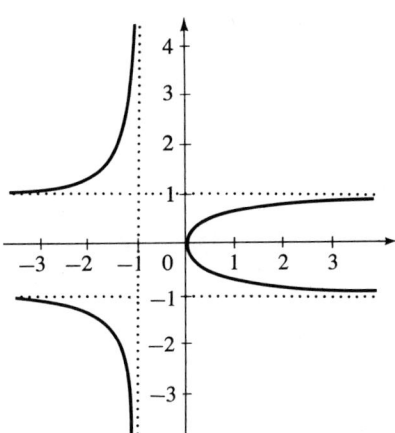

9.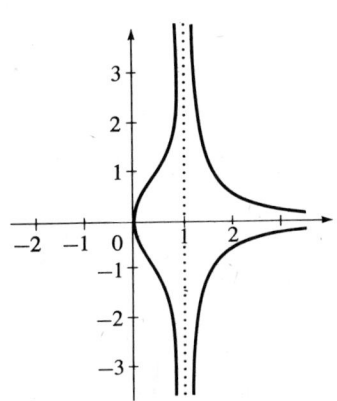

11.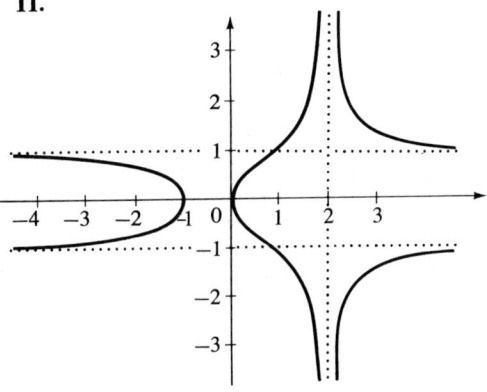

13.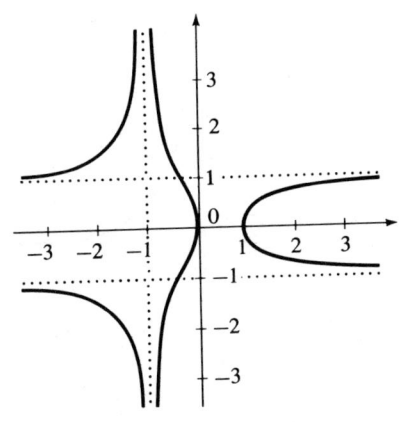

15.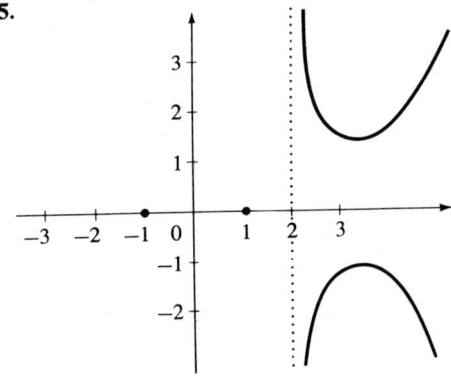

270 Answers

17.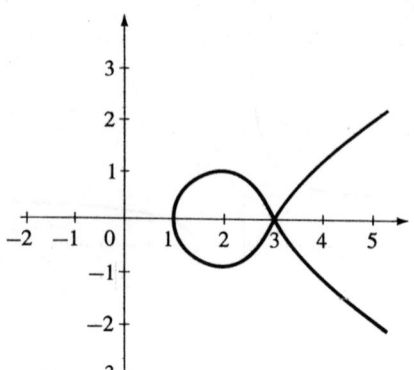

19.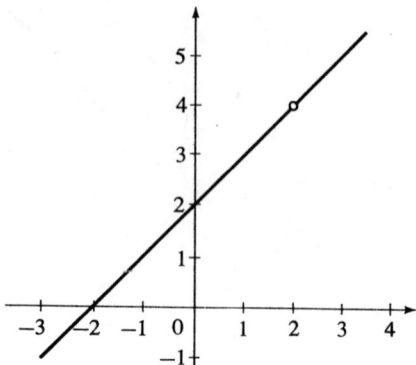

21.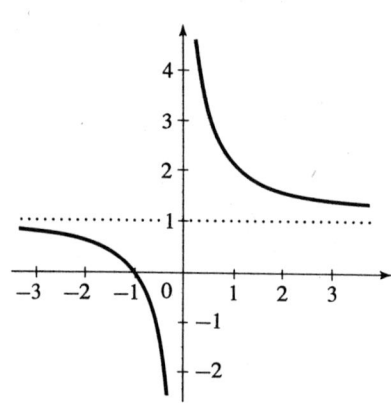

23.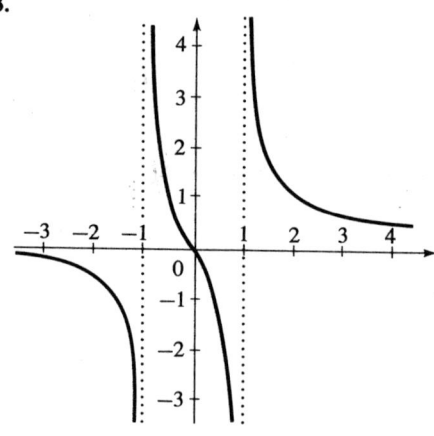

25.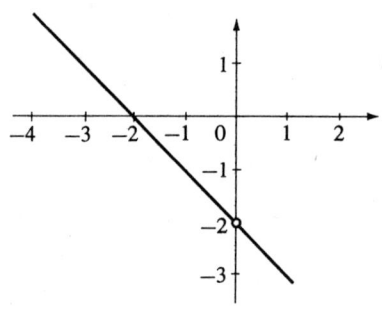

27.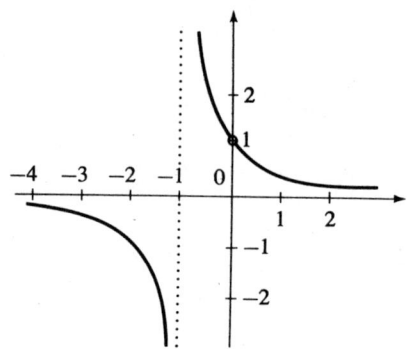

Section 7.4, page 152

1.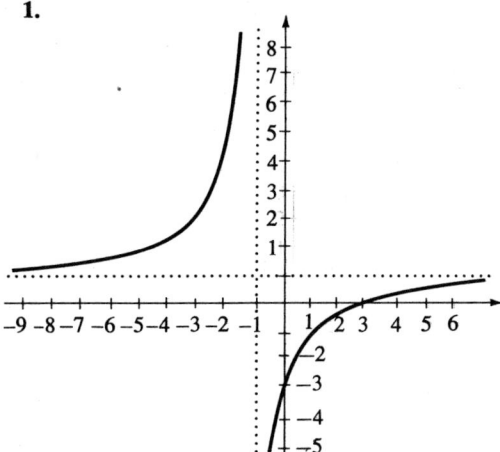

3.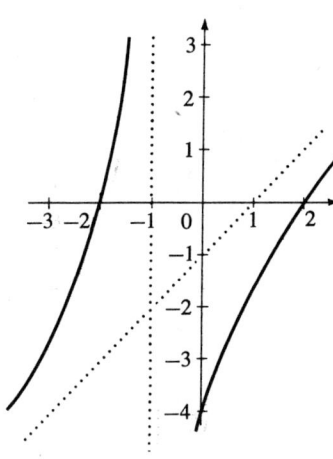

5.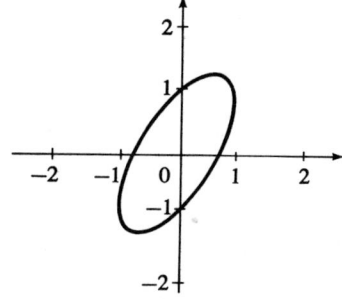

7.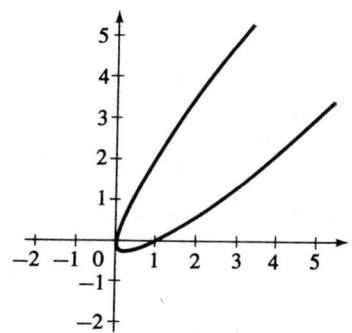

9.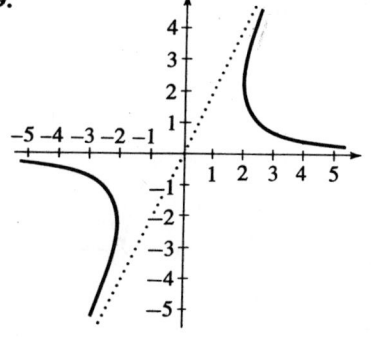

11.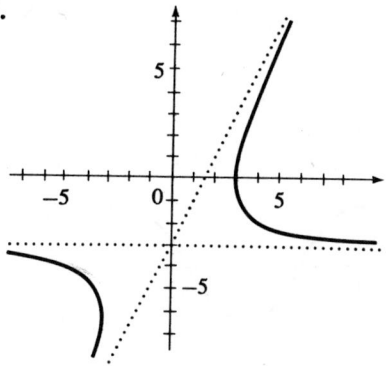

13.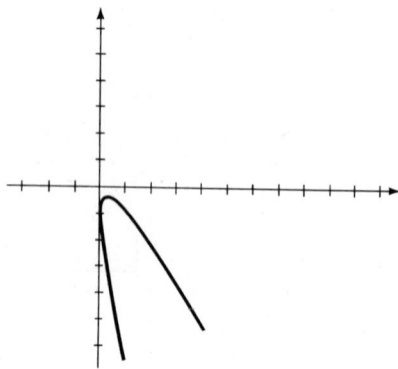

15.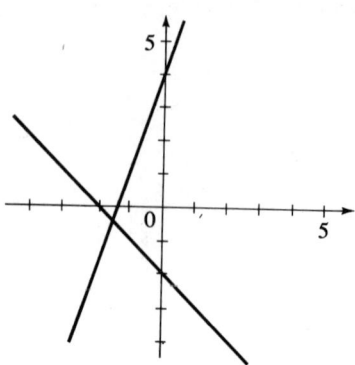

17.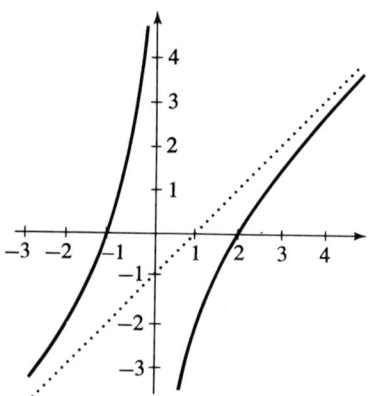

19.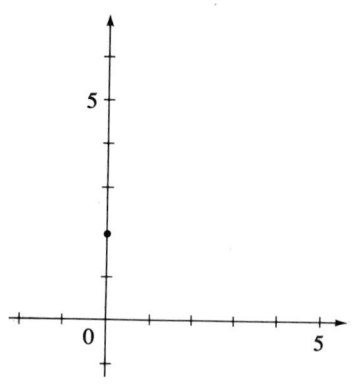

Chapter 8

Section 8.1, page 155

1. $\pi/4$, $-7\pi/6$, $3\pi/2$, $\pi/6$.   **2.** $60°$, $180°$, $135°$, $-90°$.

3.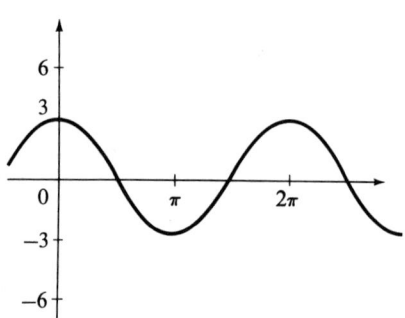

5.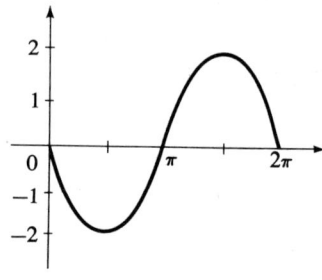

7.

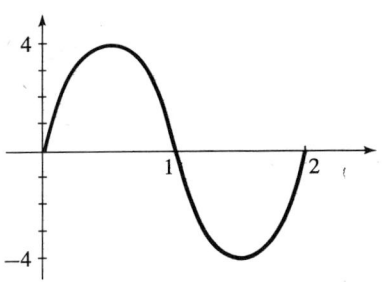

9.

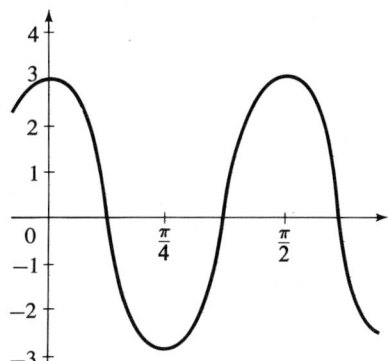

11.

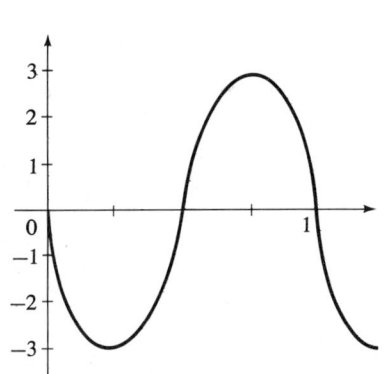

13.

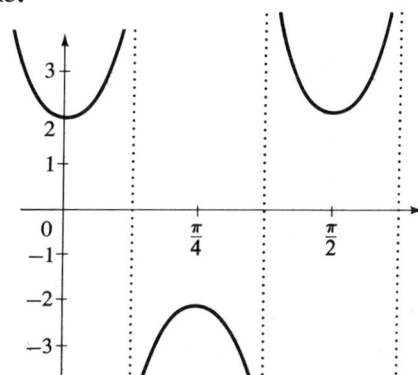

15.

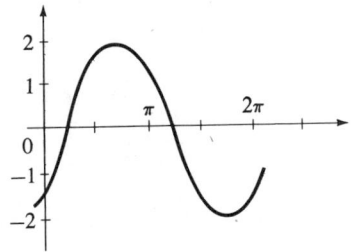

17.

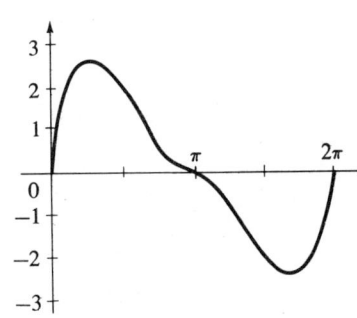

19.

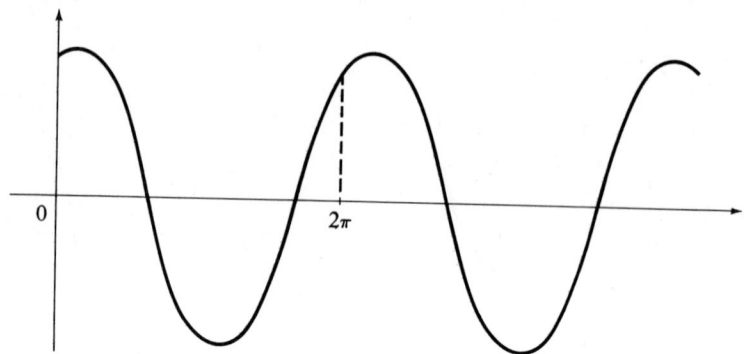

21.

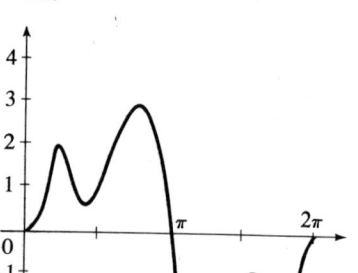

23.

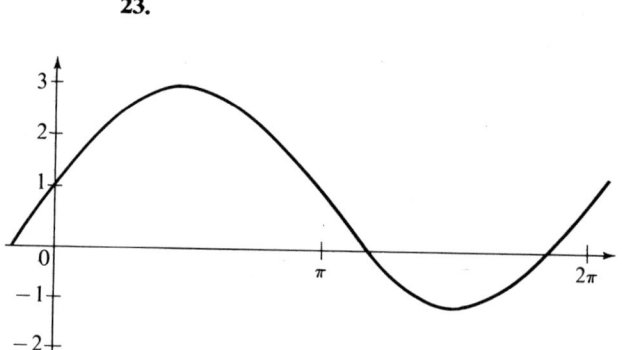

25.

27.

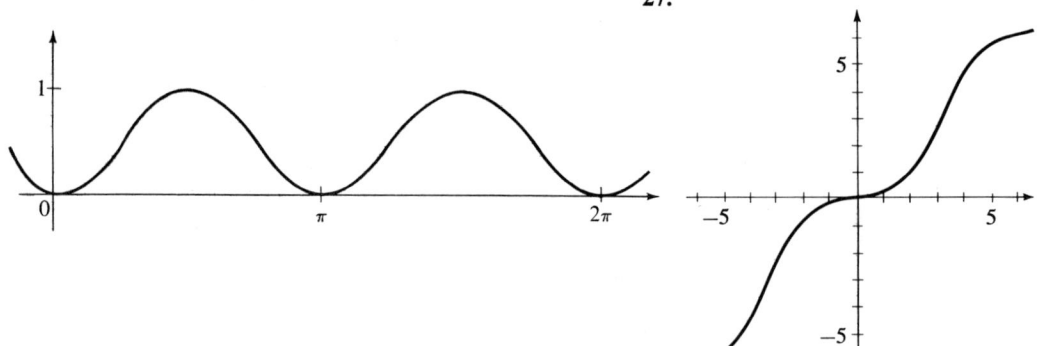

Answers 275

29.

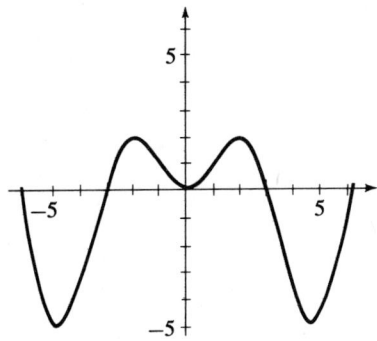

31.

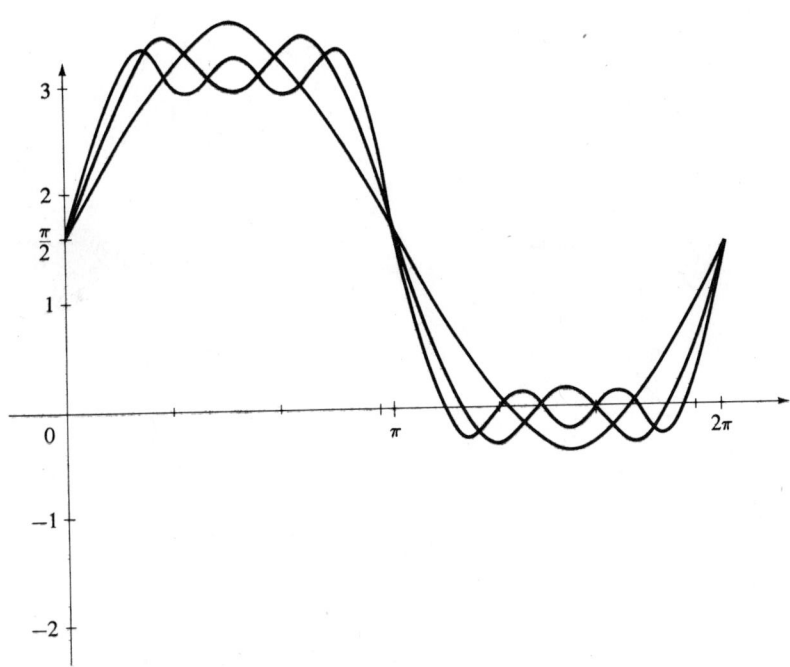

Section 8.2, page 159

1.

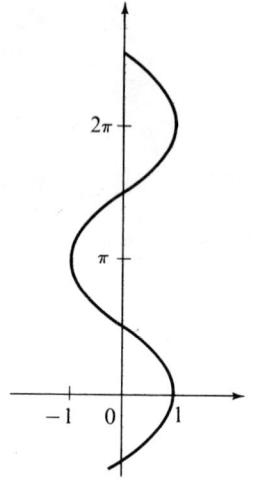

3.

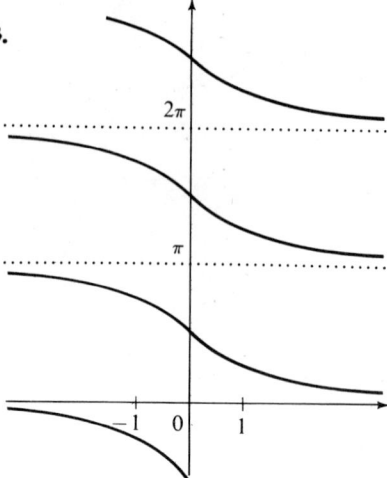

5.

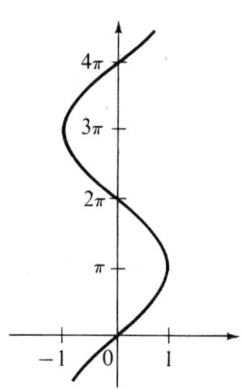

7.

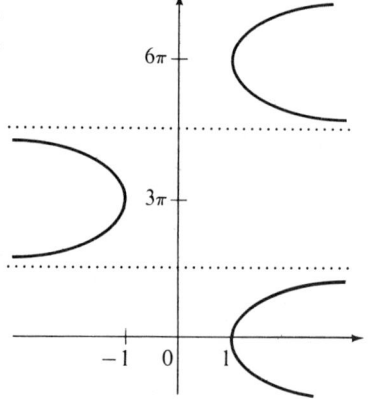

9.

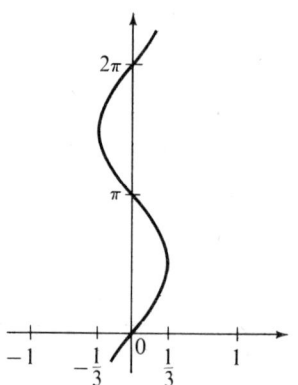

11.

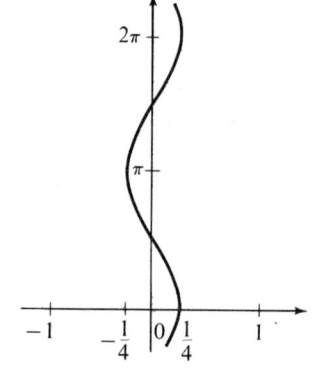

Answers 277

13.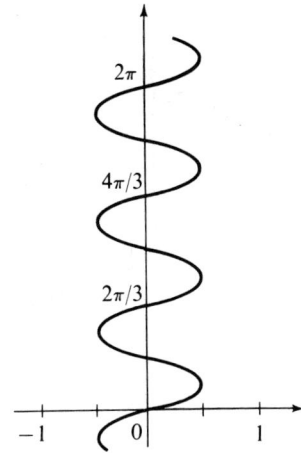

15.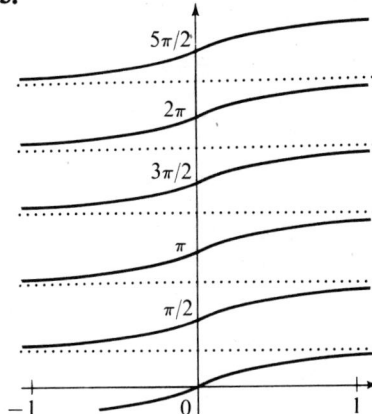

17.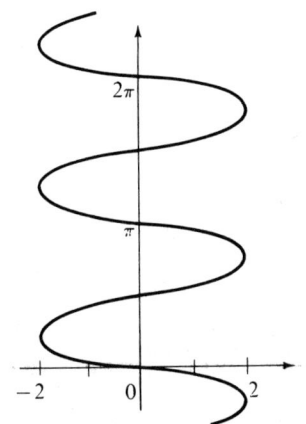

19.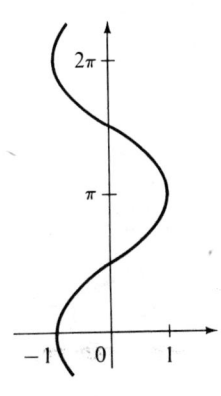

21.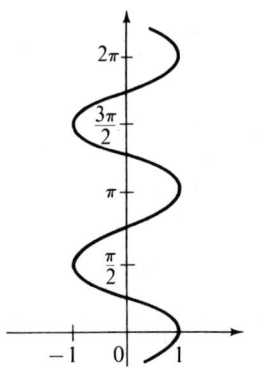

23.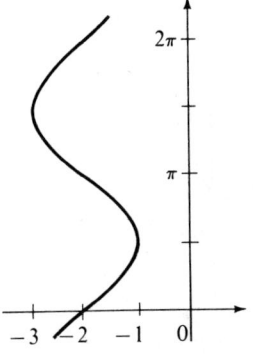

278 Answers

25.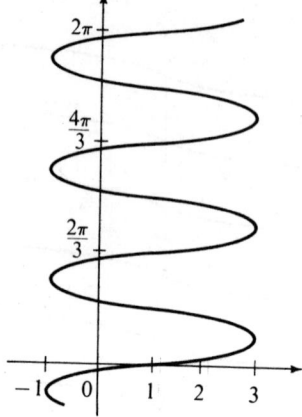

27.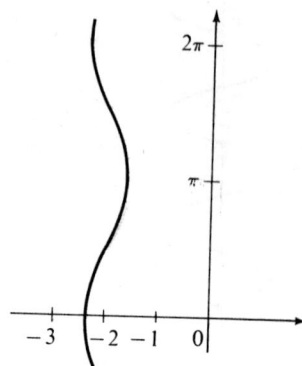

29.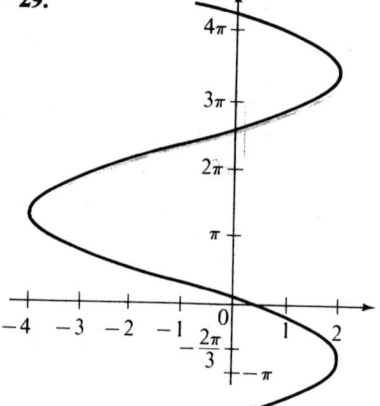

Section 8.3, page 163

1.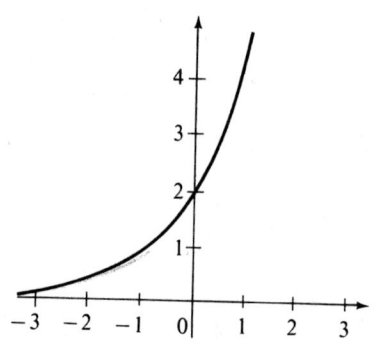

3.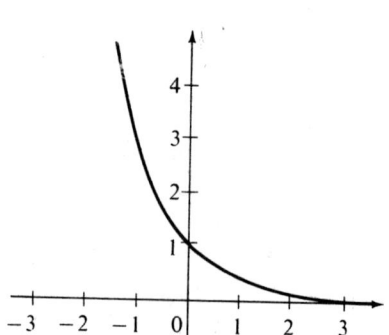

Answers 279

5.

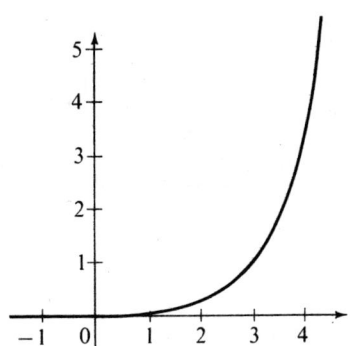

7.

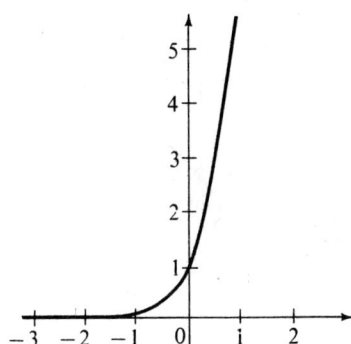

9.

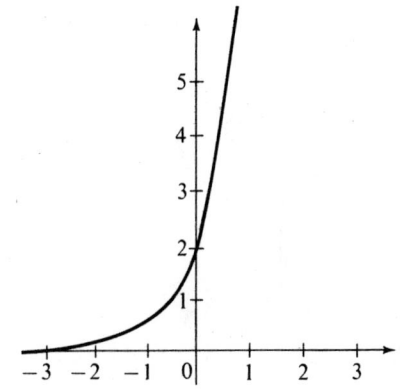

11.

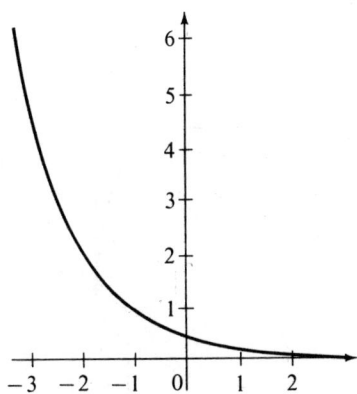

13.

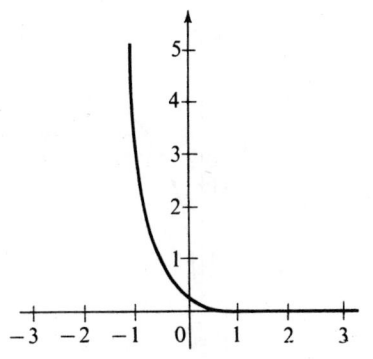

15.

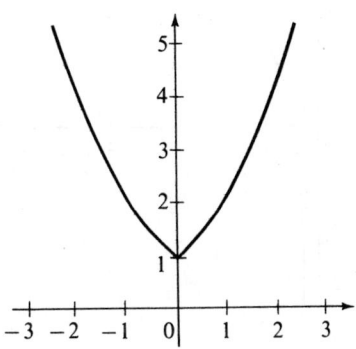

17.

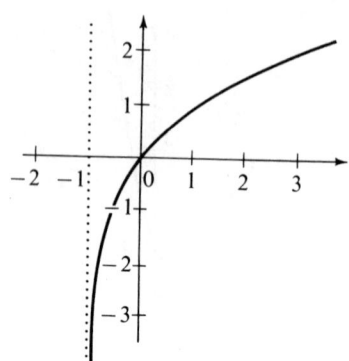

19.

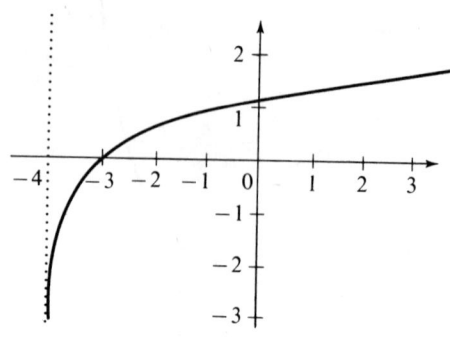

21.

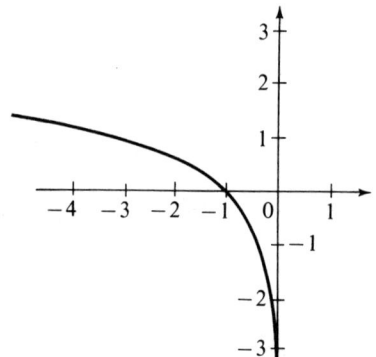

23.

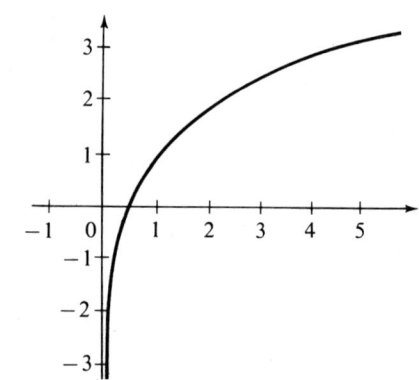

25.

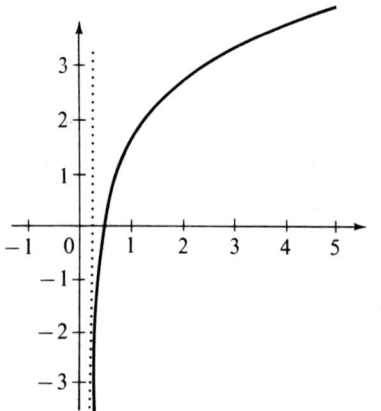

27.

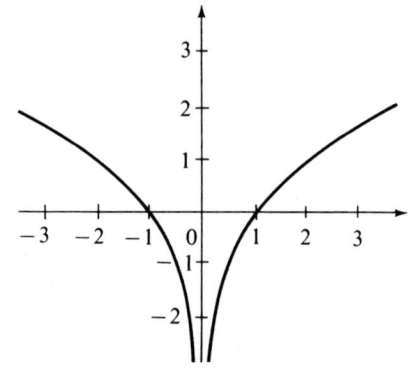

29.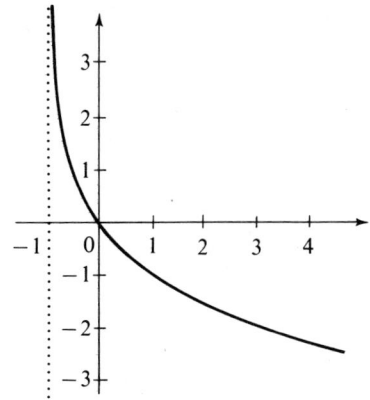

Section 8.4, page 166

1.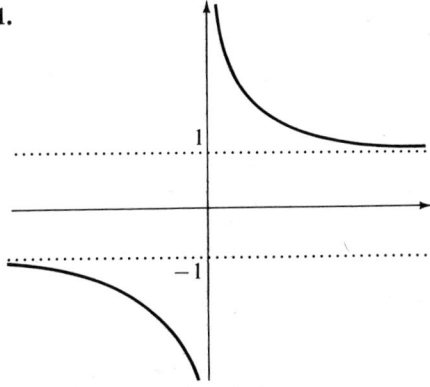

3.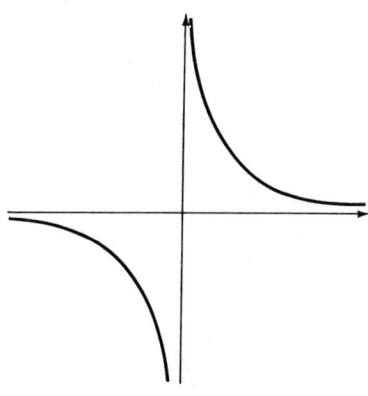

5.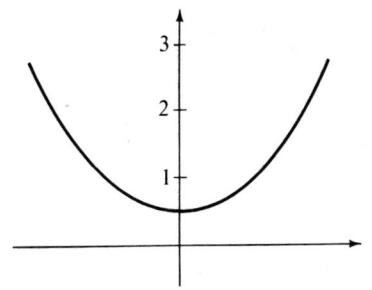

7.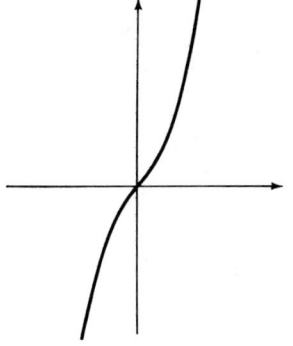

282 Answers

9.
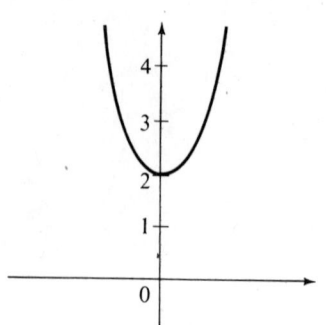

11.
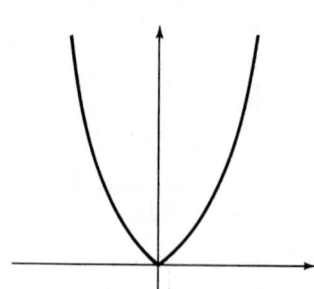

13.
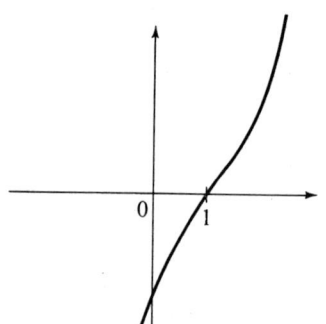

15.
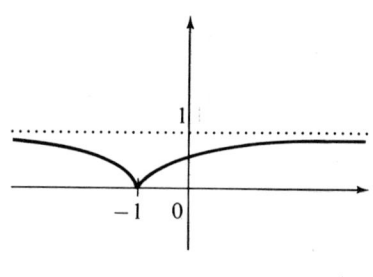

17.
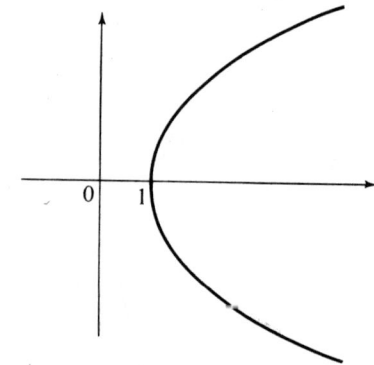

19.
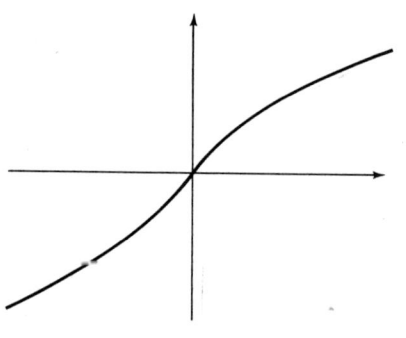

Chapter 9

Section 9.2, page 172

2. $(-4, 150°), (-2, 60°), (-1, 30°)$.

3. $(4, 300°), (3, 300°), (0, 0°)$.

5.
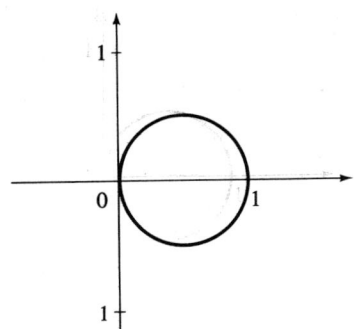

7.
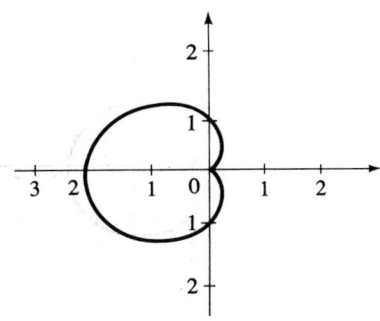

9.
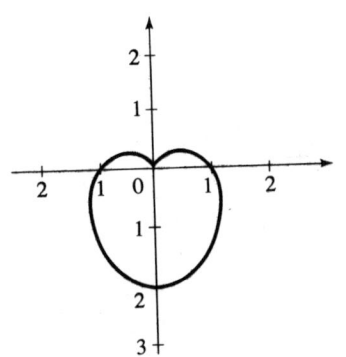

11.
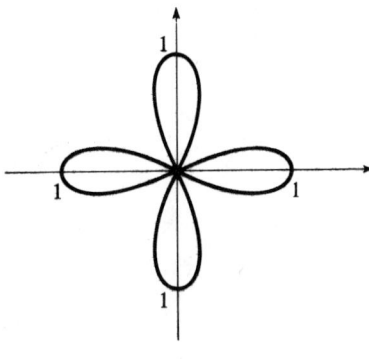

13.
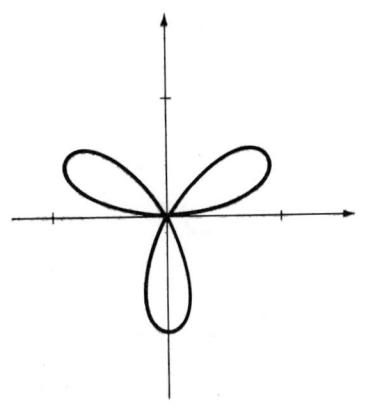

15.
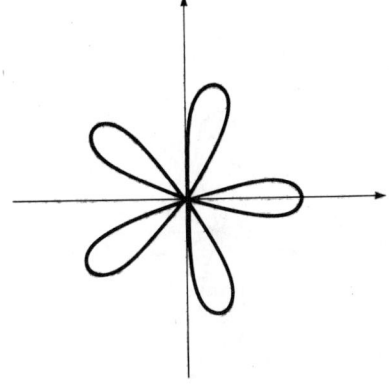

284 Answers

17.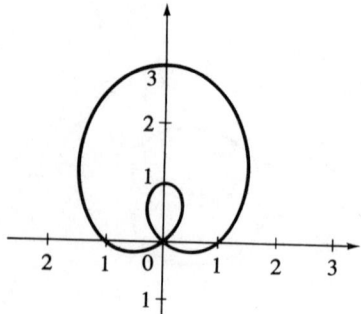

19.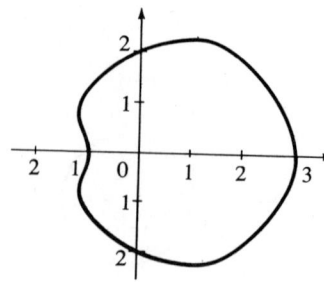

21.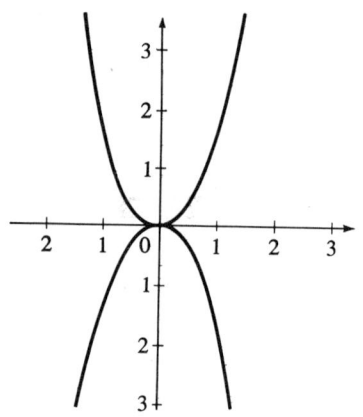

23.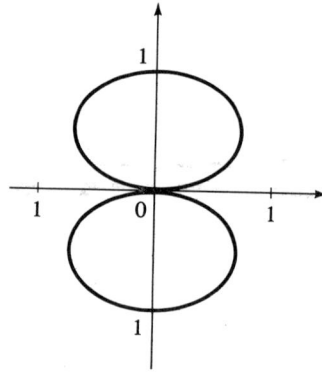

25.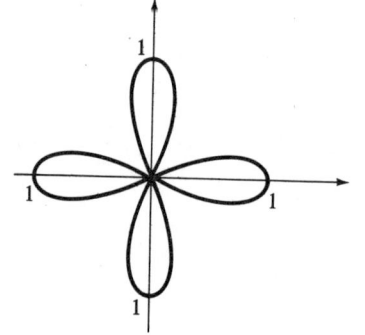

27.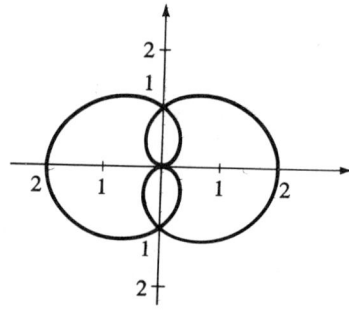

Answers 285

29.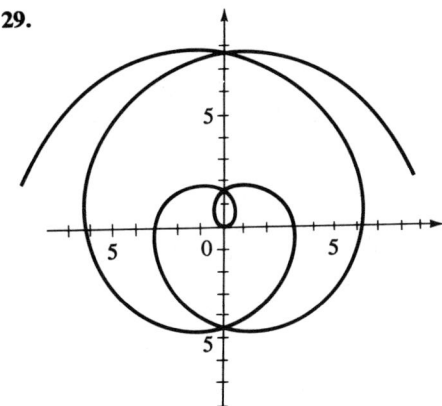

31.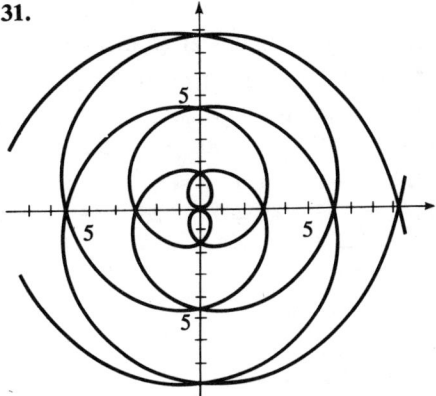

33.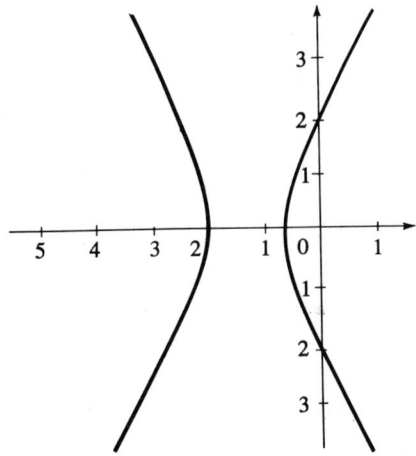

Section 9.3, page 175

1. $(\sqrt{2}, 45°), (\sqrt{2}, 315°)$. 3. $(2, 0°), (2, 180°)$.
5. $(1/2, 60°), (1/2, 300°), (0, 90°) = (0, 0°)$. 7. $(\sqrt{3}/2, 60°), (-\sqrt{3}/2, 300°), (0, 0°)$.
9. $(\sqrt{2}, 45°)$. 11. $(3, 109.5°), (3, 250.5°)$.
13. $(2+\sqrt{2}, 45°), (2-\sqrt{2}, 135°), (2-\sqrt{2}, 225°), (2+\sqrt{2}, 315°)$.
15. $(1 - 1/\sqrt{2}, 45°), (1 + 1/\sqrt{2}, 225°), (0, 90°) = (0, 0°)$. 17. $(1, 0°), (-1, 0°)$.
19. $(0, 0°), (1, 90°)$.

Section 9.4, page 177

1. $(-1, 0)$, $(\sqrt{3}/2, 3/2)$, $(1, 0)$, $(-1, 1)$.
2. $(2, 7\pi/4)$, $(2, 2\pi/3)$, $(4, 0)$, $(\sqrt{2}, 5\pi/4)$, $(2, 3\pi/2)$.   3. $r = 2 \sec \theta$.   5. $r = 1$.
7. $r = \sec \theta \tan \theta$.   9. $r = (\cos \theta - \sin \theta)/(1 + 2 \sin \theta \cos \theta)$.   11. $\tan \theta = 3$.
13. $r = 4/(2 \sin \theta + \cos \theta)$.   15. $r^2 - 2r(\sin \theta + \cos \theta) + 1 = 0$.
17. $r^2 = \sec \theta \csc \theta$.   19. $x^2 + y^2 = a^2$.   21. $\sqrt{3}x - y = 0$.
23. $x^2 + y^2 - 4x = 0$.   25. $(x^2 + y^2)^3 = (x^2 - y^2)^2$.
27. $x^2 + y^2 - 4y - 9 + 4y^2/(x^2 + y^2) = 0$ (note that the origin is not a point of the given curve).
29. $(x^2 + y^2)(x^2 + y^2 - 1)^2 = y^2$.   31. $y^2 = 2x + 1$.   33. $x^2 + y^2 = 3x + 2y$.

Section 9.5, page 181

1. hyperbola, $(0, 0)$, $x = 2, 2$.
3. ellipse, $(0, 0)$, $y = 2, 2/3$.
5. parabola, $(0, 0)$, $y = 3, 1$.
7. ellipse, $(0, 0)$, $y = 3, 2/3$.

9.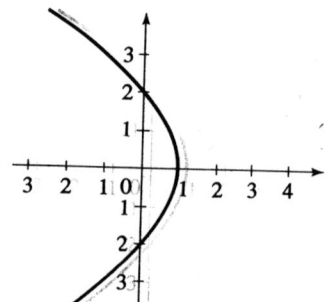

11.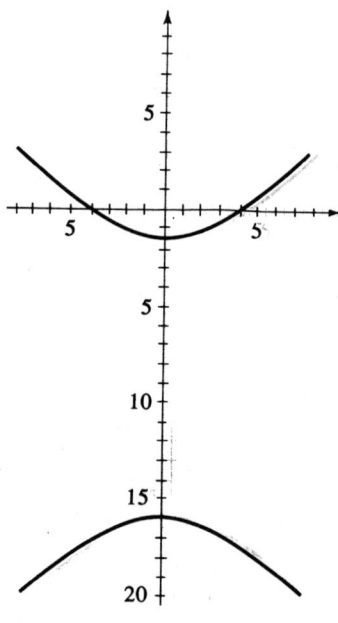

13.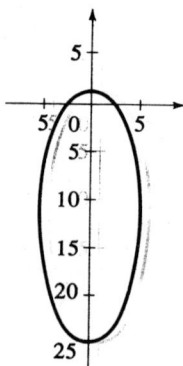

Answers 287

15. $r = 10/(3 + 2\cos\theta)$.
17. $r = 2/(1 + \sin\theta)$.
19. $r = 25/(4 + 5\cos\theta)$.

21.
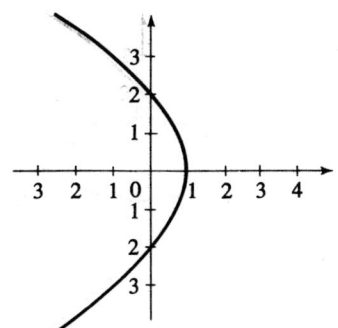

Chapter 10

Section 10.1, page 187

1. $y^2 - x + 2y - 2 = 0$.

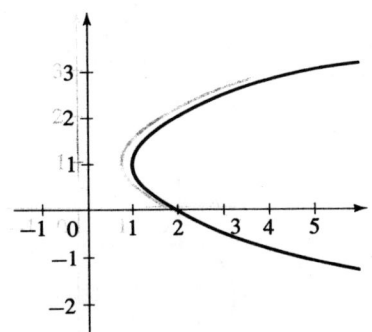

3. $y = x^2 - 1$.

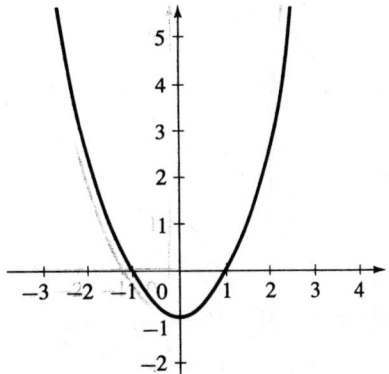

5. $x^2 - 2xy + y^2 - 2x - 2y = 0$.

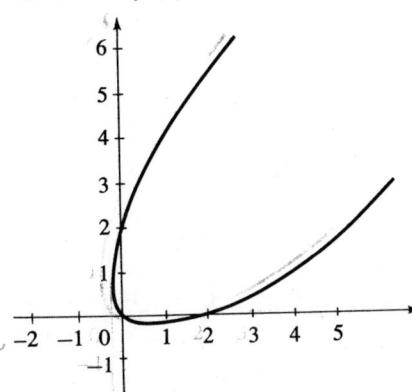

7. $y^3 = x^2$.

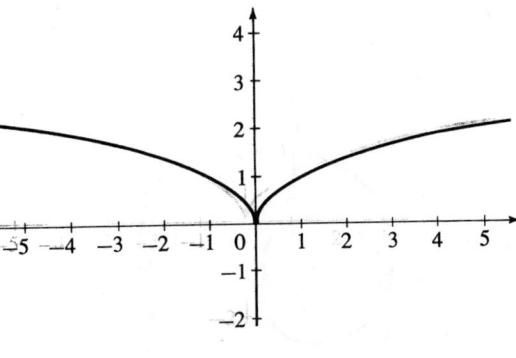

288 Answers

9. $x^2/a^2 + y^2/b^2 = 1$.

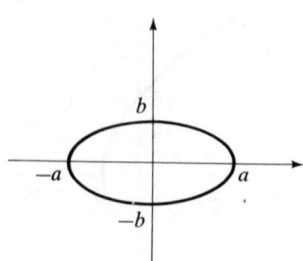

11. $(x-2)^2 + (y+1)^2 = 1$.

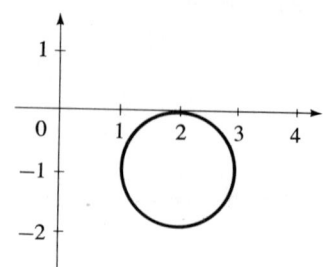

13. $(x-3)^2 - (y-2)^2 = 1$.

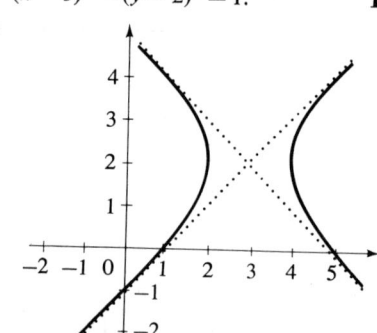

15. $y = (x+1)^2$.

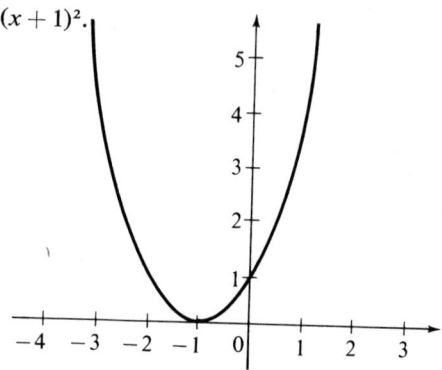

17. $y^2 = x^3$.

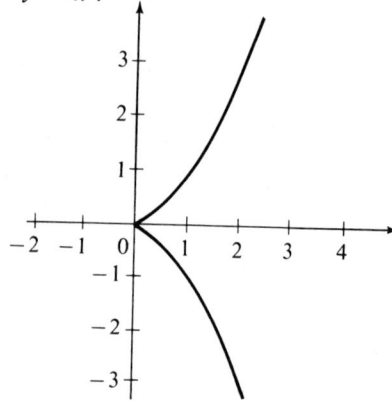

19. $x^2 + y^2 = 1$.

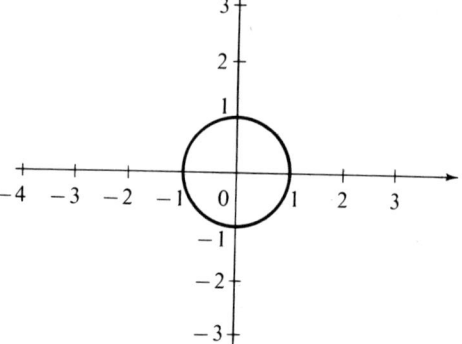

21.

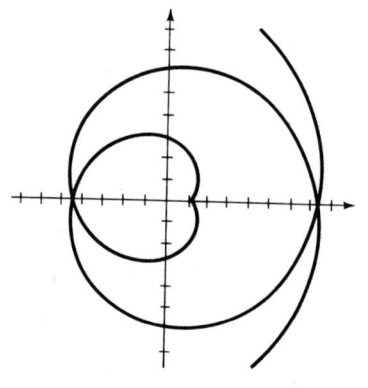

23.

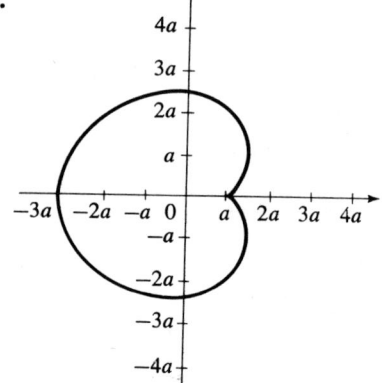

25.

(a)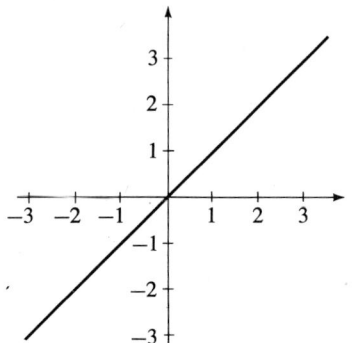

(b)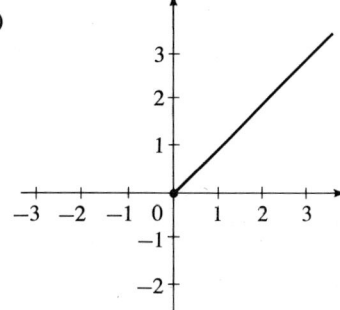

(c)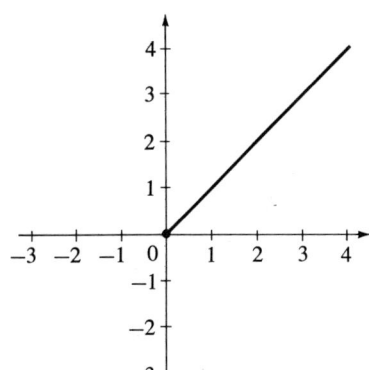

(d)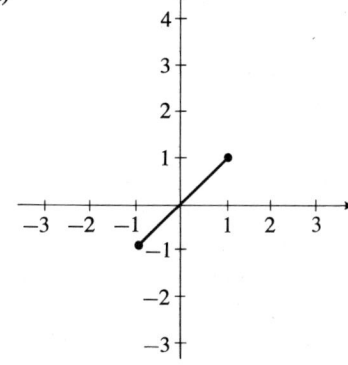

(e)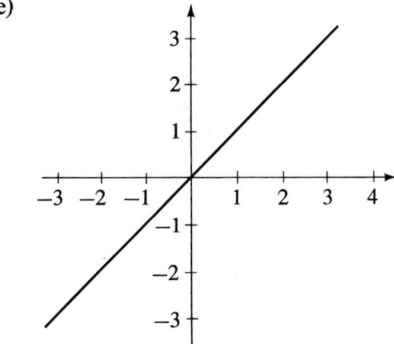

(f)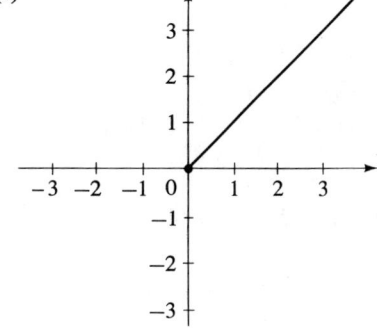

Section 10.2, page 189

1. $x = \dfrac{v_0\sqrt{3}}{2}\, t, \quad y = -16t^2 + \dfrac{v_0}{2}\, t + 10.$
3. $x = \dfrac{v_0}{2}\, t, \quad y = -16t^2 + \dfrac{v_0\sqrt{3}}{2}\, t + 10.$

4. (a) $\dfrac{v_0^2 + v_0\sqrt{v_0^2 + 1280}}{64}.$
5. $x = a \sec\theta, \quad y = b \tan\theta; \quad x^2/a^2 - y^2/b^2 = 1.$

290 Answers

7. $x = a\theta - b \sin \theta$, $y = a - b \cos \theta$.   **9.** $x = a(1 \pm \sin \theta)$, $y = a(1 \pm \sin \theta \tan \theta)$.
11. $x = a \pm b \cos \theta$, $y = a \tan \theta \pm b \sin \theta$.
13.

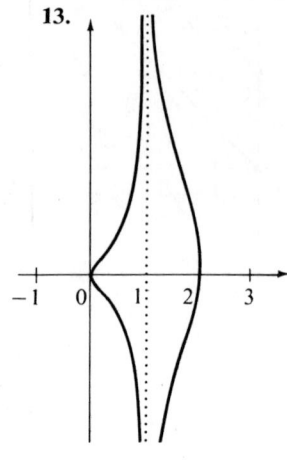

(a)

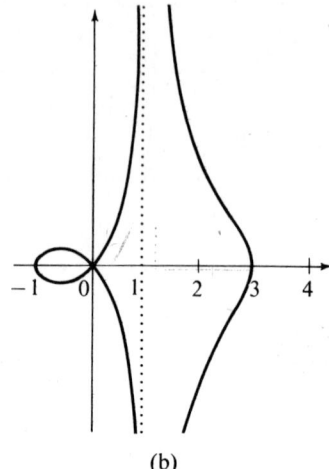

(b)

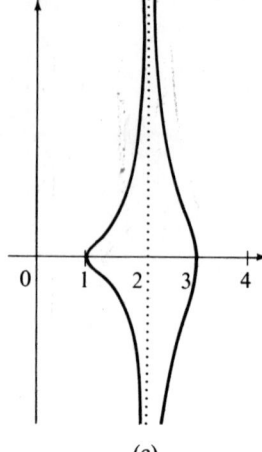

(c)

15. $x = 2a(\csc \theta - \sin \theta)\cos \theta$, $y = 2a(\csc \theta - \sin \theta)\sin \theta$.
17. $x = 3a \cos \theta + a \cos 3\theta$, $y = 3a \sin \theta - a \sin 3\theta$.
19. $x = 2a \cos \theta - a \cos 2\theta$, $y = 2a \sin \theta - a \sin 2\theta$.

Chapter 11

Section 11.1, page 195

1. $5\sqrt{2}$.   **3.** 5.   **5.** $\sqrt{131}$.   **7.** 5.   **9.** 9.   **11.** $(-2, 1, 2)$.
13. $(1, 2, 2)$.   **15.** $(-9, 7, -1)$.   **17.** $(1, 1, 5)$.   **19.** $(1, 3, 3/2)$.
21. $(8, 4, -6)$.   **23.** $(-2, 4, -3)$.   **25.** $5, -3$.   **27.** ± 5.   **29.** $-11, -5$.

Section 11.2, page 199

1. $-6\mathbf{i} - 2\mathbf{j} + 7\mathbf{k}$.   **3.** $-3\mathbf{i} - 4\mathbf{j} + 3\mathbf{k}$.   **5.** $\dfrac{4}{\sqrt{21}}\mathbf{i} + \dfrac{1}{\sqrt{21}}\mathbf{j} - \dfrac{2}{\sqrt{21}}\mathbf{k}$.

7. $\dfrac{1}{\sqrt{35}}\mathbf{i} + \dfrac{5}{\sqrt{35}}\mathbf{j} - \dfrac{3}{\sqrt{35}}\mathbf{k}$.   **9.** $B = (4, 0, 8)$.   **11.** $A = (3, 1, 3)$.
13. $A = (0, 6, -3/2)$, $B = (4, 4, -1/2)$.   **15.** $60°$.   **17.** $77°$.
19. $3\mathbf{i} + 2\mathbf{j} + 6\mathbf{k}$, $-\mathbf{i} - 6\mathbf{j} + 4\mathbf{k}$, -1, not orthogonal.
21. $3\mathbf{i} - 5\mathbf{j} + 5\mathbf{k}$, $\mathbf{i} + 3\mathbf{j} + 7\mathbf{k}$, 0, orthogonal.   **23.** $\dfrac{4}{3}\mathbf{i} + \dfrac{4}{3}\mathbf{j} + \dfrac{4}{3}\mathbf{k}$.

25. $\dfrac{7}{6}\mathbf{i} - \dfrac{7}{3}\mathbf{j} + \dfrac{7}{6}\mathbf{k}$.

Section 11.3, page 203

1. $84°, 64°, 27°$.   3. $103°, 116°, 29°$.   5. $35°, 35°, 35°$.   7. $\{-4, 2, 4\}$.
9. $\{2, 2, -2\}$.   11. $\{5, 1, -2\}$.   13. $(3, 8, 1), (4, 13, 3)$.   15. $(3, 4, 5), (5, 5, 7)$.
17. $(5, 3, -2), (9, 3, -3)$.   19. x axis: $\{0°, 90°, 90°\}, \{1, 0, 0\}$.   21. Parallel.
23. Perpendicular.   25. Coincident.   27. None.   29. Perpendicular.

Section 11.4, page 208

1. $x = 5 + 3t, y = 1 - 2t, z = 3 + 4t; \dfrac{x-5}{3} = \dfrac{y-1}{-2} = \dfrac{z-3}{4}$.

3. $x = 5 + 4t, y = -2 + t, z = 1 - 2t; \dfrac{x-5}{4} = \dfrac{y+2}{1} = \dfrac{z-1}{-2}$.

5. $x = 1 + 2t, y = 1, z = 1 + t; \dfrac{x-1}{2} = \dfrac{z-1}{1}, y = 1$.

7. $x = 4, y = 4, z = 1 + t; x = 4, y = 4$.

9. $x = 4 + 2t, y = -3t, z = 5 + 4t; \dfrac{x-4}{2} = \dfrac{y}{-3} = \dfrac{z-5}{4}$.

11. $x = 8 + 5t, y = 4 + 2t, z = 1 - t; \dfrac{x-8}{5} = \dfrac{y-4}{2} = \dfrac{z-1}{-1}$.

13. $x = 5, y = 1 + t, z = 3 + t; x = 5, y - 1 = z - 3$.
15. $x = 1, y = -2 + t, z = 3; x = 1, z = 3$.   17. Do not intersect.
19. Do not intersect.   21. The lines are identical.   23. $(2, 1, -1)$.
25. Perpendicular.   27. None.   29. Parallel.

Section 11.5, page 213

1. $-5\mathbf{i} + 5\mathbf{j} + 5\mathbf{k}$.   3. $2\mathbf{i} + \mathbf{j} + 7\mathbf{k}$.   5. $-\mathbf{i} - \mathbf{j} + 3\mathbf{k}$.   7. $\{1, -1, 0\}$.
9. $\{1, 1, -1\}$.   11. $\{8, 5, -9\}$.   13. $x = 3 + t, y = 2 - 3t, z = 1 - 5t$.
15. $x = 2 + t, y = 3 - 2t, z = 1 + 4t$.   17. $x = 2 + 4t, y = -2t, z = 5 - 7t$.
19. $24/\sqrt{30}$.   21. $113/\sqrt{542}$.   23. $\sqrt{26}$.   27. $\sqrt{893}/2$.   29. $13/2$.

Section 11.6, page 217

1. $6/\sqrt{14}$.   3. $4/\sqrt{21}$.

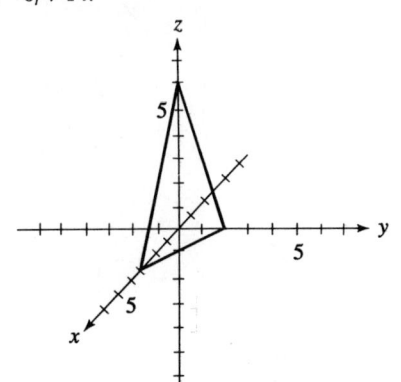
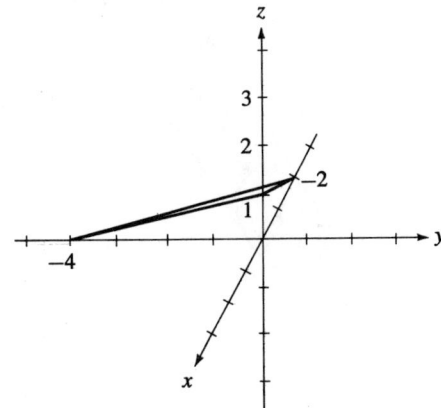

292 Answers

5. $3/\sqrt{5}$.

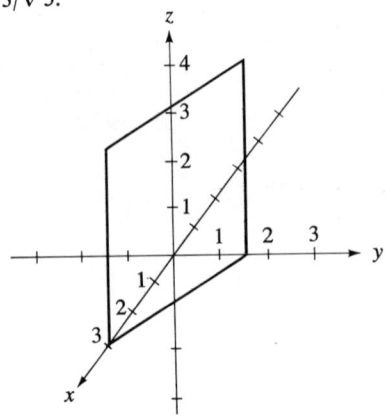

7. 7.
9. $11/2\sqrt{26}$.
11. $\sqrt{5}$.
13. $3x - y - 2z - 17 = 0$.
15. $3x - 4y + 2z + 9 = 0$.
17. $3x + y - 4z \pm 4 = 0$.
19. $2x + y - z - 4 \pm 2\sqrt{6} = 0$.
21. $x + y - 5 = 0$.
23. $3x + 5y + z - 23 = 0$.
25. $2x + y - 9 = 0$.
27. $x + 5y + 3z - 26 = 0$.
29. $3x - 7y - 5z + 22 = 0$.
31. $\sqrt{173}$.
33. $5\sqrt{2}$.

Section 11.7, page 222

1. 5/2. **3.** 0. **5.** 0. **7.** $1/\sqrt{2}$. **9.** 2, −5/2. **11.** 52°. **13.** 80°.
15. 34°. **17.** 48°. **19.** None. **21.** Perpendicular. **23.** None.
25. Parallel. **27.** Perpendicular.

Section 11.8, page 226

1.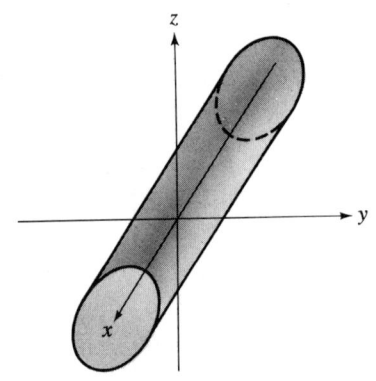

3.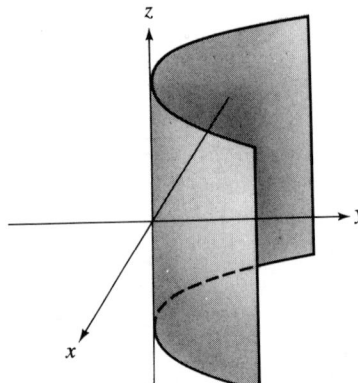

Answers 293

5.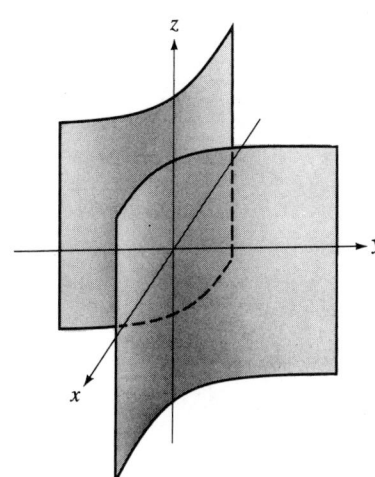

7.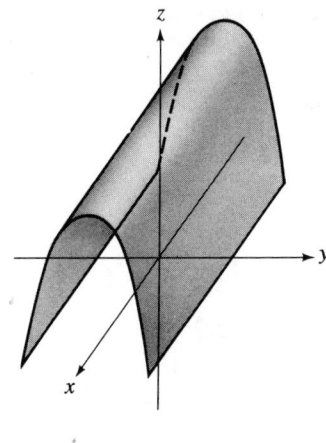

9.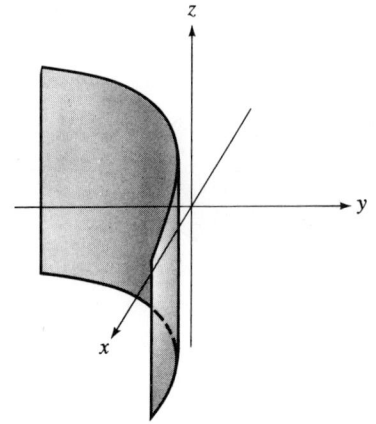

11. Sphere: $(1, 0, -2)$, 3.
13. No locus.
15. Sphere: $(-1/2, 3/2, 1)$, 2.
17. Sphere: $(1/3, -1/3, -2/3)$, $2\sqrt{2}/3$.
19. Point: $(1, 1/2, -2)$.
21. $x^2 + y^2 + z^2 - 8x - 2y + 4z + 12 = 0$.
23. $x^2 + y^2 + z^2 - 4x - 8y - 14z + 65 = 0$.

25. $x^2 + y^2 + z^2 + (-6k + 8l - 6)x - 2ky - 2lz + 10k^2 - 24kl + 17l^2 + 18k - 24l - 17 = 0$ (for any choice of k and l). **27.** $x^2 + y^2 + z^2 - 27x + 35y - 62z - 28 = 0$.

Section 11.9, page 230

1. Ellipsoid.

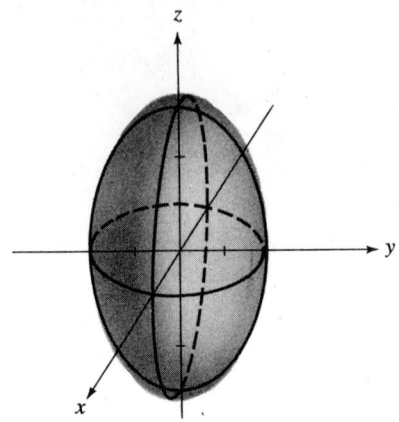

3. Circular paraboloid.

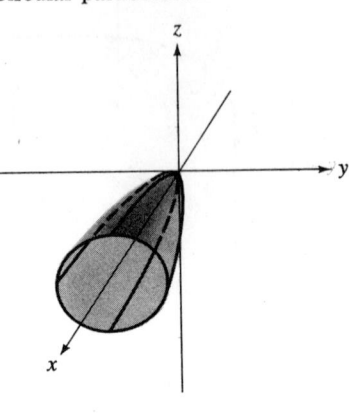

5. Circular cone.

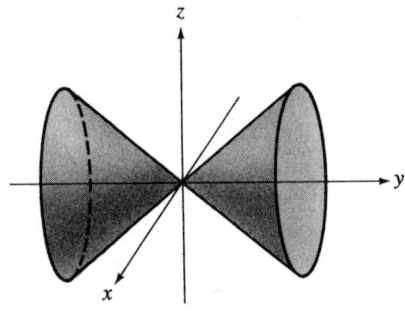

7. Circular paraboloid.

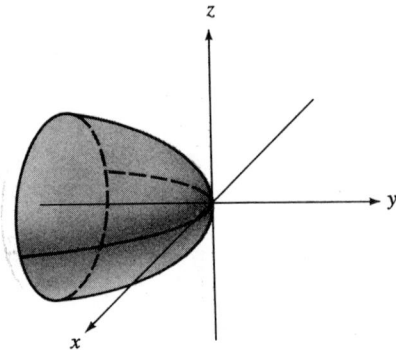

9. Hyperboloid of two sheets.

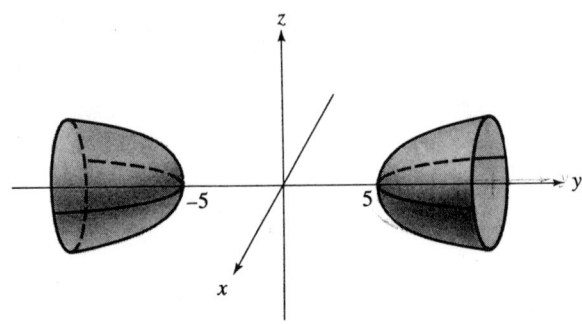

Answers 295

11. Hyperbolic paraboloid.

13. Hyperboloid of two sheets.

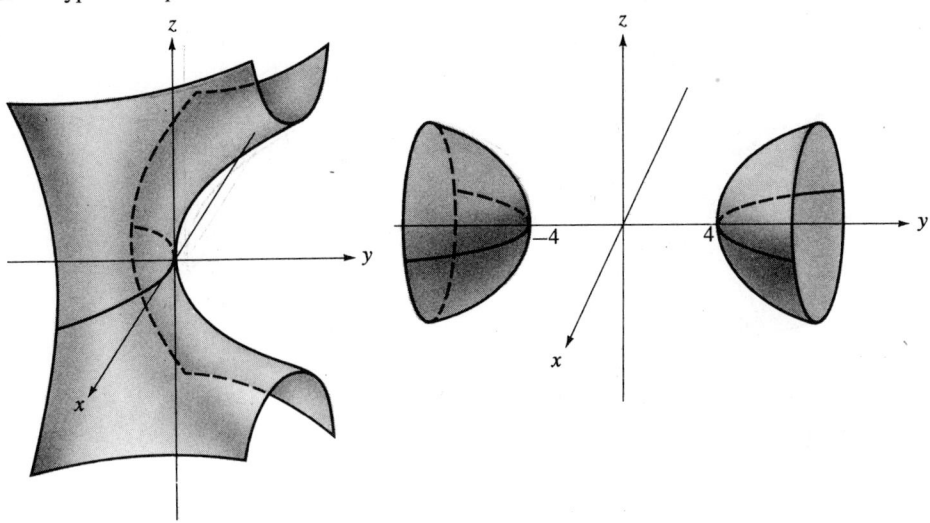

15. Circular cone.

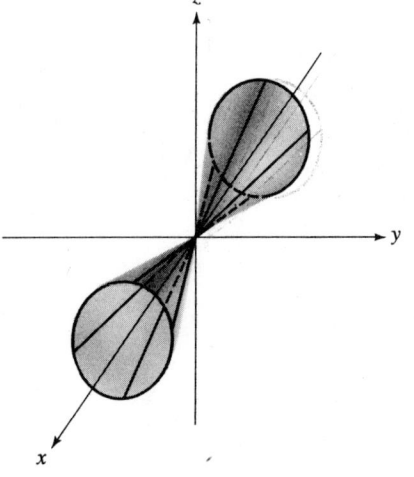

17. Hyperboloid of two sheets.

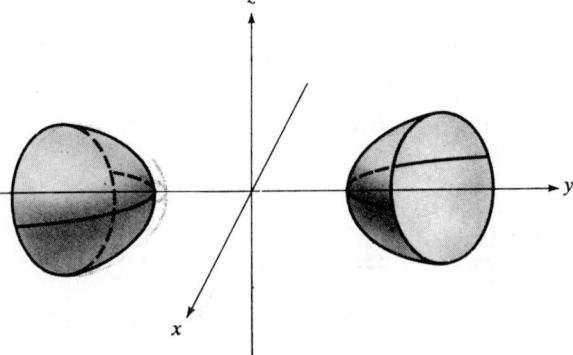

296 Answers

19. Hyperboloid of one sheet.

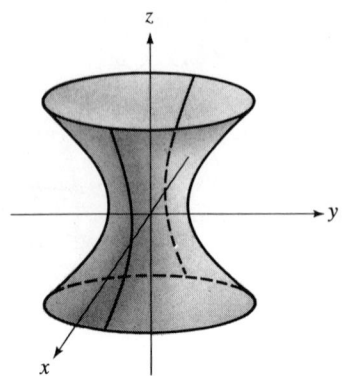

21. Hyperbolic paraboloid.

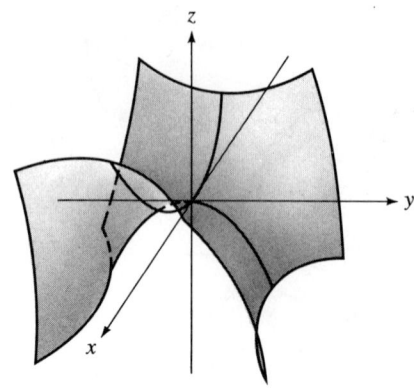

23. Circular cone.

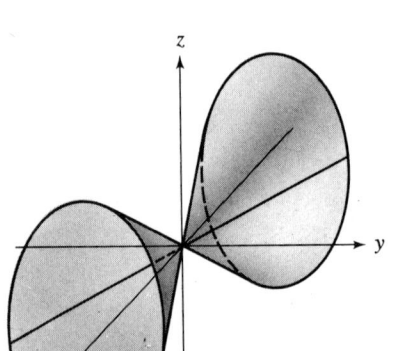

25. Hyperboloid of one sheet.

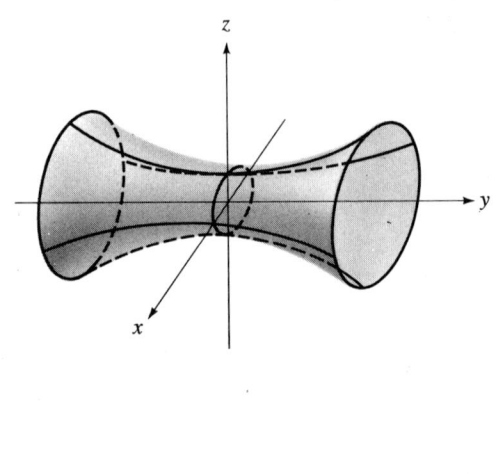

Section 11.10, page 234

1. (a) $(\sqrt{2}, \sqrt{2}, 1)$, (b) $(-3/2, 3\sqrt{3}/2, -2)$.　　2. (a) $(\sqrt{2}, 45°, 3)$, (b) $(2, 90°, -2)$.
3. (a) $(3/2\sqrt{2}, 3/2\sqrt{2}, 3\sqrt{3}/2)$, (b) $(0, 0, 1)$.
4. (a) $(2\sqrt{2}, 45°, 90°)$, (b) $(3, \text{Arccos}(2/\sqrt{5}), \text{Arccos}(-2/3))$.
5. (a) $(5, 30°, \text{Arccos}(4/5))$, (b) $(2\sqrt{2}, \pi/4, 3\pi/4)$.
6. (a) $(2, 45°, 2\sqrt{3})$, (b) $(2, 2\pi/3, 0)$.　　7. $r = 2, \rho \sin \varphi = 2$.
9. $r^2 = z, \rho = \csc \varphi \cot \varphi$.　　11. $r^2 \cos 2\theta - z^2 = 1$, $\rho^2(\sin^2 \varphi \cos 2\theta - \cos^2 \varphi) = 1$.
13. $r^2 - z^2 = 1, \rho^2 = -\sec 2\theta$.　　15. $z = 2xy$.　　17. $z = x^2 + y^2$.　　19. $xy = z$.
21. $x^2 + y^2 + z^2 - x = 0$.

Index

Abscissa, 3
Absolute value, of a vector, 31
Addition of ordinates, 150–151, 155
Addition of vectors, 30
Amplitude, 154
Analytic proofs, 19–22
Angle:
  between two lines, 16–18, 220
  between two planes, 221
  between two vectors, 35, 42, 44, 197, 212, 220
Arcsin, etc., *see* Inverse trigonometric functions
Asymptote:
  horizontal, 135–137
  of a hyperbola, 107
  odd and even, 140
  slant, 149–150
  vertical, 135
Averages, method of, 67–69
Axes:
  rectangular, 2–3, 192
  rotation of, 124–131
  translation of, 119–124
Axis:
  conjugate, of a hyperbola, 107
  major, of an ellipse, 97
  minor, of an ellipse, 97
  of a parabola, 90
  polar, 168
  transverse, of a hyperbola, 107

Basis vectors, 32, 196

Cardioid, 191
Center:
  of a circle, 72
  of an ellipse, 97
  of a hyperbola, 107
  of a sphere, 224
Circle(s), 72–87, 181–182
  definition of, 72
  degenerate cases, 75
  equation of:
    general form, 73
    standard form, 72
  families of, 83–87
Cissoid of Diocles, 191
Completing the square, 74
Component of a vector, 32, 196
Conchoid, 190
Cone, equation of, 229
Conic sections, 88–118, 127–131, 148–152
  in polar coordinates, 178–182
  and a right circular cone, 115–118
Conjugate axis of a hyperbola, 107
Conjugate hyperbolas, 110
Coordinate line, 2
Coordinate plane, 192
Coordinates:
  cylindrical, 231–232
  polar, 168–182
  rectangular, 1–3, 192
  relationships between polar and rectangular, 175–178
  spherical, 232–234
Cosecant, *see* Trigonometric functions
Cosine, *see* Trigonometric functions
Cotangent, *see* Trigonometric functions
Covertex of an ellipse, 97
Cross product, 209
Curve fitting, 65–71
Cycloid, 189
Cylinder, 223–224
  directrix of, 223
  generatrix of, 223
Cylindrical coordinates, 231–232

298 Index

Degenerate cases:
  of a circle, 75
  of an ellipse, 104, 119
  of a hyperbola, 114, 119
  of a parabola, 95, 119
Directed line segment, 29–30
  equivalence, 29
  head, 29
  tail, 29
Direction angles:
  of a line, 202
  of a vector, 35, 200
Direction cosines:
  of a line, 202
  of a vector, 36, 200
Direction numbers:
  of a line, 202
  of a vector, 38, 201
Directrix:
  of a cylinder, 223
  of an ellipse, 100
  of a hyperbola, 111
  of a parabola, 88
Distance:
  between parallel lines, 60
  from a point to a line, 56, 182
  from a point to a plane, 219
  between skew lines, 212
  between two points, 4–6, 193
Domain of an equation, 144–147
Dot product, 43, 197

e (base of the natural logarithm), 160
Eccentricity:
  of an ellipse, 100, 179
  of a hyperbola, 111, 179
  of a parabola, 101, 179
Ellipse, 96–105, 178–181
  definition of, 96
  degenerate case, 104, 119
  equation of:
    general form, 104
    standard form, 98, 102–103
  in polar coordinates, 178–181
Ellipsoid, 227
Elliptic cone, 229
Elliptic paraboloid, 228
Epicycloid, 191
Equation of a locus, 25–28
Equilibrium of forces, 41
Equivalent directed line segments, 29
Exponential functions, 160–163

Families:
  of circles, 83–87
  of lines, 61–65
Focus:
  of an ellipse, 96
  of a hyperbola, 106
  of a parabola, 88
Forces:
  in equilibrium, 41
  represented by vectors, 39–40

Generatrix of a cylinder, 223
Graph:
  of an equation, 23–25
  of a number, 2
  of parametric equations, 183–187
  in polar coordinates, 169–173
  of vector functions, 185–187

Head of a directed line segment, 29
Hyperbola, 106–115, 178–181
  conjugate, 110
  definition, 106
  degenerate case, 114, 119
  equation:
    general form, 113
    standard form, 107, 108, 112
  in polar coordinates, 178–181
Hyperbolic functions, 164–167
  definition of, 164
  graphs of, 164–167
Hyperbolic paraboloid, 228–229
Hyperboloid:
  of one sheet, 227–228
  of two sheets, 227–228
Hypocycloid, 191

Inclination of a line, 12
Inner product, 43, 197
Intercepts:
  of a curve, 134
  of a line, 52
  odd and even, 140–141
Inverse hyperbolic functions, 166–167
  definition of, 166
Inverse trigonometric functions, 156–159
Involute of a circle, 191

Latus rectum:
  of an ellipse, 98

Index

of a hyperbola, 108
of a parabola, 90
Left-hand coordinate system, 192–193
Length of a vector, 31
Line(s):
  equation of, 48–71, 182, 204–209
    from empirical data, 65–71
    general form, 54
    intercept form, 52
    normal form, 61, 182
    point-slope form, 48
    slope-intercept form, 52
    two-point form, 50
    vertical, 49
  families of, 61–65
  perpendicular, 14–15, 56
  in space, 204–209
Logarithm, 162–164
  tables, 238–239

Major axis of an ellipse, 97
Midpoint of a segment, 10, 194
Minor axis of an ellipse, 97
Multiplication of vectors:
  cross product, 209
  dot product, 43, 197
  scalar multiple, 31

Natural logarithm, see Logarithm
Nappe of a cone, 115
Normal form:
  of an equation of a line, 61, 182
  of an equation of a plane, 215
Normal line to a plane, 214–215

Oblate spheroid, 227
Octant, 192
Ordinate(s), 3
  addition of, 150–151, 155
Origin, 1, 192
Orthogonal vectors, 44, 197
Outer product, 209

Parabola, 88–96
  definition of, 88
  degenerate cases, 95, 119
  equation of:
    general form, 93
    standard form, 89, 90, 93

Paraboloid:
  elliptic, 228
  hyperbolic, 228–229
Parallel lines, 14, 60, 209
Parallel planes, 222
Parameter, 61, 183–191
Parametric equations, 183–191
  for a line in space, 205
Period, 154
Perpendicular lines, 15, 60, 221
Perpendicular planes, 222
Perpendicular vectors, 44, 197
Plane, 214–218
  normal form, 215
Point-of-division formulas, 7–11, 194
Point of intersection, 24, 173–175
Polar axis, 168
Polar coordinates, 168–182
  conic sections, 178–181
  graphs, 169–173
  relationships between polar and rectangular coordinates, 175–178
Pole, 168
Prime factors, table of, 240
Product:
  cross, 209
  dot, 43, 197
  inner, 43, 197
  outer, 209
  scalar, 43, 197
  vector, 209
Projection of one vector upon another, 45–46, 197–198
Prolate spheroid, 227

Quadradic Equation, 18
Quadrant, 3
Quadric surfaces, 227–231

Radian, 153
Radical axis, 84
Radius:
  of a circle, 72
  of a sphere, 224
Rectangular coordinates:
  relationships between rectangular and polar coordinates, 175–178
Representative of a vector, 30
Residual, 67
Right-hand coordinate system, 192–193
Rotation, 124–131
  equations of, 125

Scalar, 31
Scalar multiple, 31
Scalar product, 43, 197
Secant, *see* Trigonometric functions
Selected points, method of, 65–66, 68
Sine, *see* Trigonometric functions
Slope of a line, 13–14
Sphere, 224–226
   equation of:
      general form, 224
      standard form, 224
Spherical coordinates, 232–235
Spheroid:
   oblate, 227
   prolate, 227
Square roots, table of, 240
Squares, table of, 240
Strophoid, 190
Sum of two vectors, 30
Symmetry:
   about a line, 138–139
   about a point, 138–139

Tail of a directed line segment, 29
Tangent, *see* Trigonometric functions
Transverse axis of a hyperbola, 107
Transcendental curves, 153–167

Translation, 119–124
   equations of, 120
Trigonometric functions, 153–156
   tables of, 237

Unit vector, 32

Vector(s), 29–47, 196–204, 209–214
   definition of, 30
   directed along a line, 202
   in the plane, 29–47
   representative of, 30
   in space, 196–204, 209–214
Vector functions, 185–187
Vector product, 209
Vertex:
   of an ellipse, 97
   of a hyperbola, 107
   of a parabola, 90

Witch of Agnesi, 191

Zero vector, 30